DYNAMIC RELATIONALITY THEORY OF CREATIVE TRANSFORMATION

DYNAMIC RELATIONALITY THEORY OF CREATIVE TRANSFORMATION

GROUNDING MACHINIC ECOSYSTEMS IN LIFE-EXPERIENCES

Kerimcan Ozcan
School of Business and Global Innovation, Marywood University, Scranton, PA, United States

Venkat Ramaswamy
Ross School of Business, University of Michigan, Ann Arbor, MI, United States

ELSEVIER

Elsevier
Radarweg 29, PO Box 211, 1000 AE Amsterdam, Netherlands
125 London Wall, London EC2Y 5AS, United Kingdom
50 Hampshire Street, 5th Floor, Cambridge, MA 02139, United States

Copyright © 2025 Elsevier Inc. All rights are reserved, including those for text and data mining, AI training, and similar technologies.

Publisher's note: Elsevier takes a neutral position with respect to territorial disputes or jurisdictional claims in its published content, including in maps and institutional affiliations.

No part of this publication may be reproduced or transmitted in any form or by any means, electronic or mechanical, including photocopying, recording, or any information storage and retrieval system, without permission in writing from the publisher. Details on how to seek permission, further information about the Publisher's permissions policies and our arrangements with organizations such as the Copyright Clearance Center and the Copyright Licensing Agency, can be found at our website: www.elsevier.com/permissions.

This book and the individual contributions contained in it are protected under copyright by the Publisher (other than as may be noted herein).

Notices

Knowledge and best practice in this field are constantly changing. As new research and experience broaden our understanding, changes in research methods, professional practices, or medical treatment may become necessary.

Practitioners and researchers must always rely on their own experience and knowledge in evaluating and using any information, methods, compounds, or experiments described herein. In using such information or methods they should be mindful of their own safety and the safety of others, including parties for whom they have a professional responsibility.

To the fullest extent of the law, neither the Publisher nor the authors, contributors, or editors, assume any liability for any injury and/or damage to persons or property as a matter of products liability, negligence or otherwise, or from any use or operation of any methods, products, instructions, or ideas contained in the material herein.

ISBN: 978-0-443-30159-9

> For information on all Elsevier publications visit our website at
> https://www.elsevier.com/books-and-journals

Publisher: Kathryn Eryilmaz
Acquisitions Editor: Kathryn Eryilmaz
Editorial Project Manager: Himani Dwivedi
Production Project Manager: Gomathi Sugumar
Cover Designer: Christian Bilbow

Typeset by TNQ Technologies

Dedication

To Pana, Dora, and Jülide.

In abiding remembrance of Gürbüz Özcan and Todorka Barova.

— Kerimcan Ozcan

To Brahman and Swami Krishnananda,

Ishvara and Shivashankar,

Bhumi Devi and Soulmate Bindu,

Prakriti and Joyful Adithya and Lalitha,

Matrudevo Bhava and Pitrudevo Bhava,

And cocreators of the future everywhere.

In loving memory of my father, Shri A. V. Ramaswamy;

God bless his soul.

— Venkat Ramaswamy

Contents

About the authors xi
Acknowledgments xiii
Prologue xv

Part I
Relational dynamics

1. Theorizing relational dynamics of creative transformation

Virtual and actual dimensions of a relational universe 3
Actualization of the real and the possible 6
From virtual realities to actual possibilities 9
DRT and the four domains of the creative plane of immanence 10
Virtual and actual dynamics: Deterritorialization and discursivity 13
Differentiation as double articulation of strata via assemblages 20
Interpreting functorial domains through gauge theory 23
Seeing health ecosystems through DRT 27
Transformative potential of DRT in theory and practice 29

2. Grounding machinic ecosystems in life-experiences

Counter-actualization as the source of novelty and becoming 34
Interpenetration of actualization and counter-actualization 38
Possibilizing the virtual as experience ecosystems 41
Realizing the virtual as machinic life 48
Machinic life-experience ecosystems 55
Looking ahead 62

Part II
Dynamic relationality theory

3. Dynamic relationalities in an assemblage

Dynamic relationality #1: Within assemblages 70
Rhizomatics of machinic desire within assemblages 73
Categories and open sets as smooth manifolds and submanifolds 76
Vector fields animating the dynamics over smooth manifolds 79
Territorial escape and capture of machinic desire within assemblages 81
Deriving the plane of consistency and territories of an assemblage 83

4. Dynamic relationalities on strata

Dynamic relationality #2: Territorialization of strata between assemblages 89
Analyzing stratification across machinic and collective assemblages 93
Dynamic relationality #3: Lines of flight between assemblages across stacked strata 95
Analyzing functorial transformations across stacked strata 96
Transversality 98
Dynamic relationality of machinic desires and territories in MLXEs 103

5. Dynamic relationalities across (stacked) strata

Enriching the creative-experiencer with generative artificial intelligence 112

Dynamic relationality #4: Becoming through lines of flight   115
Dynamic relationality #5: Reterritorialization   119
Enriching the experience-verse with machinic generalized intelligence   123
Lines of flight from structures and becoming of identities   129

6. Dynamic relationalities of relational dynamics

Dynamic relationality #6: Differential transformation   138
Differential forms to measure the flow of natural transformations   140
Toward a new monadology of dynamic relationalities   143
Homotopies   150
Becoming of singularity in AGI as differential transformation   152

Part III
Organizations and creative transformation

7. Creative transformation along a molar/molecular stack

Dynamic relationality #7: Molar/molecular lines of segmentation between strata   159
Relating global transformations to local transformations   161
Transformations along stacked strata: Sheaf morphisms   166
Transversality as the pathway between global and local transformations   175

8. Diagrammatic logic of organizations in ecosystems

Category Theoretical interpretation of organizations and ecosystems   179
Enrollment and agencing in organizations and ecosystems   183
Dynamic relationality #8: Diagrammatics   185
Limit/colimit object of a system of categories   187
Higher-order categories to represent categories and functors   190
Organizations as convergence of diagrams and limit/colimit objects   192
Case: Diagrammatic logic of organizations in healthcare ecosystems   194

9. Organizational morphology and development

A category theoretical typology of organizational morphology   197
Structur*ed* corporations as limit object focused organizations with stable functor sets   201
Agenc*ed* corporations as limit object focused organizations with variable functor sets   202
Structur*ing* collectives as colimit object focused organizations with stable functor sets   204
Agenc*ing* collectives as colimit object focused organizations with variable functor sets   205
Toward a theory of organizational morphogenesis   207
Case: Morphology and morphogenesis in healthcare   212

10. Architectural transformation in an ecosystem

Extending sheaves to stacks for multilayered global to local analysis   216
Complex dynamics of order and scale in organizations   219
Dynamic relationality #9: Strategic architecturing as transformations of diagrams   228
Designing life-experience ecosystems via pragmatic semiotics of diagrams   236
Case: Healthcare application of strategic architecturing as transformation of diagrams   239

Part IV
Complex transformative emergence and evolution

11. Emergent transformation in a machinic life-experience ecosystem

Immanence within a machinic life-experience ecosystem   246

Monadic structure of differentiation into four domains 248
Case: Healthcare system and COVID-19 254
Differentiation of ecosystem sectors via double articulation 258
Case: Healthcare system and COVID-19 (Continued) 265

12. Evolutionary transformation in a machinic life-experience ecosystem

Grounding social-behavioral phenomena with gauge theory 269
From global immanence to well-being, wealth, empowerment, and welfare 274
MLXE evolution as differentiation from virtual realities to actual possibilities 279
Dynamic relationality #10: Immanence of emergence and evolution 287
From analysis and modeling to strategy development 289

Epilogue 293
Appendix—Category theory notational reference table 309
Appendix—Healthcare case: Details of category theory analysis 313
Appendix—Applying gauge theory to social—behavioral phenomena 323
References 341
Glossary of terms 351
Index 361

About the authors

Dr. Kerimcan Ozcan is an Associate Professor of Marketing at the School of Business and Global Innovation, Marywood University, USA. His expertise encompasses co-creation, interactive platforms, digitalization, strategy, and various facets of marketing such as branding, customer service, and industrial/B2B marketing. His current work emphasizes the fusion of human–AI interactions and their transformative impact on organizational and societal systems. This research highlights how the dynamic relationalities of machinic ecosystems and life-experiences coalesce into strategic organizational transformations, helping organizations navigate their evolving ecosystems. Forthcoming contributions extend the current work into real-world applications, with projects that bridge theoretical advancements with hands-on strategies for organizational growth, including strategic engagement with AI, co-innovation platforms, and ecosystem-based value creation. Dr. Ozcan's prior scholarly work includes *The Co-Creation Paradigm* co-authored with Dr. Ramaswamy, published by Stanford University Press, along with contributions to the *Journal of Marketing*, *Journal of Business Research*, *International Journal of Research in Marketing*, and *Harvard Business Review*. Previously, Dr. Ozcan taught at several other universities (including the University of Michigan and the International University of Japan) and held engineering roles in various industries. His research has garnered support from entities such as the Japanese Ministry of Education, Science, and Technology, and the University of Michigan Tauber Manufacturing Institute. Dr. Ozcan received his PhD in Marketing, an MA in Applied Economics from the University of Michigan (Ann Arbor), an MS in Management from the Georgia Institute of Technology, and a BS in Electrical and Electronics Engineering from Boğaziçi University.

Dr. Venkat Ramaswamy serves as a Professor of Marketing at the Ross School of Business, University of Michigan, Ann Arbor, USA, contributing to innovation, strategy, marketing, branding, IT, operations, and organizational behavior. His scholarly work and practical implementation of ideas have earned him global recognition. In 2004, Ramaswamy, alongside C.K. Prahalad, introduced the concept of Co-Creation in *The Future of Competition* (Harvard Business School Press) offering a new perspective on value generation through experiential environments and collaborative practices. This concept was further explored in influential articles in *Harvard Business Review* and *Sloan Management Review*, focusing on interactive value creation and innovation from an individual experience standpoint. From 2005 to 2010, Ramaswamy's research addressed the impact of digital and social media, technological convergence, and IT-enabled services on organizational engagement platforms, emphasizing the importance of using lived experiences to enhance interactions, as discussed in "Building the Co-creative Enterprise" (Harvard Business Review). In 2010, *The Power of Co-Creation* (Free Press) co-authored with F. Gouillart, examined co-creation platforms across various industry sectors, highlighting the transition from traditional product offerings to interactive, value-creating platforms. His 2014 book, *The Co-Creation Paradigm* (Stanford University Press) co-authored with K. Ozcan, advocated for a shift in individual and institutional engagement to foster a collaborative economy and society. Dr. Ramaswamy's ongoing research focuses on systemic transformations through digitalized societal ecosystems, aiming to enhance well-being, wealth, and welfare. His mentorship extends globally, guiding enterprises in adopting co-creative practices and building management capabilities. Dr. Ramaswamy holds a PhD in Marketing from the Wharton School of the University of Pennsylvania and a BTech in Mechanical Engineering from the Indian Institute of Technology.

Acknowledgments

Kerimcan Ozcan would like to extend his gratitude to Marywood University for a generous sabbatical, which provided the much-needed focus and time for the initial development of this project. Venkat Ramaswamy thanks the Ross School of Business for a flexible sabbatical, which allowed for ongoing creative collaboration on this project.

We greatly appreciate the early encouragement and support we received from Steve Hardman. We thank the following individuals who read and provided very helpful comments on parts of our manuscript: Jason Anderson, Mehmet Başer, Richard Gyrd-Jones, Oriol Iglesias, Sema Salur, Philip Sugai, and Julian Wichmann.

We are very grateful to our editor at Elsevier, Kathryn Eryılmaz, as well as Himani Dwivedi, Gomathi Sugumar, and the editorial team for their enthusiastic and timely guidance and support in publishing this work.

Finally, we would like to acknowledge scientists, scholars, and researchers, too numerous to name here, on whose shoulders this book rests—introducing Dynamic Relationality Theory as an organizing framework of how cointelligent modeling systems can be used across various disciplines and contexts. We see a coevolution of actioning systems with advances in quantum computing enabling unified gauge theoretic modeling of machinic life-experience ecosystems, such that organizations can more proactively engage in its creative transformation to amplify valuable sustainable well-being impacts and more effectively balancing health-wealth, education-empowerment, and mobility welfare in societal life. In researching and developing this book, we used generative AI both as a research tool in developing category theoretical and differential topological translations, computations, and inferences of our own conceptual models expressed in natural language and as an editorial tool to enhance the readability and language of the work for better scientific communication of the complex and wide-ranging ideas presented throughout. All the while, we have directly and independently engaged with all the references we cite in the book, even as we have supervised for semantic and factual accuracy in curating our ongoing research and analyses to the best of our abilities.

Thus, we hope this book demonstrates through its very example of research and development, a new kind of human—artificial intelligence (AI) cocreation entailing cointelligent reasoning. The purposeful and systematic involvement of generative AI in our research and development is not only integral to the conceptual and applied thesis of our book but also serves as an empirical demonstration of how cointelligence and human-AI cocreation can productively work in the real world, that is, when accompanied by formal modeling of objects and relations using the vocabulary and grammar of category theory. It is with a sincere conviction that AI can only morph into a machinic generalized intelligence successfully *if and when* both human and machine intelligences are proactively reasoning interactively and iteratively, using a framework of abstraction that is formal and rigorous, that we present this artifact to the scientific community.

Prologue

The digital revolution has ushered in an era of unprecedented interactivity. No longer are systems isolated; they are interconnected, driven by digitalized flows that redefine our understanding of networks.[1] Artificial Intelligence (AI), with its predictive prowess and learning capabilities, infuses these networks with intelligence, making them responsive and adaptive. This Human—Artificial Intelligence (HAI) interactivity is the catalyst for *creative transformation*, urging us to reimagine networks not as static constructs but as dynamic relational ecosystems.[2] Moreover, the digitalization of our world has blurred traditional industry boundaries, facilitating cross-sectoral engagement interactions. This dissolution of boundaries is not just confined to industries but extends to broader collaborations across government, civil society, and the private sector.[3] Such engagements underscore the shift toward societal impact systems, emphasizing the importance of collective intelligence in our interconnected world.

In this transformative landscape, the concept of "cocreative power" takes center stage. It is not just about individual innovation but about the symbiotic potential of combined intelligences, *cointelligent creation,* where Human Intelligence (HI) and AI converge in Machinic Generalized Intelligence (MGI).[4] This coalescence births "cocreativity," a paradigm where collaborative efforts between humans and AI drive creative transformation.[5] This transformation is characterized by the creative power of humans in coexistence with artificial intelligence, further amplified by cross-sectoral collaborations, fostering a richer array of insights and solutions, toward richer, more holistic *Life-Experiences*. As we delve deeper into this dynamic landscape, we encounter the concept of "living systems." These are not mere networks but vibrant ecosystems shaping our Life-Experiences.[6] Every interaction, every connection, is alive, pulsating with energy and information. In our rapidly evolving digitalized landscape, living systems emerge as more than just networks; they represent collective evolutions and transformations, mediated by technology. This intimate interplay between the human and the technical not only shapes our individual experiences but also paves the way for our collective becoming in an interconnected world. This shift is pivotal, for it underscores the need for a language that can encapsulate these dynamic life-experiences.

Building on the philosophy of Deleuze and Guattari, Dynamic Relationality Theory (DRT) presents a new framework for understanding the intricate systems of emergent life-experiences, especially within digitalized ecosystems of increasing *"phenomenotechnicity."*[7,8] This theory emphasizes the fluidity, contingency, and interconnectedness of relations, offering a fresh perspective on the dynamics of complex systems. Its core tenets are:

Relation (Ontological Core): Ontology delves into the nature of existence. Central to DRT is the concept of "Relation." It asserts that entities gain their essence not in isolation but through their interconnections with others. These interactional relations are not mere external links but are foundational to the entities' identities and very existence.[9]

-al- (Epistemological Core): Epistemology concerns the nature and scope of knowledge. The "-al-" in DRT underscores the evolving and generative nature of relations and thereby continuous shared knowledge creation. Relations are not static but are ceaselessly crafted and reshaped through interactional knowledge. This evolving nature is pivotal for grasping the emergent characteristics of systems.

-ity (Axiological Core): Axiology studies the nature of values. The "-ity" in DRT accentuates the primary status of relations, suggesting they are not mere derivatives but are founded in values. These values shape the valorization and evaluation of systemic impacts.

Dynamic (Pragmatic Core): Pragmatics focuses on the practical implications of concepts, knowing the world as inseparable from agency within it. DRT accentuates that relations are not just fluctuating but are intrinsically transformative to practices.[10] This dynamism is not just temporal change but involves the ceaseless creation, metamorphosis, and dissolution of relations founded in values that define the existence of entities.

Dynamic Relationality Theory (DRT), thus, presents a holistic "ethico-epistem-ontological" framework that perceives systems as fluid *"agencements"* or *agencial assemblages* of interconnected entities, acting upon (and being acted upon) relational phenomena.[11,12] They are rhizomatic arrangements of heterogeneous and self-subsistent components, with constitutional roles of content and expression, and "virtual" capacity to interact in variably emergent and recurrent processes of destabilization and stabilization. For ease of exposition, we will just use the term "assemblage" in the rest of this book. While existing frameworks like Assemblage Theory offer valuable insights, they often fall short in capturing the continuous dynamics inherent in living systems.[13] Within assemblages, entities are not static but continuously evolve, influenced by both internal and external dynamics.[14] This is where Category Theory comes into play, capturing this evolution through morphisms, representing the dynamic influences and interactions between entities.[15] Beyond individual assemblages, DRT emphasizes the intricate relationships between different assemblages, visualized through categories and their interconnections via "functors," a mathematical construct that maps objects and their relationships (morphisms) from one category to another while preserving their intrinsic structure and relational dynamics. The concept of morphisms provides the tools to understand the complex synergy of relationships of living systems and its ecosystem dynamics. This is where Differential Topology comes in, offering a lens to capture the continuous dynamics inherent in these morphisms, ensuring a comprehensive understanding of the transformational landscape.[16]

As we journey through the realms of DRT, we will encounter diverse terrains—from the foundational concepts of transformation to practical applications in sectors like healthcare. Here, the relational dynamics of ecosystems come alive, with AI-infused interactivity offering transformative insights. For instance, physicians can harness the power of DRT to understand the intricate web of factors influencing a patient's health, leading to personalized, holistic solutions.[17] This is not just theoretical; it's a profound shift with tangible, real-world implications.

DRT underscores the interplay between local (molecular) and global (molar) dynamics, revealing how overarching socio-cultural-political forces shape broader structures, while microtransformations capture nuances at the granular level. Central to DRT is the concept of "becoming" a continuous transformation that challenges established identities, visualized through natural transformations between functors.[18] The creative transformation of

"**Machinic Ecosystems**" grounded in "**Life-Experiences**" (and hence Machinic "**Life-Experience**" **Ecosystems or MLXEs for short**) stands as an expression of the limitless potential of entities, differentiating into subcategories that capture unique facets of identity, transformation, conceptual forces, and interconnectedness. This differentiation is a coordination of potentialities and actualities, guided by mathematical constructs like gauge transformations and Lagrangian dynamics.[19,20] MLXEs draw attention to an actualization of the virtual that is grounded in life-experience, while virtualizing potentialities from the actual, through MGI cocreativity.

DRT offers a toolkit for transformations, aiding strategic architecturing of MLXEs in organizational journeys. Whether it's harnessing AI-infused interactivity or leveraging the principles of cocreativity, DRT provides the tools to navigate the complexities of the modern enterprise landscape in a cointelligent world. DRT introduces a diagrammatic perspective, offering visual representations of complex ecosystems, where organizations navigate between current and aspirational states. Its architectural foundation lies in the category of "Platform" (entailing "*shared digitalized infrastructure*"), which can be public or private, open or closed, influencing the ecosystem's entities and their transformations.[21] As organizations evolve as ecosystemic assemblages themselves, they strategically architect their position within Techno-Natural Econo-Societal (or "NEST," as a mnemonic) assemblages, bringing together nature, technology, society, and economy, while balancing adaptability and stability, steering MLXEs toward envisioned goals.

In essence, the philosophy of Deleuze and Guattari, combined with category theory, offers a comprehensive understanding of dynamic relationality in a complex world of NEST-networked MLXEs. Assemblages, representing networks of diverse heterogeneous components, are at the heart of this understanding. These assemblages are driven by inherent desires and objectives in MLXEs, interacting in nonlinear ways, leading to various states or configurations of life-experiences in a universe of relational phenomena.[22] Transformations between these states offer insights into system reconfigurations, influenced by both external and internal factors. Each point in its topological space represents a unique state of the system. Transformations between these points capture dynamic shifts within MLXEs.

The theoretical framework presented integrates categories, each defining a topological space of assemblages, connected through functors and their natural transformations. These categories, when paired, represent molar or molecular strata, and are linked by functors of territorialization. The framework employs 'sheaf analysis' between these strata, which contextualizes local information within a broader global framework, and connects them via functors acting as "lines of flight" or pathways between the virtual and the actual, through which these systems influence and transform each other.[23] This approach aligns with Guattari's functorial domains, which however lacks the dynamism of DRT if it is only confined within the topological spaces of categories and their connecting functors. Differential topology offers a dynamic lens to view transformations within and between categories. Combining differential topology with category theory offers a comprehensive approach to view dynamic relationalities. This integration of differential topology provides a deeper, dynamic perspective on category theory.[24]

In sum, DRT amalgamates insights from diverse disciplines, offering a holistic understanding of the exponentially increasing phenomenotechnicity of digitalized ecosystems, grounded in life-experiences. By emphasizing the interconnected, evolving, foundational, and transformative aspects of entities and systems, DRT paves the way for innovative research trajectories. The integration of category theory and differential topology further deepens this understanding, providing rigorous tools to explore the world of relational dynamics of MLXEs and their cocreative transformation.[25]

Since the theory and methodology of DRT is in its early stages of development, and this book serves as a progress report rather than a definitive codification of it, just like all scientific theoretical endeavors described by Thomas Kuhn, it is open to revisions, retractions, and further developments. Therefore, we invite readers to visit www.drtbook.com to engage further with the content, provide feedback and suggestions in a spirit of collective and cocreative scientific enterprise, and check for any revisions or refinements to update the contents of this book as a living artifact. Furthermore, some of the more extensive details that could not be accommodated in a lengthy Appendix can be found in the same online resource.

In Part 1, we will dissect and weave together the multifarious facets of life-experiences, machinic ecosystems, and their interdependencies. Chapter 1 immerses us in a philosophical exploration, guided by Deleuze and Guattari, as it grapples with the complexities of Life-Experience within the Virtual Dimension. Here, the tangible intertwines with the intangible, as personal histories, emotional landscapes, and external stimuli converge with technology, fostering a reciprocity that shapes experiences. The Actual Dimension, delineated through Machinic Ecosystems, reveals the tangible realities of existence, emphasizing adaptability and foresight. This chapter intricately describes the transition from virtual potentialities to actualized realities, while also integrating relational dynamics with Guattari's four domains of existence. The discussion navigates through molar and molecular aspects of existence, concluding with a conceptual framework grounded in a unified gauge theory, underscoring its relevance across various sectors.

In Chapter 2, we introduce the concept of counter-actualization in creative transformation of Machinic Ecosystems, showcasing the bidirectional influence between the Actual and the Virtual through the lens of rhizomes and transformative events. Here, the interplay between technological capabilities and human engagement is emphasized, highlighting the transformative potential of human–machinic interactions. The chapter elucidates the strategic pathways for both conventional and modern tech organizations, advocating for a balanced integration of technological platforms with lived experiences and existential territories. The MLXE (Machinic Life-Experience Ecosystems) framework of relational dynamics in creative transformation of machinic ecosystems *grounded in life-experience* is unveiled, offering a comprehensive view that integrates various theories, while emphasizing the transformative impact of technology on interactive life-experiences.

Thus, Part 1 serves as a foundational exploration, detailing the nuances of creative transformation of machinic ecosystems, while setting a firm groundwork for the subsequent deep dive into Dynamic Relationality Theory (DRT).

Endnotes

1. Manovich's (2002) pioneering work on new media provided early insights into the interconnected, digitalized world.

2. For a recent exploration of human and machine collaboration, highlighting the adaptive capabilities of AI and its role in transforming industries and work, see Daugherty and Wilson (2018).
3. Benkler's (2006) study of networked information economy provides clues to interconnected systems and collaborative engagements across various sectors.
4. See Malone (2018) for his concept of "supermind" to emphasize the enhanced capabilities arising from the synergy of human and machine intelligence.
5. The ability for AI to understand and adapt within collaborative endeavors might ultimately depend on AI systems to operate with common sense (Brachman and Levesque, 2022).
6. To capture the energetic interconnections and transformations within ecosystems shaped by human—technical interplay, Bennett (2010) proposes the notion of "vibrant matter."
7. Deleuze and Guattari's work laid the groundwork for understanding relationality and interconnectedness through their exploration of complex systems and philosophical concepts (Deleuze 1990, 1994; Deleuze and Guattari 1987).
8. The concepts of "system—environment hybrids" (Hansen 2009) and "feed-forward" (Hansen, 2015) are helpful in understanding emergent life-experiences and knowledge creation, respectively, while the research stream on Gaian systems (Clarke, 2020) explores the interconnectedness of a multiplicity of dynamic ecosystems.
9. Here, we concur with the core tenet of Actor-Network Theory, which emphasizes the centrality of relations in the constitution of entities and the coconstruction of knowledge (Latour 2005, 2013).
10. Barad's exploration of matter and meaning through quantum physics resonates with DRT's pragmatic core, emphasizing the transformative nature of relations within interconnected systems (Barad, 2007).
11. An agencial assemblage acting as a system—environment hybrid entails interactional flows with other agencial assemblages (Ramaswamy and Ozcan, 2022), enacting interactional creation in MGI-infused cocreative transformation of dynamic relationalities. The term "agencial" emphasizes an agency view of creative transformation, in contrast to a conventional agential focus on just agents (i.e., actors) in the system (Ramaswamy and Ozcan, 2018b). Agencial assemblages call attention to a focus on interactive agency as defined from the relations among entities (Callon, 2021).
12. DRT can be seen as connecting the work of Deleuze (1994) and Guattari (2013) with the work of Barad (2007) via Simondon's individuation (2020), that is, from intraaction to individuation of entities engaged in an entanglement of matter and meaning-making.
13. For instance, DeLanda's "neo" assemblage theory (2016) pays less attention to Deleuze's desire (Buchanan, 2021) and incorporeal strata of living systems (Deleuze, 1990). In DRT, individuated entities as creative-experiencers imply Active Inference (Parr et al., 2022) on the one hand, but also a process of *experiencial* becoming, following the works of Heidegger (1996) and Merleau-Ponty (1962), emphasizing active creative experience that goes beyond conventional "experiential" notions of individuals as users of offerings. Experiencial agencial assemblages bring affective and creative experiences into the fold of a complex systems view of life (Capra and Luisi, 2014) and that of actor-network theory (Latour, 2005; Law, 2009).
14. We see assemblages as interactive system—environments, that is, acting as system—environment hybrids (Hansen, 2009) entailing interactional flows with other assemblages (Ramaswamy and Ozcan, 2022).
15. Rosen's (1991) challenging text, aptly titled *Life Itself*, intricately connects living systems' organizational nature and closure to efficient causation, employing category theory for comprehensive analysis of these complex dynamics. Regarding Category Theory: for an accessible introduction, see Roman (2017); for an intermediate treatment, see Leinster (2014); for a rigorous and standard reference, see Mac Lane (1998); for contemporary applications across a variety of disciplines, see Fong and Spivak (2019); and finally, for a less technical but philosophical application closest to our approach, see Gangle (2016).
16. Regarding Differential Topology, for an accessible introduction, see Dundas (2018); for a rigorous treatment of the topic, see Wall (2016); and for a less technical but philosophical application closest to our approach, see Sha (2013).
17. Recent emergence of patient-centric approaches in medicine provides an accessible context for studying relational dynamics (Veatch, 2009; Stewart et al., 2014).
18. The common thread running across Deleuze and Guattari's two-volume *Capitalism and Schizophrenia* project is the interplay between local (molecular) and global (molar) forces in society, providing a framework to understand transformations at every level (Deleuze and Guattari, 1987, 1977).

19. Gauge theory provides a mathematical framework for understanding the complex interplay of physical fields that permeate the universe, influencing the behavior of particles and governing the fundamental forces of nature. For a good introduction to gauge theory, see Baez and Muniain (1994); for deeper exposition of the techniques, see Naber (2011a, 2011b); and for a more general and superb treatment of mathematical physics, see Penrose (2004). Quantum concepts can be applied to social and behavioral phenomena, using "quantum-like" modeling to interpret complex dynamics (Haven and Khrennikov, 2013; Wendt, 2015). Gauge theory has recently been deployed to develop a general mechanics of self-organizing and complex adaptive systems with broader application potential in biological, cognitive, and social sciences (Sengupta et al., 2016; Ramstead et al., 2018; Ramstead et al., 2023). In what follows, we only use the most basic and necessary formalization of state definitions and transformation matrices, with the primary goal of developing intuition and qualitative predictions.

20. In physics, the Lagrangian L is a function that describes the dynamics of a system, typically as the difference between kinetic energy T and potential energy V, that is, $L=T-V$. For our societal system, we can define analogous "energies" based on Deleuze and Guattari's concepts. The latter's concept of "desiring production" involves flows and intensities, which can be seen as a form of psychic energy with roots in Freud's psychoanalytic theory. Before turning to id, ego, and superego as structuring the flows of libido, Freud had initially attempted to link psychological processes with neurological activity, but abandoned this approach due to the limitations of neuroscience at the time (Northoff, 2012). Today's neuroscientific ideas of free energy, prediction error, hierarchical inference, and precision modulation latent in Freud's thought (Carhart-Harris and Friston, 2010; Solms, 2020) can help further bridge the gap between Freudian psychoanalysis and poststructuralist philosophy. As for "social production," Tonkonoff (2017) documents how Deleuze and Guattari's emphasis on energy, flows, and intensities can be traced back to Gabriel Tarde's focus on the microlevel and his view of society as emerging from the aggregation of countless small interactions driven by imitations and innovations. Crucially, Deleuze and Guattari can be said to follow Freud and Tarde's early intuitions from 1890s.

21. We use 'platform' in the most generalized sense as a configured machinic assemblage that is not necessarily 'digital' (Ramaswamy and Ozcan, 2014). That said, with increasing phenomenotechnicity and digitalization of interactive experiences, platforms of digitalized interactivity are embedded in relational phenomena (Ramaswamy and Ozcan, 2018a).

22. Guattari (1995) delves into the intricacies of subjectivity and machinic interactions, elucidating how assemblages in ecosystems are driven by desires and objectives, leading to transformative states and insights.

23. A sheaf is a mathematical concept that provides a way to organize and relate local information within a global context. It offers a framework for understanding how local systems relate to and are influenced by global systems. Functors, when acting as "lines of flight," further establish connections between these strata. In Deleuzian terms, a line of flight represents a movement of deterritorialization and transformation that disrupts and destabilizes existing structures and systems, leading to the creation of new assemblages and territories. For an excellent coverage of sheaf theory we use in Chapter 7, see Rosiak (2022).

24. As we will discuss, DRT builds on Guattari's (2013) work on functorial domains, visualizing each category as a smooth manifold, with points symbolizing configurations of objects and morphisms. Vector fields on these manifolds depict the direction of transformations, while Lie derivatives measure changes in these vector fields, capturing symmetries and transformations. Differential forms gauge the intensity of transformations, and de Rham cohomology, a topological tool, assesses the structure and evolution of categories. For intercategory dynamics, functors act as maps between manifolds, enabling modeling of continuous transformations between categories. Concepts like transversality and the differential of functors provide insights into intersections and transformation rates between categories. Monads bundle manifold, vector field, and differential form, offering a comprehensive view of intracategory dynamics. Transversality and homotopy theory, focusing on nontangential intersections and continuous deformation, respectively, enrich the understanding of intercategory transformations.

25. It is critical to state at the outset that DRT is a set of concepts, statements, tools, and a methodology that allows one to develop theoretical diagrams, intuitions, and qualitative/directional predictions to capture the massive ontological and empirical complexities involved in most practical ecosystems that implicate humans, artifacts, institutions, and natural phenomena. In that respect, DRT can be compared to both Actor-Network Theory (ANT) in sociology and the neoclassical school in microeconomics. ANT captures the qualitative

detail and complexity of massively networked entities in different ontological registers via ethnographic methods. Neoclassical economics starts with a set of axioms and uses mathematics of calculus and topology to explain and predict microeconomic phenomena. Neither ANT nor neoclassical economics aim for precise empirical operationalization, although both can be productively used to develop rich intuitions and explanations for the phenomena under study. As a metamodeling framework, DRT can (and does) accommodate ANT and neoclassical insights, among others. While ANT might never, and in principle even aims to, achieve quantitative precision, the epistemological state of DRT at this point is no different from that of the marginalist economics in 1870s, developed by Jevons, Menger, and Walras, who placed utility at the center of their conceptual and methodological innovations, without truly measuring or quantifying it, and applied mathematics to derive results. It was only a century later, with the advent of experimental and behavioral economics, that many of the established theories built on top the marginalist foundation, that is, what was later to become the reigning economic paradigm of neoclassical synthesis, began to be empirically tested. Considering the exponential strides in big data analytics, machine learning, and the synergistic ecosystem of AI, IoT, and cloud technologies, however, DRT might receive empirical testing and validation at a significantly accelerated rate compared to marginalist economics.

Relational dynamics

CHAPTER 1

Theorizing relational dynamics of creative transformation

Dynamic Relationality Theory (DRT) is a holistic "ethico-epistem-ontological" framework that perceives systems as fluid assemblages of interconnected entities—implying a combination of ethical, epistemological, and ontological perspectives on phenomena. DRT prioritizes relations (ontology) that are understood through knowledge (epistemology) and are grounded in values (ethics). The "dynamic" prefix in DRT implies that relationality is not just about a change in time but involves the emergence and transformation of relations that define the very existence of entities. DRT is thus a theory of ongoing creative transformation of the world in its becoming.[1]

DRT advocates for a view of entities as defined by their dynamic relationality rather than viewing them as isolated units. It is not only fundamental to understanding the complexities of exponential change in our interconnected world but also provides novel perspectives on creative transformation of life-experiences in a computationally tech-intensive world.

Virtual and actual dimensions of a relational universe

To get started with our theorization of dynamic relationalities, we draw on the philosophical masterworks of two modern philosophers, Gilles Deleuze and Félix Guattari, and their conceptualizations of **virtual and actual dynamics**, central to DRT. Deleuze and Guattari, though frequent collaborators, had distinct philosophical projects. However, their works often resonate with and complement each other.

In Deleuze's book on "*Difference and Repetition*" (Deleuze, 1994), the **virtual** is a realm of multiplicities, potentialities, and singularities. It is not unreal but is a dimension of reality

---

[1] Guattari (2013, p. 52) explains the *transformational* approach thus: "For the conception of an unconscious founded on an economy of drive quantities and a dynamic of conflictual representations, I would thus substitute a *transformational modelling* such that, under certain conditions, Territories of the Self, Universes of alterity, Complexions of material Flows, machines of desire, semiotic, iconic Assemblages, Assemblages of intellection, etc. can engender one another. Also, it is no longer a matter here of sticking to the form of instances, but of acceding to the transmutations, the transductions of their substance." [italics ours]

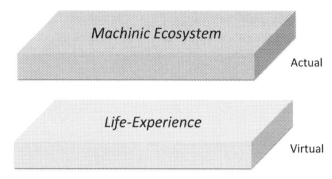

FIGURE 1.1 Virtual and actual dimensions.

that has not been actualized yet. The **actual**, on the other hand, is what has been realized from the virtual. It is the concrete, tangible manifestation of the virtual (see Fig. 1.1).

Life-experience as the virtual: The nature of life-experience is multidimensional, encompassing both the tangible and intangible elements of our existence, framed by realms of real potentialities and abstract possibilities.[2] Within these realms, one finds dynamic territories that are continually evolving, influenced by memories, desires, and external forces. These territories, an embodiment of "Territorial Spacetime," are the confluence of our immediate surroundings, personal histories, emotions, and external stimuli that shape our sense of self and existence. As entities traverse this spacetime, they are not just anchored in the physical realm but are also contextualized within a broader socio-behavioral ecosystem, making their journey multifaceted.[3]

Our engagement with technology forms a crucial axis in this ecosystem. Technologies present possibilities, creating a myriad of avenues for interactivity.[4] How we, as individuals, engage with technological possibilities leads to the creation of unique experiences and innovations. This dynamic interplay, grounded in the capacities and capabilities that technology offers, evolves into a dynamic of value creation, where both technology and the individual's capacities and capabilities reciprocally transform each other. This dynamic is further enriched by the inherent characteristics of technology and its interplay with individuals and their environment. Through a harmonious convergence, they coalesce, fostering individuality while also spawning collectives, thereby reflecting the intrinsic interconnectedness of life.

---

[2] Deleuze's entire output intricately can be seen as weaving a philosophy of life-experience. In one of his earliest works, Deleuze delves into multiplicities, difference, and becoming, drawing on Bergson's fluid temporality (Deleuze, 1988). Toward the end of his life, his exploration culminates in "Immanence: A Life," presenting life as a plane of interconnected intensities, foregrounding experience's transformative nature and ceaseless becoming (Deleuze, 2001).

[3] Tironi and Lisboa (2023) illustrate a real-world application of how AI in environmental governance shapes our understanding and interaction with our surroundings, thereby influencing the "Territorial Spacetime" within which individuals and societies operate.

[4] See Prahalad and Ramaswamy (2004) for an early exploration of how technology is fostering individual and collective experiences, where interaction with technological platforms leads to innovative, value-creating outcomes.

Parallel to these territories are incorporeal landscapes imbued with meaning, value, and possibilities. They guide our journey, acting as navigational compasses filled with aspirations and desires. These abstract realms hold the principles, goals, and values that sculpt our experiences. Living systems, in their quest to harmonize internal models with sensory data, actively seek information, thereby evolving their internal landscapes. The very act of sense-making, a drive to understand and give meaning to our experiences, delineates the intricate relationship between virtual and emergent experiences. This process not only establishes connections but also fosters the creation of engaging and meaningful encounters, blending materiality and immateriality.[5]

In synthesizing these elements, the concept of life-experience emerges as a rich canvas, weaving together the realms of real potentialities and abstract possibilities with tangible realities and technical possibilities. It captures the fluid dynamics between the real and the possible in the virtuality of life-experience anchored by our interactions and perceptions. The framework unravels the complexities of human experience, bridging the tangible territories of existence with the intangible universes of meaning, guiding our journey through the vast landscapes of life.

Machinic ecosystem as the actual: Machinic ecosystems represent a synthesis of the dynamic processes that shape our lived reality and the potential paths that these realities can take. In the heart of machinic ecosystems lies the interplay of concrete processes and forces that influence our existential territories. This realm is characterized by tangible currents—be it in the form of economic processes, media influences, or biological needs—that crisscross and mold our experiencial domain. The shaping of this space emerges from the synthesis of complex interactions among humans and nonhumans, nested within intricate social networks. Simultaneously, individuals and organizational entities adaptively structure their interactions with technological landscapes, laying the foundation for niche environments that cater to evolving demands and possibilities. As technologies change, so too does the manner, in which these entities engage with their environments, a reflection of the continuous co-evolution of practices and mechanisms in these systems.[6]

Yet, this intricate meshing of adaptation and structuration is not limited to the present moment. Machinic ecosystems also encompass the prospective trajectories that our existential territories might follow. These potentialities are not abstract fantasies; they are grounded in our current realm, representing the conceivable lines of transformation or evolution entities may pursue. Our ecosystem has myriad potentialities attached to each present state. These are the possible futures, the pathways that are anchored in our current reality yet pointing toward novel horizons. Collaboration emerges as a critical driver in this ecosystem, empowering entities to share knowledge and cocreate value.[7] This is facilitated by foundational

---

[5] Enzo Paci's critical social phenomenology reconciles lifeworld's dependence on environmental interaction with technology as a means to satisfy human needs in specific social contexts so as to highlight the critical role of technology in shaping human experiences and the evolving dynamics of human-environment-technology interactions (Gunderson, 2020).

[6] Through interconnected and adaptive co-evolution of entities and technologies, machinic ecosystems embody dynamic processes, diversity, and continuous transformations that shape existential territories, echoing Guattari's emphasis on machinic heterogenesis, multiplicity, and autopoietic transformation (Guattari, 1995).

[7] Tekic and Füller (2023) illustrate how AI fosters collaborative innovation, aligning with the role of technology in facilitating shared knowledge and cocreation within ecosystems.

infrastructures, systems, and networks that are essential for the ecosystem's functionality. Technological platforms, both as conceptual tools and spatial constructs, and with its increasing digitalization, play a pivotal role enabling the production, exchange, and consumption of offerings of value to individuals. Through these platforms, a plethora of services is rendered, each tailored to meet specific user demands, and each contributing to the grand ensemble of the machinic ecosystem.

In essence, machinic ecosystems capture the duality of our present actualities and the myriad potentialities they can birth.[8] By interweaving the tangible forces shaping our realities with the conceivable trajectories they can initiate, we arrive at a conceptual framework that foregrounds the importance of adaptability, collaboration, and foresight in navigating and shaping the machinic ecosystems we inhabit.

Actualization of the real and the possible

The transition from the virtual—a realm teeming with multiplicities of real potentialities and abstract possibilities—to the actual—the tangible manifestation of this realm—is a complex interplay of difference and repetition. The virtual, contrary to being an unreal domain, represents unactualized possibilities. During actualization, difference emerges inherently, as the virtual's differential nature does not translate into static realities. Instead, the actual materializes through divergence from its virtual origins. Complementing this, repetition, as conceptualized by Deleuze and expanded upon with Guattari, is not a mere recurrence but a dynamic instantiation process that births variations with each iteration. This repetitive process does not reproduce identities but fosters creative differentiation.[9] Therefore, actualization is not a static culmination but an ongoing "becoming"—a continuous evolution influenced by the virtual's interplay. This Deleuzian-Guattarian perspective posits reality as ever-evolving, shaped by the forces of difference and repetition.

The actualization process between the multifaceted nature of "life-experience" as the virtual and the intricate dynamics of the "machinic ecosystem" as the actual is an intricate dialectical movement. Within the virtual realm of life-experiences, individual, and collective consciousness navigate territorial spacetime, shaped by intangible potentialities such as memories, desires, and values. These potentialities offer a plethora of avenues for interactivity, particularly in their convergence with technology, where new horizons of experiences and innovations unfold. In contrast, the machinic ecosystem as the actual is the realized matrix, where tangible processes, from economic flows to technological engagements, intersect. This ecosystem is the concrete manifestation of the virtual potentialities, grounded in present actualities while also gesturing toward numerous prospective trajectories. It embodies the practical, everyday interplays among humans, technologies, and their environment.

---

[8] Dai and Hao (2018) explore transcending techno-utopianism and dystopianism, offering insight into the nuanced duality of technology's impact, aligning with the adaptability and potentialities in machinic ecosystems.

[9] Differentiation, in Deleuze's philosophy, refers to the process of actualization of the virtual, the process through which something new and unique emerges from the plane of consistency. Differenciation, on the other hand, refers to the process of specification and individuation of these differentiated entities in actual experience.

During the actualization process, the virtual's differential multiplicities do not simply transpose into static elements within the machinic ecosystem. Instead, they undergo a process of differentiation, molding the virtual's fluid potentialities into the tangible processes and structures of the actual. This differentiation is where the dynamics of difference play a role. The virtual, with its broad spectrum of experiences and aspirations, encounters the machinic ecosystem, resulting in divergence and variations that do not simply mirror the virtual but transform it. Concurrently, repetition emerges as an inherent mechanism in this actualization. Each instantiation in the machinic ecosystem, from technological innovations to socio-economic structures, does not merely reproduce the virtual's potentialities. Instead, through the process of creative repetition, each iteration engenders unique variations that enrich the ecosystem, rendering it vibrant and ever evolving.

Guattari's understanding of the virtual and the actual is similar to Deleuze's. The virtual is a realm of potentialities, while the actual is the realized, concrete manifestation of these potentialities. However, for Guattari, in his book, *"Schizoanalytic Cartographies"* the **real** is the tangible present, whereas the **possible** is a projection of the real into potential future states.[10] It is what could potentially come into existence based on the current state of the real.

Virtual-actual (engagement) in the real: *Engagements* as lines of flight between the virtual "life-experience" and the actual "machinic ecosystem" exemplify the actualization of the virtual in the real. It encapsulates both potentialities, elements yet to be realized, and tangible processes shaping our current reality. As previously discussed, this duality comprises personal nuances, memories, emotions, and environments, as well as collective tangible forces.

Central to our existence is a multidimensional space, not just based on physical location but inclusive of unique conditions and histories. Within this space, technology offers action possibilities, leading to individual engagements. As these engagements progress, behaviors evolve, yielding innovative outcomes and amplifying the ecosystem's potentialities. The properties of technology play a pivotal role in this evolution, intertwining personal experiences, technological tools, and the broader environment.[11] Through continuous interaction, both individual identities and collective definitions emerge, optimizing technological interfaces based on needs. Simultaneously, this realm is influenced by tangible processes and forces that dictate its trajectory. These forces, comprising both material and symbolic elements, are dynamic, embodying the socio-behavioral intricacies of our world. The interplay between individuals, organizations, and technological entities leads to an adaptive

---

[10] Note how Guattari (2013, p. 69) explains his choice to employ four ontological realms (after using two in his collaboration with Deleuze, and three in his early solo writings): "Axiomatics with two terms (of the Being/Nothingness type) necessarily result in a 'depotentialized' representation and an inaccessible 'grund', whilst dialectics with three terms lead to pyramidal, arborescent determinations … It is only with 3 + n entities that one can establish: (1) a trans-entitarian (matricial) generativity, without any essential priority of one essence over another (without the infrastructure - superstructure relation, for example); (2) a principle of self-affirmation, auto-retroaction, a self-transcending (Jean-Pierre Dupuy) or auto-poietic (Francisco Varela) foundation."

[11] Schubert (2015) enhances this understanding by illustrating how technology, particularly through computer simulations, actively shapes societal futures. His exploration of simulations as epistemic tools that transform indeterminate situations into determinate ones reflects the role of technology in evolving personal experiences within multidimensional spaces, emphasizing its impact on societal dynamics and the emergence of new practices and thinking patterns.

structuration of the ecosystem. Here, the coshaping and mutual evolution of technological and organizational practices is evident.[12]

Understanding this ecosystem necessitates a grasp of the transformative journeys connecting the potential and the tangible. This transition, indicative of the intricate movement from the virtual to the actual, exemplifies the materialization of potentialities into concrete experiences. Moreover, as entities navigate this ecosystem, they not only experience the interplay between personal and collective domains but also contribute to processes that reconfigure the ecosystem's structures in its coevolutionary becoming.

Virtual-actual (enactment) of the possible: *Enactments* through lines of flight at the nexus of our internal meanings, values, and the potential pathways of our existential territories provide a lens through which we can understand the dialectical movement from the virtual to the actual.

Drawing from our earlier discussion, the intangible landscapes of meaning and value play pivotal roles in guiding the transformation processes within these experiences. Analogous to forces in gauge theory, these intangible elements influence the dynamics within the entities. Principally, living systems, in their inherent nature, are predisposed to refine their internal models by consistently assimilating information. Technology, especially digital forms, amplifies this predisposition, reshaping human perceptions and experiences. As these perceptions evolve, there's a clear inclination to synthesize the virtual, thereby molding connections between potential and actual emergent experiences. The culmination of this synthesis is a balance between tangible and intangible components, enriching overall quality of experiences.[13]

Concurrently, the tangible trajectories our existential territories might adopt are anchored in our present realities. These trajectories, based on our earlier delineation of machinic ecosystems, signify potential lines of transformation emanating from our current existential states. One can conceptualize such trajectories as mathematical entities, providing values across a comprehensive landscape. Key to deciphering these trajectories are collaboration, co-design, and open innovation. The collective drive to innovate together amplifies the generation of shared value, leading to open innovation strategies that transcend traditional boundaries. Here, shared digitalized infrastructure acts as a keystone, enabling value-oriented platforms that meet user demands.

Crucially, the transition from abstract potentials to tangible trajectories illustrates the dialectical play between difference and repetition. Actualizing the virtual involves iterative

---

[12] Tim Ingold's ideas on lines, meshworks, and continuous co-evolution, highlighting dynamic interactions and growth within interconnected spaces, and his emphasis on materiality and participatory engagement provides a nuanced framework for understanding the life ecosystem's adaptive structuration and the materialization of potentialities into lived experience (Ingold, 2015).

[13] Hansen's concept of "feed-forward" contributes to our understanding of machinic experiences, emphasizing proactive engagement with futures, and illustrating how digital technology's preemptive nature dynamically synthesizes the virtual and the actual. This interaction, mediated by new forms of perception, enhances experience quality, while iterative repetition ensures continuous adaptation and divergence in existential trajectories (Hansen, 2015).

repetition, each cycle producing new divergences and unexpected outcomes. This process of iteration and divergence becomes evident as collaborative efforts bring forth new ideas and practices, echoing our earlier point about the dynamic nature of the virtual's actualization.[14]

From virtual realities to actual possibilities

Deleuze's exploration of the dynamic relation between the virtual and the actual in *"Difference and Repetition"* can be seen as corresponding to and complementing Guattari's more complex interplay of the Virtual, Actual, Real, and Possible in *"Schizoanalytic Cartographies"*. The concepts of Repetition in Deleuze's work and the Real and Possible in Guattari's work, while not directly analogous, can be understood in relation to one another when considering the broader philosophical frameworks of both thinkers.

Deleuze distinguishes between the virtual, a dimension of unmanifested potentialities, and the actual, the tangible realization of these potentials. This actualization is not mere replication but a dynamic process generating differences with each iteration, termed repetition. Each repetition introduces variations, ensuring that the virtual, when actualized, brings forth something new and unexpected. In "Difference and Repetition" Deleuze posits that repetition is not about the recurrence of the same but rather about the instantiation of difference. Repetition is not a matter of identical recurrences but of producing variations and differences with each iteration. Repetition, for Deleuze, is bound up with the process of actualization. As the virtual is actualized, it does not simply produce identical copies but generates differences. This is because the virtual is a realm of differential multiplicities, and its actualization inherently involves divergence and differentiation. Thus, repetition is a creative force that brings forth the new, the different, and the unexpected.[15]

While Deleuze focuses on the process of actualization, Guattari delves deeper, exploring potential future trajectories, transitioning from the real (the present) to the possible (potential futures). This shift is not predetermined but is a dynamic journey that can lead to unforeseen outcomes. The **real** is the tangible, present reality for Guattari. It is what exists here and now. The **possible**, on the other hand, represents potential future trajectories based on the current state of the real. It is not just a mirror image of the real but a projection of what could come into being. The transition from the real to the possible involves processes of transformation, change, and becoming. It is not a deterministic or predictable transition but one that can take multiple paths and produce unexpected outcomes.

Deleuze's concept of repetition, generating difference, mirrors Guattari's transition from the Real to the Possible, broadening the horizon of potentialities. Deleuze's concept of repetition, as a force that produces difference and variation, can be related to Guattari's transition

---

[14] Briggle and Mitcham's (2009) discussion of "embedding and networking" in technosocieties and how the interplay of technology and society brings about an "experiential gap" underscores the complexity of actualizing virtual potentialities into tangible outcomes.

[15] Batayeh et al. (2018) exemplify Deleuze's concept of actualization, demonstrating how innovation in healthcare does not merely replicate existing models but evolves them. Their analysis of socially responsible healthcare practices reveals a "cycle of actualization," where each iteration leads to new, patient-centered approaches, resonating with Deleuze's emphasis on creative and divergent actualization.

from the real to the possible. Just as repetition brings forth the new and the different, the movement from the real to the possible involves the emergence of new potentials and trajectories. Repetition, in Deleuze's sense, can be seen as one of the mechanisms through which the real gives rise to the Possible in Guattari's framework. As reality undergoes repetitions, it does not just reproduce itself identically but generates new possibilities and potentialities. In other words, the creative and differential nature of repetition in Deleuze's philosophy can be seen as resonating with the dynamic interplay between the real and the possible in Guattari's thought. Repetition introduces variations and differences into the real, thereby expanding the realm of the possible.

In summary, while Deleuze's concept of repetition and Guattari's distinction between the real and the possible are distinct philosophical ideas, they can be understood in relation to one another when considering their shared emphasis on difference, transformation, and the emergence of the new. Repetition, as a force of difference, plays a role in the unfolding of the real into its manifold possibilities.[16]

DRT and the four domains of the creative plane of immanence

The Creative Plane of Immanence is the foundational or primordial category, a virtual multiplicity teeming with potential. The objects in this category are the entities of the world, and the morphisms are the transformations that these entities can undergo. This category is characterized by its potential for transformation and change, reflecting DRT's emphasis on the fluidity and dynamism of existence.[17] It is described as a plane of consistency or immanence, which is a virtual field of potentialities from which actual entities emerge through processes of differen*t*iation and differen*c*iation. DRT builds on Guattari's fourfold schema to understand the intricate dynamics shaping reality and subjectivity in this Creative Plane of Immanence (**P**), characterized along the dimensions of virtual, actual, real, and possible, as shown in Fig. 1.2.[18]

---

[16] Blockchain technology, through its iterative cryptographic processes and game-theoretical incentives, epitomizes Deleuze's concept of repetition as a force of difference. Each transaction within a blockchain introduces variations, fostering a shift from traditional trust to a new form of procedural confidence, thus unfolding new possibilities in trust dynamics (De Filippi et al., 2020).

[17] Guattari (2013, p. 52) describes this "psychophysics" as: "'Prior' to the establishing of a matter and extension that can be located in the energetico-spatio-temporal dimensions of the physical world, it will begin with transformations that establish themselves 'straddling' the most heterogeneous of domains conceivable. It will presuppose diverse modalities of 'transversality' between: (1) Flows of matter and energy; (2) the abstract machinic Phyla that preside over objective laws and changes; (3) existential Territories, considered from the angle of their self-enjoyment (their 'for itself') and, finally, (4) incorporeal Universes, which escape from the energetic, legal, evolutionary and existential coordinates of the three preceding domains."

[18] Guattari denotes existential territories, incorporeal universes, machinic phyla, and energetic-signaletic flows as T, U, Φ, and F, respectively. We use "L," "X," "M," and "E," respectively to bring attention to emergent life-experiences from machinic ecosystems in a creative plane of immanence.

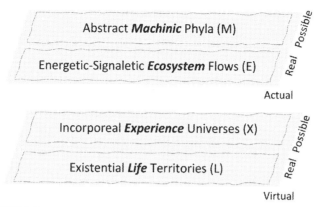

FIGURE 1.2 DRT and the four domains of the creative plane of immanence.

Existential **Life** Territories (**L**), as personal, subjective realms, constantly evolving due to memories, desires, and external influences, represent the *Virtual Real*, the unmanifested potentials of existence.[19] Energetic-Signaletic **Ecosystem** Flows (**E**), as the tangible currents, both material and semiotic, that shape our reality, symbolize the *Actual Real*, the processes that mold our existential territories in stakeholder ecosystems.[20] Incorporeal **Experience** Universes (**X**), as abstract realms that influence our territories of subjective life-experiences, signify the *Virtual Possible*, the intangible forces guiding our aspirations following Deleuze's Logic of Sense.[21] Abstract **Machinic** Phyla (**M**) symbolize concrete evolutionary paths, representing potential phenomenotechnicity futures based on our current reality, the *Actual Possible*.[22] The dynamics between these dimensions can be understood using Deleuze's concepts. For **L** to **E**, subjective life-experiences (**L**) actualize into tangible ecosystemic flows (**E**),

---

[19] Misra and Stokols (2012) illustrate how personal and subjective realms are shaped within hybrid settings combining physical and digital elements, where individual identity and socio-emotional needs are continually evolving in response to a blend of real and virtual influences.

[20] To see how tangible currents (re-)shape social and material landscapes of stakeholder ecosystems in the context of technological system integration, please see Mulder and Kaijser (2014).

[21] The intangible forces in **X**, shaping and guiding our aspirations and subjective experiences, constitute the systemic and functional aspects of innovation systems (Reale, 2019).

[22] Sony and Naik (2020) show how technological advancements in Industry 4.0, i.e., potential future paths, require comprehensive integration—vertical, horizontal, and end-to-end—considering the socio-technical impact on people, infrastructure, and processes, i.e., Actual Possible.

I. Relational dynamics

similar to the Virtual becoming Actual.[23] For **E** to **X**, material and semiotic flows (**E**) influence our understanding, leading to the creation of abstract meanings and potentials (**X**).[24] This mirrors Deleuze's idea of repetition producing variations. For **X** to **M**, abstract potentials (**X**) actualize into potential evolutionary trajectories (**M**).[25] For **M** to **L**, potential futures (**M**) interact with and reshape existential territories (**L**), introducing new potentialities.[26]

In practical terms, consider a groundbreaking medical device for breast cancer screening using AI. The introduction of this device, the continuous refinement of its algorithms, and the feedback from patients and medical professionals can be understood using Guattari's four dimensions and the dynamics of differentiation and repetition. The device's introduction, its adoption, and its impact on medical practices and patient experiences represent the intricate convergence between the virtual and the actual, the real and the possible.

- The groundbreaking introduction of the AI-driven medical device for breast cancer screening signifies a direct emergence of difference, revolutionizing traditional mammography and initiating new cycles of repetition, transitioning the landscape of healthcare from Existential Life Territories (**L**) to the dynamic realm of Energetic-Signaletic Ecosystem Flows (**E**), reshaping patient experiences with novel screening contexts.
- Behind the technological forefront, dedicated researchers tirelessly refine the AI algorithms, embodying hidden processes of differentiation that redirect the trajectory of medical diagnostics, transitioning the diagnostic paradigm from Energetic-Signaletic Ecosystem Flows (**E**) to the more abstract and nuanced realm of Incorporeal Experience Universes (**X**), marking a qualitative transformation in the repetition of diagnostic procedures.
- As the AI's prowess in early cancer detection becomes evident, a palpable tension arises between the established norm of biannual screenings and the potential for less frequent AI-driven screenings, emphasizing the distinction between the virtual promises of technology and its tangible benefits, transitioning the screening frequency debate from Incorporeal Experience Universes (**X**) to the realm of Abstract Machinic Phyla (**M**), reshaping the temporal rhythm of patient screenings.
- Echoing the voices of patients, feedback indicates a profound desire for a more holistic treatment approach, signifying a transformative shift in perspective and inducing an inversion in differentiation, steering the medical community from potential futures (**M**) back to the intimate realm of lived experiences and subjective territories (**L**), challenging the traditional treatment paradigms.

---

[23] Huang (2023) study of how local sociotechnical affordances, driven by community dialog and public ownership, influenced the development of global internet infrastructure, illustrates how localized, subjective experiences (**L**) can actualize into broader, tangible realities (**E**).

[24] The process of material and semiotic flows influencing the creation of abstract meanings and potentials in stakeholder ecosystems can be seen in the way collaborative dynamics crucially mediate the capabilities of AI systems and their impacts (de Neufville and Baum, 2021).

[25] Umbrello et al.'s (2023) work on Anticipatory Ethics for Emerging Technologies (ATE) is concerned with the transformation of abstract potentials into concrete evolutionary paths.

[26] For example, in regional innovation systems, potential future innovations dynamically interact with and reshape current existential territories (Doloreux and Parto, 2005).

- As the AI-driven device gains traction across hospitals, there's a noticeable change in the character and quality of radiologist training protocols, indicating a profound shift in perspective that modulates hidden processes of differentiation, transitioning the professional development landscape from Existential Life Territories (**L**) to the dynamic realm of Energetic-Signaletic Ecosystem Flows (**E**), emphasizing the evolving nature of medical training in the age of AI.
- The increasing standardization of the AI-driven device across medical institutions marks a temporal shift in repetition, influencing the distinction and tension between hospitals rapidly adopting the technology and those lagging behind, transitioning the healthcare landscape from the tangible, material flows of technology in stakeholder ecosystems (**E**) to the more abstract realms of reputation and potential (**X**), highlighting the disparities in healthcare technology adoption.
- The integration of the AI-driven device into radiology marks a transition from abstract potentials (**X**) to concrete professional development trajectories (**M**), with a direct emergence of difference as radiologists evolve from passive AI output recipients to active diagnostic participants, and a shift in repetition as training modules transition from annual to quarterly formats.
- The AI device's trajectory in radiology leads to an increased emphasis on patient feedback postdiagnosis, signifying a phase shift in repetition away from potential futures (**M**) and a direct emergence of difference as medical institutions prioritize subjective patient life-experiences (**L**), refining AI algorithms for a more holistic diagnosis and treatment approach.

As this stylized example demonstrates, the interplay between virtual realities and actual possibilities, as explored by Deleuze and Guattari, offers profound insights into the dynamic evolution of ecosystems.

Virtual and actual dynamics: Deterritorialization and discursivity

Molar versus **molecular**: These terms indicate the nature of organization and process within systems—whether they are more structured and uniform (molar) or more fluid and diverse (molecular). This distinction is not strictly about scale (like global/local) or the specific form of organization (like strata/assemblage).[27]

---

[27] Global structures or phenomena can be likened to Molar systems due to their large-scale, structured nature. Conversely, local phenomena, with their specific and potentially more fluid characteristics, might be seen as akin to Molecular processes. However, one can also encounter global molecular processes (like worldwide cultural shifts) or local molar structures (like a local institution or tradition), as we will see in Chapter 7. Similarly, molar systems often manifest as Strata—rigid and structured layers, while molecular processes are frequently akin to assemblages—dynamic and interconnected networks. Still, the two are not entirely synonymous; molecular processes can disrupt strata, and molar systems can incorporate elements of assemblages.

In the philosophy of Deleuze and Guattari, "molar" refers to structured, stable systems or processes characterized by organization, uniformity, and homogeneity. These encompass recognizable, established structures, norms, and practices prevalent in societal institutions and traditional beliefs.

- Within breast cancer care, molar encompasses set guidelines for mammogram screenings, standard protocols for staging the cancer, and established treatment regimens including chemotherapy and mastectomies—reflecting structured, uniform approaches across prevention, diagnosis, and treatment.

In contrast, "molecular" denotes dynamic, fluid processes that exist within or alongside more structured systems. Emphasizing diversity, variability, and transformative potential, the molecular encompasses deviations, disruptions, and minor variations that challenge or diversify established norms.

- Molecular in breast cancer care embodies fluidity: personalized genetic-based prevention strategies, adaptive algorithms improving diagnostic accuracy, and precision medicine tailoring treatments to a tumor's unique genetic profile. This adaptive approach continuously evolves, responding to individual needs and advancing research.

While actual states can be both molar (structured like established institutions) or molecular (dynamic like emergent social movements), and virtual potentials can influence or disrupt both molar (like major societal shifts) and molecular systems (like subtle transformation of a minor cultural trend), most molar systems do however correlate with the "actual" because they are tangible, established, and present, while molecular processes, being more fluid and dynamic, align with the "virtual" in their potential for transformation.[28] Existential Territories (**L**) and Incorporeal Universes (**X**) align with molecular characteristics due to their emphasis on fluidity and dynamic potentials. **L** represents personal, evolving subjective life realms influenced by memories and desires, embodying the transformative nature of molecular processes. **X**, being abstract and intangible, represents underlying potentials and latent forces, further echoing molecular traits. Conversely, Energetic-Signaletic Flows (**E**) and Abstract Machinic Phyla (**M**) resonate with molar attributes, emphasizing structured

---

[28] The interaction between these concepts is not just epistemologically apparent but also ontologically relevant, due to the inherent properties and interrelationships within the molar/molecular and actual/virtual dichotomies.
- Perception and Categorization: Our cognitive frameworks often bind the perceptible and established (molar) with the currently existent (actual) since they are readily identifiable and stable. Micro-changes or subtleties (molecular) correlate with potential futures (virtual) because they represent undercurrents that might give rise to new realities.
- Nature of Existence and Transformation Pathways: Molar entities possess fixed attributes, giving them an "actual" existence, while molecular entities, being dynamic, align with the "virtual." Molar structures and molecular processes are intrinsically linked; stability arises from dynamism and vice-versa. This interplay often mirrors the transformation between actual states and their latent potentials.

systems and tangible trajectories. **E** denotes tangible currents that mold reality, embodying the molar's concrete, structured nature. **M**, though addressing potential futures, is grounded in current realities, suggesting a fixed, structured trajectory typical of molar frameworks. In summary, while **L** and **X** highlight dynamic possibilities and transformations, **E** and **M** underscore stability and structured pathways.[29]

Structure of the creative plane of immanence: The creative plane of immanence, a vast field of interconnected potentials, is intricately structured through *cutouts*, *complexions*, *constellations*, and *rhizomes* (Fig. 1.3). These structures give form to Existential Life Territories (**L**), regulate Energetic-Signaletic Ecosystem Flows (**E**), conceptualize Incorporeal Experience Universes (**X**), and depict dynamic networks within Abstract Machinic Phyla (**M**), respectively, providing a holistic framework to understand life's diverse nuances and interconnected complexities. Together, they encapsulate the essence of life's potential, ecosystemic flows, experiential realms, and intricate machinic networks, providing a comprehensive framework for understanding their vast complexities and inherent potentialities.

Cutout, as a specific configuration that delineates Existential Life Territories (**L**), represents a distinct snapshot of untapped possibilities within life's vast landscape. These snapshots or territories define the structure of life's experiential fabric, carving out specific zones of potential experiences. Each cutout carries unique latent possibilities that, when engaged, reveal diverse outcomes, reshaping and enriching our understanding of life.[30] The inherent boundaries within cutouts allow us to discern between different experiences, much like defining territories on a map or open sets in a topological space.[31] They encapsulate specific moments or

---

[29] The concepts of molar and molecular manifestations are pivotal in understanding the underlying assemblages within various categories. An "individual," in a molecular context, represents a singular, dynamic entity, encapsulating personal experiences, memories, and desires. As a "stakeholder", this same entity manifests in a molar context, where it is recognized as a social actor, and becomes part of a structured, more homogenized group or system, with collective identity. A "network", in a molecular sense, is a fluid, ever-changing collection of connections and relationships, comprising narratives and discourses created by individuals to make sense of their experiences with technology. In a broader, more structured perspective of similar relational dynamics, an "ecosystem" encompasses comprehensive systems, policies, and organisms evolving together. The "creative-experiencer", in a molecular context, embodies the potential for innovation and latent creativity in each entity. In a molar context, it is channeled through "coinnovation", where the collective potential of multiple entities is harnessed to foster an environment conducive to collaborative and transformative innovation. The "experience-verse" in its molecular form represents the diverse, constantly evolving, individualized experiences and interactions within the digital ecosystem. In its molar manifestation as a "platform," these individual experiences coalesce into the technological and policy infrastructure that shapes and guides the collective experience. The transition from molecular to molar reflects a shift in perspective or analysis level rather than positing different entities or a literal transformation of one entity into another. It's a conceptual shift from focusing on the individual, dynamic interactions (molecular) to observing the organized, systemic patterns and structures that emerge from these interactions (molar).

[30] Tania et al.'s (2022) use of a multicriteria approach to assess innovation in different territories of Spain highlights the complexity and uniqueness of each territory, reflecting the diverse potentialities and outcomes inherent in each "cutout."

[31] In a topological space, the key concept is that of "nearness" or "closeness" without necessarily having a precise notion of distance. This aligns well with the idea of Existential Life Territories (**L**), which are about subjective experiences and personal identities, concepts that are inherently fuzzy and not easily quantifiable.

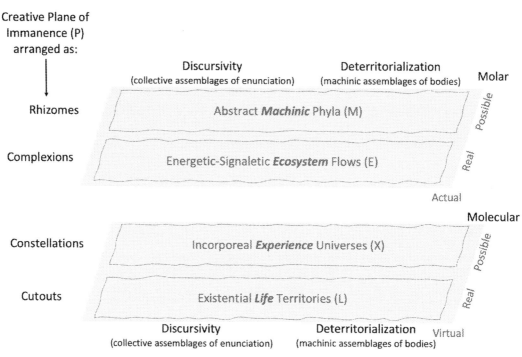

FIGURE 1.3 Further aspects of the creative plane of immanence (P).

durations, preserving them as discrete segments of potentiality. In essence, cutouts are the foundational building blocks that capture, contain, and present life's myriad potentials awaiting exploration and realization.

- Consider the myriad of experiences, from initial screenings to posttreatment care in breast cancer prevention, diagnosis, and treatment. Within the expansive realm of breast cancer care, each significant stage or experience can be seen as a cutout. For instance, a woman entering a clinic for her annual mammogram is one such snapshot. Similarly, a subsequent biopsy, a consultation about the results, the decision to undergo surgery or chemotherapy, and posttreatment follow-ups each represent individual cutouts. These cutouts capture specific moments and their accompanying interactions—be it between the patient and the physician, the devices used, or the data interpreted. Collectively, these cutouts map out the multifaceted territory of experiences in the broader journey of breast cancer care.

Complexion represents the intricate interplay and evolution of factors shaping the present state of Energetic-Signaletic Ecosystem Flows (**E**). It captures the tangible forces, processes, and interactions within boundaries that define a system's current configuration.[32] While rooted in the now, complexion acknowledges the system's history and potential futures. Bridging potentialities with reality, complexion is pivotal in translating possibilities into tangible outcomes. Like regions in a manifold that display uniform behavior yet form part of a larger, more complex structure, complexion captures the nuanced flows that, while continuous, are bound by certain constraints.[33] These constraints, whether physical or situational, guide the rhythm and direction of the flows, ensuring ecosystems remain both adaptive and coherent in their evolution.

- Consider the dynamic environment of a healthcare facility where various elements (human and nonhuman) interact in real time. The interactions between physicians, nurses, patients, devices, and data form the complexion of breast cancer care. For example, when a patient's mammogram shows an anomaly, the physician, with the help of data from past screenings, other medical records, and the input from technicians, determines the next steps. This interplay, informed by past cases (history) and cutting-edge medical practices (potential futures), reflects the intricate complexion of diagnosis and decision-making.

Constellation represents an assemblage of abstract ideas that converge to form significant patterns, encompassing vast durations and weaving the intricate web of Incorporeal Experience Universes (**X**). Though intangible, these interconnected notions shape and guide transformations across various domains.[34] Acting as a foundational blueprint, constellations determine potential trajectories of events based on internal patterns and relationships.[35] While setting the stage for specific outcomes, constellations remain adaptable, ensuring that transformations stay dynamic and attuned to emerging insights.

- Consider the amalgamation of knowledge, insights, and patterns in breast cancer care. Envision the diverse range of data and knowledge about breast cancer—symptom recognition, risk factors, genetic predispositions, to the effectiveness of various treatments—as distinct nodes of information. When these nodes are interlinked based on their inherent relationships and patterns, they form a "constellation." This interconnected map provides physicians a comprehensive understanding, guiding their decisions in patient care. As new knowledge emerges, this constellation can adapt,

---

[32] Luna-Ochoa et al. (2016) illustrate how technological agglomerations emerge from a blend of endogenous and exogenous factors, akin to the "complexion" in energetic-signaletic ecosystem flows. Their analysis of Mexico's aerospace and nanotechnology clusters underscores the interplay of local and global dynamics in shaping such ecosystems, mirroring the intricate interactions within "complexion."

[33] A smooth manifold is a space that is locally similar to Euclidean space, and smooth functions on the manifold can represent smooth transformations or flows. This aligns with the idea of Energetic-Signaletic Ecosystem Flows (**E**), which are about processes of change and transformation.

[34] Networks of personal and emotional communication established through digital technologies, or phatic systems, serve as modern constellations of interaction, that is, symbolic tokens in the context of precise, communal information exchange (Wang and Tucker, 2016).

[35] Much like a constellation, a vector field assigns a vector to each point in a space, which can be seen as representing the "force" or "direction" at that point. This aligns with the idea of Incorporeal Experience Universes (**X**), which are about abstract concepts and ideas that guide and shape our actions.

integrating new nodes and altering connections, ensuring that the medical approach is always informed by the latest insights.

Rhizome is a structure epitomizing decentralized, nonlinear systems, where each component can freely connect with another, which reflects the endless possibilities of the Abstract and Concrete Phyla of Machines (**M**), always in a state of flux and ceaselessly adapting to encountered intricacies. Retaining their unique attributes, each element collaborates in this network, adjusting to complexities without needing structural redefinition, granting adaptability and robustness. With no obligatory sequences, elements in a rhizome have the liberty for myriad interactions, ensuring system flexibility.[36] Emphasizing continuous change, the rhizomatic system evolves and adapts, avoiding rigid categorizations and embracing ever-shifting configurations. Much like intertwined spaces in a fiber bundle, rhizomes thrive without hierarchical constraints and navigate diverse temporalities.[37]

- Consider the interwoven and interconnected nature of various components of breast cancer care. The healthcare ecosystem for breast cancer care can be visualized as a rhizomatic structure. A patient diagnosed with breast cancer does not follow a linear journey. She might consult multiple physicians, undergo various tests administered by different technicians, and use several devices for monitoring and treatment, all while being part of support groups, rehabilitation programs, and more. The data generated flows in multiple directions - from the devices to physicians, from nurses to medical databases, and so forth. This nonlinear, decentralized network ensures adaptability and comprehensive care, much like a rhizome navigating the complexities of its environment.

Deterritorialization and discursivity: Navigating the realms of Existential Life Territories (**L**), Energetic-Signaletic Ecosystem Flows (**E**), Incorporeal Experience Universes (**X**), and Abstract Machinic Phyla (**M**), one encounters the dynamic interplay of deterritorialization and discursivity, intricately structured by cutouts, complexions, constellations, and rhizomes, reshaping our understanding of identity, meaning, and existence. Deterritorialization is the disruption of fixed identities, emphasizing constant change and fluidity in entities. This dislodging often leads to reterritorialization, where entities find new structures or "territories." Discursivity extends beyond mere words, encompassing semiotic systems, bodily expressions, and material flows. Meaning arises not solely from structured linguistic systems but also nondiscursive practices, highlighting the intricate interactions across diverse "plateaus" or assemblages.

Deterritorialization, in the philosophy of Deleuze and Guattari, refers to the process of dislodging entities from fixed organizational, spatial, or conceptual territories,

---

[36] Watanabe et al. (2017) elucidate the resilience of Uber's ICT-driven business model, embodying a rhizomatic structure in its decentralized, adaptive approach. Uber's global expansion, underpinned by co-evolutionary acclimatization and the ability to harness diverse regional characteristics, mirrors the flexible, nonlinear interactions and continuous adaptation characteristic of rhizomatic systems.

[37] A fiber bundle is a space that is locally a product of two spaces, but globally may have a more complex structure. The base space of the bundle can represent the Existential Life Territories (**L**), while the fibers represent the different Phyla associated with each point in the territories. This aligns with the idea of Abstract Machinic Phyla (**M**), which are about complex systems and structures that emerge from simpler components.

emphasizing fluidity and constant transformation.[38] Rejecting static identities, they argue that entities are always in flux, undergoing processes of deterritorialization and subsequent reterritorialization into new configurations. Central to this idea is the concept of the **"Machinic Assemblage of Bodies" (MAB)**. Entities are visualized not as isolated beings, but as assemblages—dynamic combinations of components that connect, disconnect, and reconnect in varying configurations. These assemblages are "machinic" in their interconnected and productive nature, always in states of desire and continual production. In essence, for Deleuze and Guattari, deterritorialization denotes the inherent changeability of things, and machinic assemblages represent the interconnected networks that entities form, perpetually evolving in response to various forces. Together, these concepts challenge traditional views of identity and existence, positing a world marked by continuous change and complex interrelations.[39]

For Deleuze and Guattari, discursivity encompasses more than just formal language; it integrates a vast array of semiotic systems, bodily expressions, and material flows.[40] It's not just about conveying information but about the production and intersection of meanings across various "plateaus" or assemblages. This complex interplay leads to their notion of the **"Collective Assemblage of Enunciation" (CAE)**. This concept posits that expressions, whether spoken or otherwise, are not singular acts. Instead, they emerge from a network of heterogeneous components: bodies, technologies, social systems, and more. Enunciations are deeply collective, drawing from shared histories, linguistic codes, and societal norms. They are not merely about producing statements but also about crafting and situating subjectivities within these multifaceted assemblages. Hence, when individuals express or communicate, they engage in a vast, interconnected web of meanings, histories, and systems. Deleuze and Guattari's reconceptualization underscores the dynamic, interconnected nature of meaning production, emphasizing it as a collective and assemblage-driven endeavor.[41]

The concept of double articulation emerges within the interplay of discursivity and deterritorialization. Firstly, discursivity extends beyond mere language to encompass a plethora of semiotic systems and material flows. Deterritorialization then disrupts these established systems, paving the way for new configurations and connections. Within this, CAE reflects the idea that expressions are born from an intricate web of bodies, technologies, histories, and societal norms. It is a collective endeavor, wherein meanings are not just conveyed but are produced and intertwined across various "plateaus." In parallel, MAB emphasizes entities

---

[38] Deleuze and Guattari (1987), especially in Chapter 11, delve into deterritorialization in the context of "Machinic Assemblages of Bodies" (MAB) and territoriality.

[39] Akin et al. (2021) illustrate deterritorialization in the sharing economy, whereby users of the Norwegian platform "Nabohjelp," seen as a fluid, transformative assemblage, engage in unexpected ways, disrupting and reshaping traditional community structures and identities.

[40] Deleuze and Guattari (1987), especially in Chapters 4 and 5, explore how language and semiotics are not isolated entities but part of broader assemblages, that is, communication extends beyond mere language, encompassing a network of varied influences and expressions, thus reinforcing the interconnected nature of discourse as a collective phenomenon.

[41] Adamsone-Fiskovica's (2015) study on public framing of science in "technoscientific futures" illustrates how public discourse on science is not just about information dissemination but about forming complex interplays of meanings, where lay narratives in science, encompassing a variety of semiotic systems and societal influences, contribute to the collective assemblage of enunciation.

as dynamic networks, always in a state of desire, production, and evolution. Double articulation thus signifies the dual-layered process: the initial layer where heterogeneous components form assemblages and a subsequent layer where these assemblages are expressed, leading to the production of meaning. This intermingling between deterritorialization and reterritorialization, between forming assemblages and enunciating them, underscores the fluid, interconnected nature of existence and meaning-making. For Deleuze and Guattari, meaning is not static but continuously crafted through complex interrelations and transformations.

By exploring the double articulation of strata of discursivity and deterritorialization within Existential Life Territories (**L**), Energetic-Signaletic Ecosystem Flows (**E**), Incorporeal Experience Universes (**X**), and Abstract Machinic Phyla (**M**), structured by cutouts, complexions, constellations, and rhizomes, we gain a holistic understanding of the nuanced processes of life, meaning-making, and transformation, as we will show next.

Differentiation as double articulation of strata via assemblages

Deleuze's process of differentiation, characterized by a "double articulation" of machinic assemblages of bodies (MABs) and collective assemblages of enunciation (CAEs), can be seen as the articulation of the four domains: Existential Life Territories (**L**), Energetic-Signaletic Ecosystem Flows (**E**), Incorporeal Experience Universes (**X**), and Abstract Machinic Phyla (**M**). Each domain is a unique combination of these assemblages, reflecting the specific dynamics and processes associated with that domain.

The Plane of Pure Immanence (**P**) is the overarching category encompassing all entities and their potential transformations. The objects in **P** are the abstract entities (MAB-CAE nexuses) of the world, dynamic processes of transformation.[42] The morphisms in **P** are the potential transformations these entities can undergo, including changes in understanding, representation, and articulation, reflecting the double articulation of substance and form of content and expression. The fracturing of **P** as a plane of consistency into the four subcategories (**L**, **E**, **X**, **M**) is a process of differentiation, where the virtual multiplicity of **P** differentiates into distinct domains. This process is a creative emergence, guided by the double articulation of substance and form of content and expression, shaping the way these entities are formed and understood (see Fig. 1.4).

---

[42] None of the "objects" in the categories are "actual entities". Moreover, any entity is a "section" across some of these objects spread across multiple categories. In a way, we are showing processes that are related to each other in particular ways. So, an entity as an assemblage of assemblages would be that entity's component processes implicated in various categories' processes (as objects) related to each other via morphisms, and thereby related to other diverse entities. Thus, we can show how gross entities affect and are affected by other gross entities, in ways in which their constitutive machinic processes (desire in Deleuze and Guattari's "Anti-Oedipus," and lines of segmentation/flight in their "A Thousand Plateaus") themselves circulate and interact with other similar machinic processes across the MLXE's four domains. In Part Three, we pull together these machinic processes/desires/lines (objects related by morphisms within categories and then related across them through functors) in organizations through diagrams of limit/colimit objects. We will also see there that the distinction of Organized Body versus Body without Organs applies equally to a physical entity and an organization of entities.

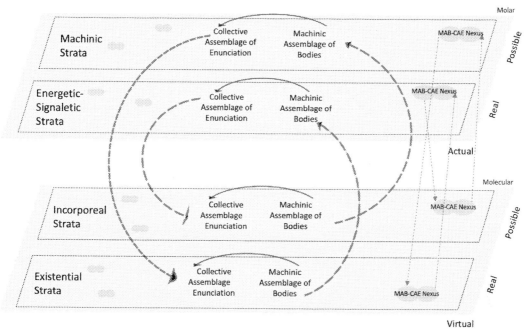

FIGURE 1.4 DRT and double articulation of strata.

As personal realms of subjective life-experiences, **L** is where individual identities and histories are constituted. Double articulation, by interlacing the collective assemblage of enunciation and machinic assemblage of bodies, continuously molds and refashions these territories. Through cutouts, these territories demarcate specific zones of potential experiences, continuously redefined by discursive practices and deterritorializing influences. The constant ebb and flow of individual subjectivities, influenced by societal norms, personal desires, and external forces, create a dynamic patchwork of evolving territories. In the Existential Life Territories (**L**), the double articulation shapes the formation of entities and their relationships. The substance of content could be the physical and emotional experiences of individuals, while the form of content could be the specific ways these experiences are structured. The substance of expression could be the personal narratives and discourses that individuals create to make sense of their experiences, and the form of expression could be the specific ways these narratives are articulated. The MABs here could be the entities themselves, and the CAEs could be the narratives and discourses they create.[43]

The tangible currents, **E**, comprise the material and semiotic forces actively shaping our reality. Double articulation operates within these flows by forming assemblages and articulating them. These assemblages, both collective and machinic, are enmeshed in intricate webs of

---

[43] Ahn (2016) shows how actor-networks, exemplified in Brain-Machine Interfaces, reflect a dynamic patchwork of interactions and adaptability, paralleling how identities and experiences are continuously molded and redefined through "double articulation" in Existential Life Territories.

meaning. The complexion of **E** is vital in translating the latent potentials of these assemblages into tangible realities. As flows intersect, deterritorialize, and reterritorialize, they constantly redefine the complexion of ecosystems, maintaining a delicate balance between adaptation and coherence. In the Energetic-Signaletic Ecosystem Flows (**E**), the double articulation shapes the transformations that these entities undergo and the paths they follow. The substance of content could be the actual processes, while the form of content could be the specific ways these processes are structured. The substance of expression could be the policies, guidelines, and protocols that guide these processes, and the form of expression could be the specific ways these policies and guidelines are articulated. The MABs here could be the entities involved in these processes, and the CAEs could be the policies, guidelines, and protocols that guide them.[44]

Within the abstract realms of **X**, double articulation casts its influence by weaving together abstract ideas with the tangible forces of the world. Constellations serve as convergence points for these ideas, allowing them to form significant patterns that guide transformations across domains. As individuals articulate their subjectivities through collective assemblages, they also navigate the vast, interconnected networks of abstract notions present in **X**. These constellations, while being influential blueprints, remain adaptable to the continuous deterritorializing and reterritorializing forces. In the Incorporeal Experience Universes (**X**), the double articulation shapes the way these abstract concepts and ideas guide the transformations. The substance of content could be the actual principles, goals, and values, while the form of content could be the specific ways these principles, goals, and values are prioritized and balanced. The substance of expression could be the public discourses, debates, and narratives that articulate these principles, goals, and values, and the form of expression could be the specific ways these discourses, debates, and narratives are conducted. The MABs here could be the entities who articulate these principles, goals, and values, and the CAEs could be the public discourses, debates, and narratives that shape them.[45]

The infinite possibilities within **M** echo the fluidity and adaptability inherent in rhizomatic structures. Double articulation in this domain emphasizes the interconnectedness of entities, as they form dynamic assemblages that defy rigid categorizations. The decentralized nature of rhizomes, coupled with the collective and machinic assemblages, ensures that the entities within **M** remain in a state of perpetual flux. Deterritorialization in this realm translates to a ceaseless adaptation of configurations, allowing for expansive evolutionary paths. In the Abstract Machinic Phyla (**M**), the double articulation shapes the way these entities interact and the connections they form. The substance of content could be the actual structures and organizations, while the form of content could be the specific ways these structures and organizations are arranged and coordinated. The substance of

---

[44] Bagis et al. (2022) demonstrate how the interaction of individual mindsets, social structures, and organizational processes, through the interplay of material and semiotic forces, shape the development of innovation capabilities in the automotive industry.

[45] Kim et al.'s (2023) study on sociotechnical challenges in computer vision discusses the interplay of abstract concepts (like algorithmic accuracy) with tangible forces (like sensor data), providing a pertinent example of double articulation in Incorporeal Experience Universes.

expression could be the administrative policies, procedures, and protocols that structure and organize the system, and the form of expression could be the specific ways these policies, procedures, and protocols are articulated. The MABs here could be the entities that structure and organize the system, and the CAEs could be the policies, procedures, and protocols that guide them.[46]

In this framework, **P** is the overarching category that encompasses all entities and their potential transformations. **L**, **E**, **X**, and **M** are subcategories of **P** that focus on specific aspects of these entities and their transformations. The objects and morphisms in **L**, **E**, **X**, and **M** are derived from the objects and morphisms in **P**, reflecting the specific focus of each subcategory. In each category, the double articulation process involves the interplay between the substance and form of content (the objects and their transformations) and the substance and form of expression (the understanding, representation, and articulation of these objects and transformations) within each subcategory. This interplay can be represented mathematically as a functor from the category to itself, which maps each object to its image under the double articulation process and each morphism to its image under the double articulation process.

The functors between the different strata and assemblages represent lines of flight, movements of deterritorialization, and transformation that disrupt and destabilize existing structures and systems, leading to the creation of new assemblages and territories. These functors capture the dynamic interplay between the existential, economic, symbolic, and technical dimensions of reality, and the constant flux and transformation between different levels and dimensions of reality. The sequence captures the complex interplay between individual experiences, societal structures, cultural narratives, and technologies.[47]

Through the process of double articulation, the MAB and CAE interact and influence each other, shaping the trajectories of entities through the Existential Life Territories, the dynamics of the Energetic-Signaletic Ecosystem Flows, the guiding principles of the Incorporeal Experience Universes, and the connections within the Abstract Machinic Phyla.

Interpreting functorial domains through gauge theory

In the comprehensive framework of terrestrial existence, the Creative Plane of Immanence (**P**) is defined as the all-encompassing domain, integrating all categories including humans,

---

[46] Human-machine alliances in hybrid intelligent systems, that embody the fluidity and adaptability of rhizomatic structures in the abstract machinic phyla, illustrate the dynamic interplay and coevolution of human creativity and machine logic (Ostheimer et al., 2021).

[47] Geels and Kemp (2007) typology of change processes in socio-technical systems—reproduction, transformation, and transition—can be used to illuminate the mechanisms underlying the transformative interactions between **P**'s subcategories (**L**, **E**, **X**, **M**) in DRT.

nonhuman organisms, artifacts, technologies, and their prospective evolutions. This conceptualization is analogous to the total space in gauge theory, a representation of the entire physical system being examined, inclusive of every conceivable state of the fields.[48]

Within this expansive realm, the Existential Life Territories (**L**) emerge as distinct configurations or cutouts, which articulate specific states or conditions of entities sourced from **P**. These territories, delineated by cutouts, furnish the requisite "background" or context, facilitating the existence and transformational trajectories of entities. Each point within these territories is tantamount to a state or condition of an entity located within the territorial life spacetime (**L**-spacetime).[49]

The dynamics within this space are governed by the Energetic-Signaletic Ecosystem Flows (**E**). These flows, marked by their complexions, are indicative of the states of a system; they embody dynamic processes occurring within the entities of **P**. Manifesting at every nexus within the existential territories, the characteristics of these flows—defined by their respective complexions—dictate the transformative pathways experienced by entities. This is resonant with fields in gauge theory, symbolic of the fluctuating physical quantities extending across Space-time.[50]

Introducing an added dimension to this schema is the Abstract Machinic Phyla (**M**). This structure, reflective of the rhizomatic nature of connections, establishes its base within the Existential Life Territories (**L**). Attached to every distinct point in this foundational space is a fiber, symbolizing the internal states of the entity existing at that particular juncture. The entities, in this context, can be understood as sections of this fiber bundle—mathematical representations that allocate a specific value in the fiber for each corresponding point in the base space. This conceptual arrangement is reminiscent of the underpinnings of gauge theory, wherein the fibers demarcate the internal "degrees of freedom" or unique "states" that particles may assume at every specific point within space-time.[51]

---

[48] In social theory, fields are structured spaces of power struggle (Bourdieu, 1977) or strategic action (Fligstein and McAdam, 2012), where agents, shaped by their resources and dispositions, interact, compete, and strive for resources or capital. Social fields can be seen as analogous to gauge fields, structuring interactions and influencing agent behavior. Agents' resources and habitus are akin to particle properties, determining their actions within fields. Transformations in social fields, driven by shifts in power or challenges to the status quo, can be related to gauge transformations, preserving certain symmetries while allowing change. By using concepts from gauge theory to model social interactions, power dynamics, and field transformations, we can formalize a rigorous approach to analyzing social phenomena, enhancing our understanding of social structures and their influence on individual and group behavior. See Part IV and the Appendix for a discussion of a Gauge Theory interpretation of social-behavioral phenomena.

[49] The dynamic interplay and transformation within Existential Life Territories (**L**), where innovations reshape the background and conditions of entities, can be seen in Raven and Verbong's (2009) study of how boundary-crossing innovations in socio-technical systems alter regime relationships, from isolated to symbiotic.

[50] Goulet's (2021) research on socio-technical transitions, particularly in the context of agricultural bio-inputs in Brazil, underscores the significance of coherent interactions between various elements (technologies, users, organizations) in dictating transformative pathways within a system, analogous to fluctuating fields in gauge theory.

[51] Applying STS theories to the unique challenges in the South, Furlong (2014) analysis of infrastructure stability and change in varied contexts highlights the importance of recognizing diverse conditions and coexistence among systems and understanding of the nuanced, context-specific "states" that entities may assume in different systems, analogous to the "degrees of freedom" in gauge theory.

To delineate the interactions and evolutionary dynamics of these entities, the framework propounds the concept of Incorporeal Experience Universes (**X**). These universes, mapped out as constellations, stand as assemblages of guiding principles, objectives, and intrinsic values. These constellations, by virtue of their internal configurations, exert influence over transformations within entities stemming from **P**. They bear similarity to the operative forces within gauge theory, signifying the impacts or resultant effects that fields exert on particles.[52]

Hence, **P** is the overarching total space that encompasses all entities and their potential transformations (see Fig. 1.5). **L**, **E**, **X**, and **M** are subspaces of **P** that focus on specific aspects of these entities and their transformations. The entities in **L**, **E**, **X**, and **M** can be seen as "sections" of **P**, and the transformations in **L**, **E**, **X**, and **M** can be seen as "connections" on **P**. This mirrors the structure of gauge theory, where the total space is a fiber bundle over the base space, with the fibers representing the internal states of the fields at each point in space time. The fields themselves are represented as sections of this fiber bundle, and the forces are represented as the curvature of a connection on the fiber bundle.

In the context of gauge theory, the food sector can be seen as correlating with the base space of Existential Life Territories (**L**). This is where the conditions or states of the system and the entities within the territorial life spacetime (L-spacetime) are defined, much like the spacetime in gauge theory. However, these states are not static; they are influenced by the dynamics of the other domains. Health as Energetic-Signaletic Ecosystem Flows (**E**) can be seen as the gauge fields that act on the base space of food. The health conditions of populations, influenced by their diet, can be seen as the values these fields take at each point in the base space. Changes in health conditions can indirectly lead to transformations in the food sector as well, such as shifts in dietary preferences leading to changes in food practices. Education as Incorporeal Experience Universes (**X**) can be seen as the forces that act on the fields. The knowledge and understanding about health (and its relationalities to food), cognized through sense-making act as forces that influence the state of the health field. This influence can be seen in how learning about nutrition and health can lead to changes in dietary habits and wellness routines, which in turn can influence food and health practices. Mobility as Abstract Machinic Phyla (**M**) can be seen as the internal degrees of freedom of the fields. The ability to move and connect, physically and virtually, can change the internal states of the health field. For instance, the spread of information about health and nutrition can lead to changes in dietary habits and without directly causing observable transformations in the health sector. However, these changes in the health field can indirectly influence the food sector by altering demand for certain types of food, which can lead to changes in farming practices and food practices.

While ecosystems are related to other ecosystems, in the rest of this book, we will focus primarily on the "Health" ecosystem for the sake of exposition of the various theoretical concepts of DRT. However, in Part Four, which concludes the book, we will discuss the importance and implications of understanding the continuous dynamics of interrelatedness of ecosystems, especially from a policy perspective. There, we will also discuss how we might

---

[52] In their study on mass production's historical evolution, Kanger and Sillak (2020) illustrate how meta-regimes, that is, aggregated rules (akin to guiding principles in **X**), guided transformations across socio-technical systems, resonating with DRT's view of influential constellations in shaping entity transformations.

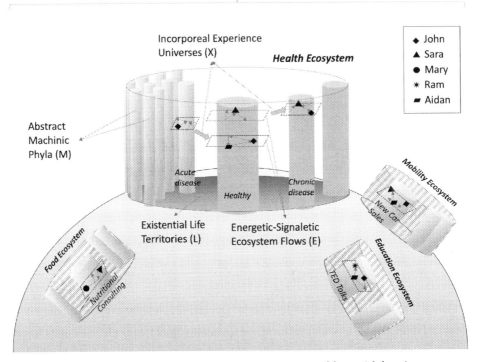

FIGURE 1.5 DRT and gauge theoretic interpretation of functorial domains.

more generally correlate food, health, education, and mobility with **L**, **E**, **X**, and **M**, respectively, which also aligns with Guattari's cartographies.

Building upon the profound interplay of health with other ecosystems, health's relationship with food, education, and mobility is intricate and multifaceted. The health of populations, conceptualized as Energetic-Signaletic Flows (**E**), is intrinsically intertwined with the state of the food sector, the Existential Territories (**L**). As the base for health fields, food sector dynamics shape dietary habits, nutrition, and, consequently, overall health. Furthermore, the Incorporeal Universes (**X**) of education disseminate critical knowledge about health, thereby influencing societal behaviors, preferences, and decisions regarding health. The paradigm shifts in health practices, driven by new knowledge, can realign dietary priorities, emphasizing nutrition-rich crops and sustainable farming. Simultaneously, the Abstract Machinic Phyla (**M**) embodied in mobility, with its profound capacity to connect and inform, alters the internal dynamics of health without necessarily changing observable health outcomes. For instance, health apps might amplify health awareness, thereby shifting dietary choices, influencing food practices, and reshaping the learning landscape. In essence, the health ecosystem, while primarily representing the energetic-signaletic aspects of life, also resonates with existential territories, incorporeal universes, and abstract machinic attributes, underscoring the intricate interconnectedness of these realms in the overall scheme of existence.

Seeing health ecosystems through DRT

In the grand tableau of New York City's healthcare landscape, visualize the Healthcare ecosystem as a vast, multidimensional Creative Plane of Immanence (**P**), which serves as an overarching domain that seamlessly integrates all entities: hospitals, clinics, healthcare professionals, patients, insurance providers, and even nonhuman elements like medical technologies and electronic health records. Just as the total space in gauge theory represents the entirety of a physical system, inclusive of every conceivable state of the fields, **P** encompasses all elements, their potential interactions, and prospective evolutions within the healthcare continuum.[53]

The Existential Life Territories (**L**) within this domain emerge as distinct configurations or cutouts, mapping the healthcare environment's nuanced layers. Each point within this L-spacetime represents a specific health condition or situation at a certain moment in time and location. It's here that the essence of life in healthcare manifests, with conditions, situations, and locations evolving over time. For instance, two points representing the same health condition at two consecutive moments, at the same location within the New York City healthcare ecosystem, would be in close proximity within this L-spacetime. These territories, carved out by the very nature of their delineation, establish the requisite "background" or context, facilitating the existence and transformative pathways of healthcare entities.

The dynamic heartbeat of this environment is represented by the Energetic-Signaletic Ecosystem Flows (**E**), the vital processes that embody the state of the healthcare system. Envision these flows, marked by their intricate complexions, as the daily activities like patient consultations, surgeries, and medical tests. These flows, symbolic of the fields in gauge theory, dictate the transformative pathways experienced by entities. As they crisscross the healthcare landscape, they represent various healthcare pathways a patient might traverse, shaping the trajectories of patients through this complex terrain.

Adding depth to this intricate weave are the Machinic Assemblages of Bodies (MAB) and Collective Assemblages of Enunciation (CAE). The MAB encompasses tangible aspects like the physical bodies of patients and healthcare providers, medical equipment, and healthcare facilities. Simultaneously, the CAE symbolizes the intangible components: Conversations, policies, public health campaigns, medical research, and the critical dialog between patients and their caregivers. Both interact and influence one another, reflecting the rhizomatic nature of connections where each component can interlink and adapt without structural redefinition.

Dive deeper, and you will encounter the Abstract Machinic Phyla (**M**). Visualize it as a 3D representation within the territorial life spacetime (L-spacetime). Each point here captures a specific MAB-CAE nexus—a unique convergence of health conditions, moments in time, physical locations, and the tangible and intangible interactions therein. This phylum, reminiscent of the fiber bundles in gauge theory, elaborates on the internal dynamics of health, reflecting the myriad possible interactions and transformations a patient might undergo.

Lastly, hovering above this intricate landscape are the Incorporeal Experience Universes (**X**), visualized as guiding vectors within this system. These universes, formed as constellations of guiding principles, influence the very fabric of healthcare in New York City. Whether

---

[53] Please see Chapter 11 for an extended discussion of Plane of Immanence, Existential Territories, Energetic-Signaletic Flows, Incorporeal Universes, and Abstract Machinic Phyla.

it is a commitment to public health or a focus on patient-centered care, these vectors, akin to forces in gauge theory, shape the transformations of all entities within the system.

Furthermore, it is pivotal to recognize health's profound interplay with other ecosystems. The health of New York's populations is inherently tied with its food sector. As the health fields, or Energetic-Signaletic Flows (**E**) of the health ecosystem, evolve, they are influenced by the dynamic shifts in the Existential Territories (**L**) of food, altering dietary habits and overall health. The Incorporeal Universes (**X**) of learning amplify awareness, molding societal behaviors, and choices about health. For example, as knowledge advances, there's a tilt toward nutrition-rich crops and sustainable farming. Concurrently, the Abstract Machinic Phyla (**M**) of mobility, through avenues like health apps, magnifies health awareness, subtly shifting the landscape of healthcare without necessarily altering observable outcomes. Together, these intricate interconnections underscore the vast complexities and inherent potentialities of New York City's healthcare system within the broader aggregate of terrestrial existence.

In the lively expanse of NYC, envision a vast, intricate tangle that weaves the tales of its denizens—John, Sara, Aidan, and a few other players. At its core, it is a grand portrayal of the Creative Plane of Immanence (**P**). This expansive canvas encapsulates everything: the bustling hospitals, quiet clinics, diligent professionals like John, and resilient patients like Sara.

John, a devoted doctor, not only navigates the complexities of healthcare but also finds himself intertwined with Energetic-Signaletic Ecosystem Flows (**E**). In our city's story, these flows symbolize the bustling activities of the healthcare world—the patient consultations, surgeries, and diagnostic tests. Each of John's interactions, every diagnosis, and treatment recommendation, is an embodiment of the dynamic pathways shaped by these flows.

Enter Mary, a dietitian in the heart of the city. Her role highlights health's profound engagement with the city's food ecosystem. As John's patient, Sara, transitions through various health states, her dietary needs evolve. Mary, leveraging the Existential Life Territories (**L**) of food, crafts dietary plans that influence Sara's overall wellbeing. The meals Sara consumes, the dietary choices she makes, are all influenced by the dynamic shifts in these territories. In our city, a bite of an apple or a sip of green tea is not just nourishment; it is an interplay of health and food, intricately connected in the grand weave of life.

Another layer of this tangle is woven into the city's schools, universities, and digital platforms. Ram, scientist, author, and TED Talks presenter, embodies Incorporeal Experience Universes (**X**) of education. His powerful talks on the significance of reducing carbon footprints and the interconnectedness of personal and environmental health have reached a wide audience, including John. As John delves into Ram's talks, he realizes the profound impact of environmental decisions on public health. Knowledge is not just power; in our city, it is a catalyst for both personal and planetary well-being, influencing health choices and lifestyles. Through Ram's influential teachings, John begins to integrate a broader understanding of wellness into his medical practice, considering not just the individual, but also the world they inhabit.

Meanwhile, Sara, apart from her health journey and her role in the mobility landscape as a car sales agent, is also an avid user of health apps. These digital tools, part of the Abstract Machinic Phyla (**M**) of mobility, subtly transform the healthcare landscape. With each step counted and calorie logged, Sara's interactions with these apps shape her health consciousness, representing the rhizomatic connections of physical activity, health awareness, and technology.

Aidan, the versatile digital assistant, adapts and transforms across these interconnected terrains. In John's clinic, it aids in managing patient records; with Sara, it logs dietary intakes and tracks physical activity; and in Ram's classroom, it assists in educational presentations on health and nutrition. Aidan embodies the machinic assemblages of bodies (MAB) and the collective assemblages of enunciation (CAE), intertwining tangible actions with intangible data flows.

One day, as John visits Sara's car dealership, guided by her sales expertise and assisted by Aidan's reminders, he is also influenced by Ram's educational campaigns on reducing carbon footprints. Perhaps John contemplates an electric vehicle, showcasing the complex interplay of health, mobility, and education.

In this city, entities like John, Sara, Mary, Ram, and Aidan coexist, each portraying dynamic roles. Their identities are not stagnant but rather interwoven threads in a vivid, ever-evolving tapestry. From healthcare and food to education and mobility, every interaction, every decision, echoes the multifaceted dynamic of transformation, showcasing the essence of life's continuous journey.

Transformative potential of DRT in theory and practice

DRT is a groundbreaking theoretical approach that reimagines our understanding of complex systems of machinic life-experiences, particularly digitalized ecosystems of increasing phenomenotechnicity, by foregrounding the fluid, contingent, and interconnected nature of relations. DRT offers a novel perspective on the dynamics of complex systems of increasingly tech-intensive networks. It integrates insights from a diverse range of disciplines and philosophies, providing a comprehensive and nuanced understanding of digitalized ecosystems. Through its emphasis on the ontological, epistemological, axiological, and pragmatic cores on the dynamic nature of relations (see Prologue), DRT opens up new avenues for research and innovation.

Organizational and economic studies: In the realm of organizational and economic studies, DRT offers a transformative lens. Organizations are no longer static entities but fluid assemblages, continuously evolving due to a myriad of internal and external dynamics. This fluidity is captured through morphisms, representing dynamic influences that shape organizational behavior and structure. Similarly, economic systems are intricate assemblages with nonlinear relationships. Transversality reveals intersections between different economic entities, and functors integrate these categories, providing a structured perspective on economic dynamics. This perspective reshapes our understanding of strategic planning in businesses, emphasizing the need to navigate complex ecosystems and strategically architect positions to influence the ecosystem's entities and their transformations.

- In the intricate web of organizational and economic systems, entities thrive through symbiotic growth. Embracing the fluidity of relations and continuous knowledge evolution, organizations should prioritize mutual benefits, navigating the dynamic ecosystem with a focus on interconnected growth and transformation.

Information and systems science: In the domain of information and systems science, DRT provides a dynamic lens for understanding transformations within and between data categories. Information systems evolve through differential transformations, enriching data analysis methodologies. Digital platforms, visualized as fluid assemblages, evolve based on underlying energy principles. These platforms undergo reterritorialization, with foundational principles governing their relational dynamics experiencing shifts. This perspective ensures that digital governance is adaptive, efficient, and citizen-centric.

- Digital platforms, defined by their relational data dynamics, evolve through differential transformations. By valuing data relationships over mere volume, the realm of information science should emphasize adaptive, citizen-centric governance, ensuring efficient data interconnectivity and transformation.

Cognitive and behavioral sciences: DRT revolutionizes our understanding of cognitive and behavioral sciences. It emphasizes the interplay between individual cognitive processes and societal influences, revealing nuances at the granular level. The continuous transformation or "becoming" of consumers challenges established market segments, reshaping our understanding of consumer behavior. Similarly, societal roles and identities are in a state of "becoming," continuously transforming and challenging established norms. This dynamic reshapes societal structures and perceptions, offering a fresh perspective on identity transformation in sociology.

- Human cognition and behavior emerge from a patchwork of interconnected experiences. Recognizing the continuous transformation of individuals and societies, the domain should appreciate the rich network of experiences, emphasizing the significance of the collective over the individual in understanding behavior.

Philosophy and design: DRT offers a philosophical perspective on understanding existence and reality. The Plane of Consistency/Immanence represents limitless potential, differentiating into subcategories that capture facets of identity, transformation, and interconnectedness. Diagrammatic representations offer insights into complex design ecosystems, emphasizing the dynamic interplay between human needs and design solutions. The concept of "lines of flight" signifies departures from established cultural norms, propelling societal deterritorialization.

- Philosophy and design intertwine, revealing limitless potential through interconnected entities. Balancing human needs with design solutions, every design choice should reflect deeper societal values, embracing continuous transformations that challenge and reshape cultural norms.

Business strategy and innovation: DRT revolutionizes business strategy and innovation. Technological innovations are guided by the concept of "lines of flight," driving departures from established tech paradigms. Organizations navigate complex ecosystems, architecting their positions to influence the ecosystem's entities and their transformations.

- In the evolving landscape of business, innovative strategies emphasize interconnected growth. Merging risk with tradition, businesses should value new ideas while respecting foundational principles, ensuring harmonious evolution in both technology and organizational dynamics.

Healthcare and technology management: In healthcare and technology management, DRT provides insights into the evolution of healthcare ecosystems. The continuous transformation in healthcare delivery ensures services are patient-centric and adaptive. IT systems undergo differential transformations, ensuring data integrity, security, and relevance.

- Healthcare ecosystems, defined by their relational dynamics, prioritize patient well-being. Emphasizing continuous transformation in delivery and data governance, healthcare practices should remain adaptive, ensuring technology serves the primary value of patient-centric care.

Public policy and administration: In the realm of public policy and administration, DRT reshapes our understanding of policy formulation. Policies are crafted considering the dynamic relationality within and between assemblages. The continuous transformation in community aspirations and infrastructural developments reshapes urban landscapes, ensuring urban planning is both adaptive and visionary.

- Public policies, crafted from dynamic relational insights, serve the heart of the community. Embracing the evolving nature of community aspirations, decisions should prioritize the public's best interest, ensuring visionary urban planning that remains adaptive to societal needs.

Environmental and developmental policy: DRT offers a fresh perspective on environmental and developmental policy. Environmental policies are crafted considering the dynamic relationality within ecosystems. The movement between potentialities and actualities shapes the trajectory of economic diversification and growth, ensuring policies are sustainable and future-proof.

- Environmental policies, rooted in the relational dynamics of ecosystems, prioritize nature's intrinsic value. Balancing potential with actual developments, policies should ensure that every decision respects the ecosystem's balance, promoting sustainable and forward-looking developmental strategies.

In the next chapter, we shall extend our inquiry, navigating the interplay between the Actual and the Virtual through the prism of counter-actualization, elucidated via transformative events and the rhizomatic structures inherent in life. This complex interplay of transformation and innovation is unveiled through the symbiosis of technological capabilities and human engagement, casting light on the bidirectional influence shaping our machinic and experiential landscapes. The chapter promises a rich framework of insights for both conventional organizations, as they navigate the digitization of experience ecosystems, and modern tech entities, as they strive to intertwine tech platforms with the very fabric of lived experience. Through the introduction of the Machinic Life-Experience Ecosystems (MLXE) framework, we will integrate disparate theories and highlight the transformative power of human-machinic synergies. The reader is thus invited to delve deeper into the complexities of these interactions, as we seek to underscore the transformative impacts of technology on the fabric of experience and interaction.

CHAPTER 2

Grounding machinic ecosystems in life-experiences

In Chapter 1, we established the significance of life-experience as the site which originates all the creative transformations that we experience as tangible, actual entities, and processes in machinic ecosystems. We discussed the process of actualization from the virtual to the actual through difference and repetition, as well as two different aspects of the real and the possible via virtual-actual (engagement *and* enactment) lines of flight. We also investigated the topological structure of the virtual and the actual as well as the double articulation of assemblages in each realm.

We discussed how the Plane of Immanence (**P**) epitomizes the limitless potential of entities, pulsating with ceaseless metamorphosis. This foundational category differentiates into four distinct subcategories: Existential Life Territories (**L**), Energetic-Signaletic Ecosystem Flows (**E**), Incorporeal Experience Universes (**X**), and Abstract Machinic Phyla (**M**). **L** serves as the crucible of identity, a topological space where entities delineate their existential territories. **E** represents the heartbeats of transformation, visualized as smooth manifolds, signifying potential states, and transformative flows. **X**, depicted as vector fields, is the abstract canvas guiding transformative trajectories with conceptual forces. **M**, visualized as fiber bundles, is the realm of interconnectedness where entities craft intricate systems. This differentiation, steered by the dual articulation of substance and form, content and expression, reflects unique dynamics in each domain. Functors guide transitions between these realms, echoing dynamic interplay. The evolution of **P** is a process of differentiation and realization, influenced by the interplay of the virtual and the actual. **L** and **X** capture the foundational layers of potentialities, while **E** and **M** trace actual evolutionary pathways. The gauge transformations and the Lagrangian dynamics elucidate the forces driving these transformations, capturing the energy and tension inherent in differentiation and actualization.

However, as we will see in this chapter, the ever-present emergent and evolutionary dynamics we encounter in Machinic Life-Experience Ecosystems cannot be explained just by unidirectional processes of actualization.

Counter-actualization as the source of novelty and becoming

The relationship between the virtual and the actual is not a mere linear transition but is, instead, marked by its dynamic reciprocity, epitomized by the concept of counteractualization. The emergence of the actual from the virtual does not terminate the realm of potentialities. Instead, the actual exerts influence upon the virtual, modulating and enriching it.

This dynamic interplay can be understood through Deleuze and Guattari's conceptualization of the *Rhizome*, as delineated in "A Thousand Plateaus."[1] Unlike hierarchically structured representations, the rhizome, with its decentralization, serves as an apt metaphor for the interrelation of the virtual and the actual. The actual does not simply emerge; it also reintroduces new conditions and pathways in the virtual, reconfiguring its landscape. This conceptualization disrupts a linear understanding of actualization, positioning it as a part of an interconnected, recursive web. Furthermore, Deleuze's notion of *Events*, as expounded in "The Logic of Sense", reinforces this dynamism.[2] Events, while being virtual entities with their roots in potentiality, exert tangible effects upon actualization. Their realization is not a culmination but introduces new determinants that recalibrate the virtual realm.

Central to this discourse is the Deleuzian Guattarian concept of *Becoming*, which is not a mere transition from one static state to another but signifies a perpetual transformation.[3] Counteractualization, in this light, facilitates this endless becoming. As the actual reshapes the virtual's potentialities, it guarantees the ceaseless evolution of reality. In essence, Deleuze and Guattari's notion of counteractualization offers an ontology where reality is persistently in flux. The actual, though emerging from the virtual, serves as both a culmination and a beginning. It recalibrates the virtual, ensuring that the cosmos of potentialities and realizations remains in dynamic interplay.

Counteractualization between the Machinic Ecosystem and Life-Experience is a dynamic reciprocity that refuses stasis. The actual does not merely arise from the virtual; it also acts back upon it, influencing its potentialities. This interaction, rooted in the conceptual scaffolding of the rhizome, events, and becoming, ensures that reality remains vibrant, interconnected, and ceaselessly evolving.

First, the rhizomatic nature of this relationship elucidates the nonlinear, nonhierarchical interplay between the Machinic Ecosystem and Life-Experience. The machinic ecosystem, as an intricate web of tangible processes, is a realized manifestation. However, post actualization, it does not remain inert. Instead, drawing from the rhizomatic structure, the ecosystem multidirectionally influences and is influenced by the virtual realm of life-experiences. Its tangible processes, while being rooted in actuality, create new pathways, intersections, and

---

[1] Deleuze and Guattari's (1987) concept of the *rhizome*, emphasizing multiplicity and nonhierarchical connections, mirrors the fluid, multidirectional interactions between the virtual and the actual, challenging traditional linear narratives of causality and development.

[2] Deleuze (1990) delves into the nature of *events* as paradoxical occurrences that belong neither solely to the realm of the virtual nor the actual, but instead on the surface between these realms, serving as catalysts that transform and redefine reality. Events act as pivotal moments that transcend mere actualization, instead actively reconfiguring and enriching both the virtual and actual domains.

[3] Deleuze and Guattari (1987) explore the intricate concept of *becoming*, emphasizing its characteristics as more than a transition but a process of transformation involving multiplicity and the anomalous. Accordingly, becoming encompasses diverse forms illustrating a complex network of transformations beyond static states.

nodes in the vast terrain of the virtual's potentialities. The flow is not unidirectional; the actual, once realized, retraces its steps, adding depth and texture to the virtual, aligning with Deleuze and Guattari's conceptualization of counteractualization.[4]

Events, in Deleuze's schema, signify those critical junctures where the virtual potentialities of life-experience transmute into tangible realities within the machinic ecosystem. These are moments of realization that hold transformative potential. Once actualized, they do not dissipate; they linger, continuously influencing the existential territories, molding potentialities, and paving the way for unforeseen actualizations. Events thus catalyze changes that reshape both the landscapes of the virtual and the actual.[5]

Lastly, becoming, as Deleuze and Guattari emphasize, is about continuous transformation. Within our context, the machinic ecosystem, even as a realized entity, is perpetually in a state of becoming. As it interfaces with the virtual realm of life-experience, it experiences transformations. Simultaneously, life-experience, too, is in flux, modified by its interactions with the actual. This continuous metamorphosis, facilitated by counter-actualization, underscores the perpetual becoming of both domains.

Interactional creation: In the intersection of technology and human interaction, a unique pattern emerges where potential futures derive from present realities. This is influenced by the multifaceted interplay of user engagement, technological capabilities, and emergent outcomes. While these trajectories have their foundations in the present, they represent a diverse array of transformation pathways that entities might adopt (see Fig. 2.1).

At the core of this discourse is the concept of "Territorial Life Spacetime," as we discussed earlier. This space, extending beyond mere physical dimensions, encapsulates the conditions and situations entities navigate. Within it, we identify undifferentiated yet real potentialities shaped by prior experiences, desires, and external factors. The foundational systems and networks in this ecosystem play a critical role. They serve as infrastructural support, enabling the creation, exchange, and consumption of value. In this context, collaboration is prominent, aided by shared platforms, leading to collective creativity and innovation. This is evident when multiple stakeholders converge, resulting in new ideas or solutions transcending traditional boundaries. These systems, by facilitating digitalized interactions, diversify the ways users can engage.[6]

Furthermore, the balance between technological possibilities and user engagement is essential for the ecosystem's dynamism. Technological affordances, or action potentials, dictate how users engage and navigate. As entities engage with these affordances, their interactions lead to unforeseen innovations, signifying the ecosystem's adaptive nature. Additionally, the inherent characteristics of technology promote interconnectivity among users,

---

[4] Herrera-Vega's (2015) analysis on the changing dynamics of technology in relation to human agency and social systems illuminates how technological advancements redefine interactions within social systems, mirroring the multidirectional and transformative influence of the actual on the virtual in the machinic ecosystem, thus underlining the nonlinear and evolving nature of these relationships.

[5] Hopster's (2021) analysis of technosocial disruption illustrates how machinic ecosystems, as critical "events," transform social relations and cognitive experiences, thereby continuously influencing and reshaping life-experiences.

[6] Pérez-Pérez et al.'s (2021) system dynamics model for examining sustainable processes in manufacturing can uncover varied strategies for future developments, mirroring the multifaceted interplay of technology, user engagement, and emergent outcomes in shaping potential futures from present realities.

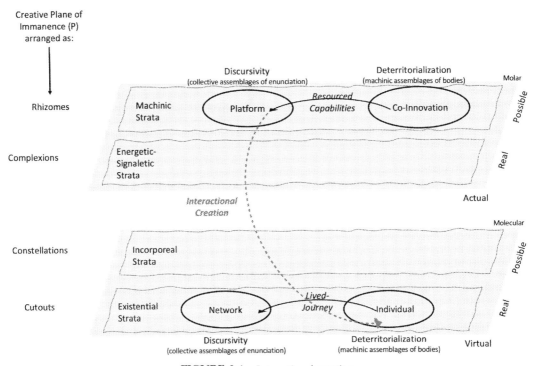

FIGURE 2.1 Interactional creation.

technology, and their environments. This relationship facilitates a state of continuous becoming for individuals, marked by their interactions, resulting in a unique process of individuation. As technology aligns with user needs, this interconnectedness deepens, leading to the emergence of collectives—groups formed from dynamic interplays.

At the convergence of these realms lies "Interactional Creation." This concept highlights the role of shared digitalized infrastructures and platforms in enhancing individual interactions and generative processes. It signifies a shift from mere collaboration to the core of individual user engagement and its potentialities, emphasizing the influence of ecosystem tools on user engagement. In conclusion, machinic life is a confluence of potential trajectories, grounded realities, and the continuous realm of becoming. Through the act of interactional creation, entities are influenced by their current contexts and can also counteractualize, shaping unforeseen futures. This highlights the rhizomatic and becoming aspects of the relationship between the virtual and the actual, ensuring the perpetual transformation of reality.

Events: Experience ecosystems encapsulate the dynamics of our lived realities, encompassing both tangible processes and intangible influences that shape our existential territories (see Fig. 2.2).

These processes, as mentioned earlier, are not fixed but are in constant flux due to various factors, such as economic shifts, media inputs, and biological imperatives. These tangible factors are intricately connected with the broader socio-technological landscape, emphasizing the interdependent relationships between humans, technologies, and their environments.

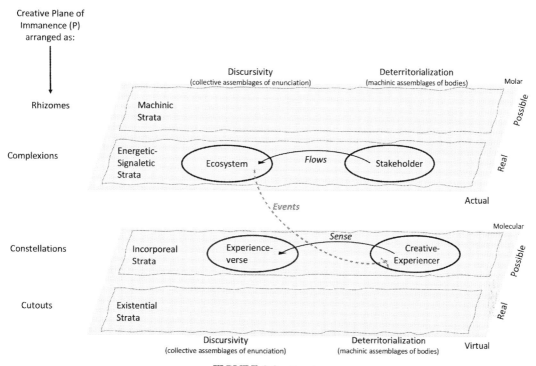

FIGURE 2.2 Events.

In this context, the interactions between individuals and technology become central. As individuals adapt to technological landscapes, it fosters the mutual evolution of both technological and human domains.

Simultaneously, there exists an intangible dimension, directing our aspirations and self-understanding. This dimension acts as a guiding compass, rich with principles, goals, and values, influencing our interactions and decisions. Much like fields in physics dictate particle behavior, these intangible realms drive the evolution of social and behavioral dynamics. Concepts like the free energy principle elucidate the mechanisms underlying these dynamics, positing that living systems aim to harmonize internal models with external sensory input. Within this paradigm, technology acts as a catalyst, altering perceptions and shaping experiences. As individuals engage with technology, they partake in active sense-making, perpetually recalibrating the relationship between potential and actual experiences.[7]

Central to the discussion here is the concept of "event," which as previously highlighted, serve as nexus points between the tangible dynamics within ecosystems and the guiding intangible principles. They are the transformative moments that bridge the gap between actual processes and the virtual guiding principles, ensuring continuous evolution in our

---

[7] Löhr (2023) emphasizes that modern technologies are conceptually disruptive, altering not just social practices but also the fundamental ways we understand ourselves and the world. By shaping our concepts, technologies like AI and social media recalibrate our sense-making processes.

I. Relational dynamics

lived realities. In summary, experience ecosystems underscore the interconnectedness of tangible realities and virtual possibilities. They highlight the intricate relationships and interactions within our environments, continually shaped by both tangible factors and intangible guiding principles, in a continuous cycle of evolution and transformation.

Interpenetration of actualization and counter-actualization

The relationship between life-experiences (the virtual) and the machinic ecosystem (the actual) is a complex interplay. In the modern era, the dynamics between these two domains seem to increasingly favor counteractualization from the machinic ecosystem back onto life-experiences. The directionality of most modern enterprises and organizations reflects this shift. Here are some considerations as to how and why this is the case:

1. **Technological determinism**: Modern societies are profoundly shaped by technological advancements. Enterprises develop technologies that not only cater to current societal needs but also shape and create new ones. Once these technologies are embedded in the machinic ecosystem, they recursively act back on the fabric of society, influencing behaviors, attitudes, and experiences. The rise of social media platforms like Facebook and Instagram offers a quintessential example. While they were initially designed to cater to the human need for connectivity, their algorithms and interfaces have now shaped and directed the way people communicate, perceive, and present themselves.[8]
2. **Technological feedback and decision systems:** Modern enterprises have seamlessly integrated data analytics and AI-driven expert systems into their core operations. The machinic ecosystem's real-time feedback, garnered through advanced analytics, enables businesses to constantly adapt and refine their products, services, and experiences. Concurrently, the use of expert systems, AI, and machine learning underscores a paradigm where decision-making across diverse sectors, from finance to healthcare, is deeply influenced by the machinic realm. Thus, the prevailing directionality emerges from the machinic ecosystem to life-experience, as enterprises leverage these technologically augmented feedback and decision mechanisms to shape societal interactions and experiences.[9]
3. **Rapid technological iteration and infrastructure rigidity**: The unprecedented pace of technological innovations means that the machinic ecosystem is constantly evolving, with frequent updates and changes that directly impact and reshape life-experiences. Concurrently, when certain technologies or systems embed themselves as cornerstones within this ecosystem, they establish path dependencies.[10] These entrenched

---

[8] Tracing digital platforms' development from HTML to HTML5, Tabarés (2021) shows how technological advancements not only fulfill but also create and direct societal needs and behaviors.

[9] Watanabe et al. (2018) underscores the transformative impact of digital solutions in the forest-based bioeconomy, exemplifying how digitalization facilitates real-time feedback and data analytics to optimize supply chains.

[10] As an example of path dependency in technological systems, Quitzau's (2007) analysis of the embedding of water-flushing toilets in Western society highlights how entrenched infrastructures, once integrated into daily practices, create a rigidity that resists the adoption of innovative alternatives.

infrastructures, like car-centric urban designs, not only illustrate the rigidity within the machinic system but also reveal how they shape and, at times, constrain societal experiences. The combination of rapid technological change and the inertia of established infrastructures means that the machinic ecosystem, in its fluidity and rigidity, holds a commanding influence over life-experiences, often bypassing the slower pace of societal norms and structures.

4. **Capitalist imperatives, global influence, and narrative control:** In the modern era, enterprises, fueled by capitalist motives, prioritize profit and market reach. When a product or service becomes central within the machinic ecosystem, businesses strategically iterate to cement its relevance, as evidenced by smartphones' evolution and their profound impact on daily experiences. This drive is magnified on a global scale, with enterprises disseminating their innovations across diverse cultures, amplifying the machinic ecosystem's pervasive influence on life across boundaries. Simultaneously, many of these enterprises command significant sway over media and narratives. Through this control, they not only dictate the products we consume but also the stories and values we internalize, intertwining the commercial and the cultural, and reinforcing the omnipresent influence of the machinic ecosystem on society.[11]

Comparatively speaking, then, while life-experiences do guide the initial creation of products, services, and technologies (actualization), the speed, scale, and strategic intent of modern enterprises have amplified the counteractualizing influences of machinic ecosystems on life-experiences. This directionality reflects the evolving power dynamics in our technologically driven society, where the machinic ecosystem, anchored by enterprise-driven innovations, has a more profound shaping influence on individual and collective experiences.

Enterprises and organizations are in the relentless pursuit of value creation, innovation, and competitive advantage. However, if they focus solely on counteractualization dynamics, via interactional creation and events, without considering the nuanced actualization processes of engagements and enactment, they risk a partial and fragmented understanding of the full potential and transformative capacity of these dynamics, which can lead to opportunity myopia on the one hand and societal backlash on the other. Here are some considerations as to how and why this is the case:

1. **Limitation of scope:** Focusing solely on interactional creation and events centers primarily on the convergence and emergent phenomena that stem from technological and human interactions. While these intersections offer a wealth of opportunities, they are fundamentally rooted in the present. By not addressing the actualization processes, organizations miss the vital transformative journeys that bridge the potential with the

---

[11] Tijmes (1999) argues that technology not only shapes societal practices and experiences (per Albert Borgmann's device paradigm) but also creates a sense of scarcity and mimetic desire, influencing consumer behavior and reinforcing technology's societal dominance.

tangible. Such an oversight limits the horizon of understanding and hampers the ability to preemptively innovate.[12]

2. **Overlooking the essence of meaningful innovation**: Engagements and enactment underscore the vital transitions between the virtual behaviors, and organizational and the actual, highlighting the intertwined evolution of technology, human practices. By not fully appreciating this co-evolutionary dynamic and the foundational drivers such as memories, emotions, and internal values, organizations risk becoming overly focused on short-term outcomes. This could lead them to produce technologically advanced solutions that, while impressive, may not resonate deeply or meaningfully with users, missing out on creating genuinely impactful and sustainable innovations.[13]

3. **Limiting innovation and collaboration potential**: The oscillation between the virtual and actual, as seen in actualization processes, underscores the significance of divergent thinking and the serendipitous innovations that can emerge from iterative explorations. Solely focusing on counteractualization dynamics might push enterprises toward a convergent mindset, sidelining the creative divergence crucial for revolutionary breakthroughs. Moreover, by not fully embracing the tenets of collaboration, co-design, and open innovation inherent in processes of enactment, organizations risk missing out on harnessing the collective genius and shared value that can redefine industry boundaries.[14]

4. **Lack of comprehensive strategy:** Interactional creation emphasizes the present trajectories stemming from current realities, while events revolve around tangible and intangible dynamics shaping our existential territories. An organization's strategy centered only on these dynamics can become reactive rather than proactive. Without recognizing the more profound transitions and transformations highlighted in the actualization processes, strategic planning can miss the foresight required to harness future opportunities fully.

5. **Societal backlash**: By neglecting the comprehensive and nuanced understanding afforded by the actualization processes of engagements and enactment, organizations might inadvertently promote or implement technologies and practices that are misaligned with societal values, norms, or expectations. This misalignment can result in deterritorializing lines of flight, where the intended positive trajectories are disrupted, leading to unintended negative impacts and a potential backlash from society. Such backlash can manifest in various forms, including loss of trust, resistance to adoption,

---

[12] Fox (2016) underscores the broader impact of virtual-social-physical convergence and its latent realities, advocating for a comprehensive view that extends beyond immediate technological—human interactions to encompass transformative potential in actualization processes, essential for holistic innovation.

[13] Kundu et al.'s (2018) study on subsurface arsenic removal technology exemplifies the risk of overlooking meaningful innovation by focusing narrowly on technical efficacy without adequately integrating social considerations, such as community buy-in and user practices, leading to innovations that fail to resonate deeply with intended users or effect meaningful change.

[14] For a more nuanced take on this, see Xiang et al. (2021), whose study on the Shaoxing textile cluster demonstrates how cultural embeddedness initially supports innovation but may later lead to a restrictive lock-in, as convergent mindset limits divergent thinking and collaborative innovation.

or active opposition, which can ultimately harm the organization's reputation, stakeholder relationships, and long-term viability.[15]

In conclusion, while interactional creation and events are pivotal for understanding the dynamic interactions of technology and human engagement, enterprises and organizations need a holistic approach. By also accounting for the intricate actualization processes, they can ensure they are not merely reacting to the present but are poised to shape and harness the full spectrum of potential futures and gain societal legitimacy.

Conventional organizations emphasize transforming intangible elements, such as values and culture, into actionable strategies, anchoring on events like product launches to bridge tangible processes with virtual principles. In contrast, more modern platform organizations prioritize the turning of potential concepts into concrete realities. These platforms maximize technological potential, allowing for value co-creation, with each interaction holding the power for unexpected innovation. The upcoming sections delve deeper into these dynamics, comparing conventional entities' emphasis on "experience ecosystems" with modern platform organizations' focus on the convergence of technology and human interaction, termed "machinic life."

Possibilizing the virtual as experience ecosystems

Conventional organizations that predominantly operate within structured frameworks, relying on established processes, hierarchies, and legacy systems to achieve goals, can focus on possibilizing the virtual as experience ecosystems. This means that they look at the intangible elements of an organization, such as its values, culture, and ethos, and attempt to mold these into actionable strategies. They can engage in "experience ecosystems" where the intangible becomes the guiding compass. This can be seen as extending their emphasis on customer relationships, branding strategies, and loyalty programs, where the aim is to align intangible brand values with tangible consumer experiences. Here, organizations view transformative events as pivotal moments that can be leveraged to bridge the gap between tangible processes and virtual guiding principles. Such events could be product launches or market expansions.

In adopting a perspective that emphasizes "experience ecosystems" as the process of possibilizing the virtual, an organization centers its operations around the dynamics of lived realities.[16] Such a perspective is keenly attuned to the interactions between individuals, technologies, and the encompassing socio-technological environment. With its strong emphasis on tangible and intangible influences, this perspective offers the organization advantages like:

---

[15] Galaz et al.'s (2021) exploration of AI's systemic risks in sustainability domains highlights how AI's rapid deployment, often without comprehensive risk assessment, can lead to issues like algorithmic bias and resilience trade-offs, echoing the potential for societal backlash against misaligned technologies.

[16] Çipi et al. (2023) underscore the necessity of integrating technological innovation within the fabric of societal and business operations.

1. **Intuitive understanding of present realities**: Such an organization is more aligned with current tangible dynamics like economic shifts and media trends, making it agile and responsive to immediate changes.
2. **Strategic use of technology**: Recognizing the mutual evolution of humans and technology, the organization will be adept at leveraging technology to enhance human experiences.
3. **Emphasis on events**: Recognizing events as transformative pivot-points allows the organization to capitalize on key moments, bridging the gap between virtual possibilities and tangible realities.
4. **Values and principle driven**: By emphasizing intangible guiding principles, the organization will likely have strong values, ensuring stakeholder trust and building reputation.

However, technology-driven experience ecosystems has its limitations:

1. **Limited recognition of personal nuances**: While the conventional framework is responsive to tangible and intangible factors, it might not deeply integrate personal memories, emotions, and unique histories into its operations and thus have a more static, less adaptive understanding of the continuously evolving lived realities.
2. **Limited digitalized engagement**: The conventional perspective might not delve deeply into the intricate nuances of machinic trajectories, which chart the course of human interaction within digital terrains.
3. **Narrow scope of intangible interactions**: The conventional framework might not fully account for the expansive possibilities that arise when intangible values and meanings permeate machinic landscapes.

Next, we discuss two new concepts of "Life-Experience Stakeholder-Ecosystems" and "Co-Innovation of Life-Experience Platforms," which go beyond the concept of Experience Ecosystems and Technology Platforms. The former delves deeper into the integration of individual nuances, emphasizing the significance of personal histories and emotions, thereby ensuring a more personalized stakeholder experience. This perspective offers dynamic adaptation to shifts in the socio-technological domain and paves the way for enhanced innovation through its comprehensive understanding of energetic-signaletic flows. On the other hand, the latter provides profound insight into our interactions within digital terrains. This perspective ensures deeper digital integration, combining the nuances of lived experiences with machinic trajectories. As a result, organizations benefit from heightened technological influence and can develop value-oriented platforms that resonate deeply with users, bridging the gap between the virtual and the actual.

Life-experience stakeholder-ecosystems emerge as intricate conceptual frames that comprehend and integrate the expansive nuances of incorporating engagements into events, where life-experience is reimagined in the context of energetic-signaletic flows of stakeholder-ecosystems.[17] This synthesized perspective centers on the relationship and

---

[17] Rodríguez et al. (2015) elucidate the impact of communication technologies in enhancing the complexity and dynamism of social interactions, not just through exchange of information but also by actively influencing the formation and evolution of engagements and events within these ecosystems.

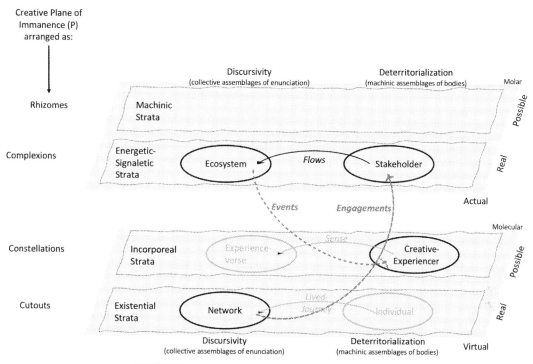

FIGURE 2.3 Life-experience stakeholder ecosystems.

dynamic transition between life-experiences and the energetic-signaletic flows within stakeholder ecosystems, providing a deeper understanding of our existential realities (see Fig. 2.3).

The tangible dynamics, influenced by myriad elements like economic shifts, media inputs, and biological imperatives, become deeply intertwined with the personal nuances, memories, emotions, and environments. This results in a unique interface where life-experiences actively guide, mold, and are simultaneously reshaped by the energetic-signaletic dynamics. As individuals navigate these ecosystems, they actively engage with technology, mediating their experiences with the ever-evolving socio-technological landscapes. This becomes a central avenue for mutual evolution, where technological affordances and human experiences collaboratively redefine each other. The guiding compass of intangible principles, goals, and values in life-experiences resonates with and influences these energetic-signaletic flows. The active recalibration and sense-making, fundamental to harmonizing internal models with external inputs, become even more paramount, enriching the overall experience by emphasizing the interconnectedness of tangible and virtual aspects of the stakeholder ecosystem.

Conversely, the conventional view of ecosystems is also enhanced by viewing energetic-signaletic flows within ecosystems through the lens of life-experiences. These flows, inherently dynamic and indicative of both potentialities and tangible processes, become intricately linked with the lived realities of individuals. Within the multidimensional space that envelops unique conditions and histories, these flows cater to individual engagements with technology, propelling innovation and amplifying potentialities. The properties of technology, pivotal in these flows, foster deep intersections with personal experiences and the broader environment. As entities engage within this domain, behaviors evolve, ensuring a cohesive co-shaping and mutual evolution of technological and organizational practices. The complex interplay between individuals, organizations, and technological entities, central to life-experiences, enhances the understanding of energetic-signaletic flows, highlighting their significance in the dynamic adaptation and structuration of the ecosystem.

Life-experience stakeholder-ecosystems, thus, propose a harmonized understanding, grounded in the rich integration of life-experiences and energetic-signaletic flows within ecosystems.[18] This synthesis underscores the importance of events as transformative moments that bridge tangible and virtual elements. By counteractualizing these events and emphasizing active engagements, these ecosystems illuminate the dynamic interdependence of our lived realities and the multifaceted environments we inhabit. Through this understanding, the depth of interactions and transformations within our environments becomes more nuanced, emphasizing the continuous cycle of evolution and transformation.

An organization rooted in a "Life-Experience Stakeholder-Ecosystem" perspective benefits from a synthesized understanding of life-experiences and energetic-signaletic flows. In serving its stakeholders, an organization oriented around this perspective offers a more integrated, holistic, and dynamic approach. It would ensure technological advancements align with deeper human needs and energetic-signaletic flows, providing richer experiences for stakeholders. Advantages of this perspective include:

1. **Deep integration of individual experiences**: The organization will place significant importance on individual nuances, histories, and emotions, ensuring a richer and more personalized stakeholder experience.
2. **Dynamic adaptation**: By emphasizing the mutual evolution of technology and human domains, the organization can continuously adapt to shifts in the socio-technological landscape.
3. **Enhanced innovation and potentialities**: With its holistic understanding of energetic-signaletic flows, the organization will be better equipped to foster innovation and amplify potentialities within its domain.

To transition from a conventional perspective on ecosystems of experiences toward a more comprehensive "Life-Experience Stakeholder-Ecosystems" view, organizations should deliberate on the deeper connotations and synergies implied by this shift. They must:

- Understand how tangible processes and intangible influences dynamically shape our existential territories, informed by economic shifts, media inputs, biological imperatives, and the broader socio-technological landscape.
- Expand the understanding of "experience" to encompass the entirety of life's nuances, memories, emotions, and unique conditions. Recognize how individuals' life-experiences intricately interface with the energetic-signaletic flows of ecosystems, redefining both.

---

[18] Sanne's (2012) study on how organizational learning shapes responses to complex, dynamic environments underscores the importance of integrating life-experiences with energetic-signaletic flows, emphasizing adaptive learning and recalibration in response to evolving socio-technological landscapes.

- Grasp the interconnectedness of tangible realities and virtual possibilities, spotlighting relationships within our environments shaped by both tangible factors and intangible guiding principles.
- Enhance the understanding of ecosystems by factoring in the transformative journeys between the virtual and the actual. Recognize the materialization of potentialities into concrete experiences and how entities not only experience the interplay of personal and collective domains but also contribute to the reconfiguration of the ecosystem's structures.
- Acknowledge the dynamics of lived realities and the actualization of experiences within conventional experience ecosystems.
- Directly emphasize the multidimensional space of existence that includes unique histories and conditions. Understand that within this space, technology offers potentials leading to engagements that continuously evolve behaviors, identities, and collective definitions.

By discerning enhanced meanings and interconnections, organizations can transition from merely grasping the intricate dynamics of experiences (experience ecosystems) to a more holistic comprehension of how life's myriad facets interweave with these dynamics, fostering richer and continually adaptive ecosystems (Life-Experience Stakeholder-Ecosystems). This broadened perspective promotes a more integrated approach, recognizing the profound interplay between life's experiences and the ecosystems they inhabit and influence.

Machinic ecosystem co-innovation: Enactment of individuals as creative-experiencers transforms the very fabric of machinic ecosystems through the infusion of Incorporeal Universes of Experience in its co-innovation.[19] This multifaceted perspective pivots on the relationship between the Incorporeal Universes of Experience and the dynamic trajectories of the machinic ecosystem in its co-innovation, offering a comprehensive understanding of our interaction within digital terrains and existential realms (see Fig. 2.4).

The dynamics of lived realities, rich in tangible processes and intangible influences, become more nuanced when approached through the prism of machinic trajectories. Within these trajectories, the technological landscape intertwines with human domains, fostering mutual evolution. As individuals navigate and adapt to this ecosystem, their interactions with digital forms become heightened, driving perceptions and interpretations grounded in the Incorporeal Universes.

The intangible dimension of aspirations, values, and self-understanding, which has traditionally influenced decision-making and interactions, now permeates the machinic landscape, influencing its very contours and potentialities. Technology in this enriched ecosystem does more than just act as a catalyst. It reshapes and redefines perceptions, serving as a bridge between the potential (virtual) and actual experiences. Through active sense-making, individuals constantly recalibrate this relationship, with events serving as transformative pivot-points, bridging tangible dynamics and guiding intangible principles.

---

[19] Sneltvedt (2018) highlights the importance of envisioning and materializing sociotechnical futures, particularly how they contribute to shaping a good society through the integration of technology with human experiences and societal values, and the relevance of public engagement in the transformation and enactment of technological ecosystems.

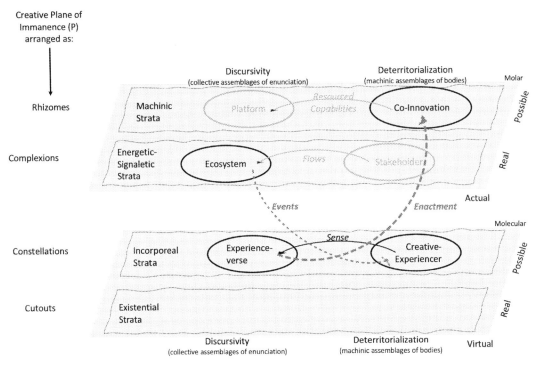

FIGURE 2.4 Machinic ecosystem co-innovation.

On the flip side, the Incorporeal Universes are re-envisioned within the confines of the machinic ecosystem. The intangible landscapes of meaning and value, which traditionally guided transformation processes, find new potentials and trajectories within this ecosystem. The dialectical movement between the virtual and the actual becomes more prominent, with technology amplifying the predisposition of living systems to assimilate and refine information. This leads to a synthesis of the virtual, molding connections between potential experiences and their tangible counterparts within the machinic ecosystem.

These tangible trajectories, which are reflections of potential transformation lines, become richly embedded with intangible values and meanings. They act as mathematical entities, charting values across a broad landscape, signifying potential paths of transformation. The infrastructure within the ecosystem enables value-oriented platforms that accommodate user demands, with open innovation strategies being a result of collaborative and co-creative endeavors. The iterative process of difference and repetition comes into play here, with each cycle generating new divergences and unforeseen outcomes.

Through creative transformation via Incorporeal Universes in co-innovation of machinic ecosystems, we discern the profound interconnectedness of digitalized terrains, existential realms, and the dynamic trajectories they chart.[20] This perspective illuminates the nuanced interactions within our environments, emphasizing the evolving cycle of interpretation and transformation.

For stakeholders, an organization following this perspective offers an enriched and technologically advanced approach. It leverages the expansive potentialities of the digital realm, intertwining them with the intricate nuances of lived experiences. An organization rooted in this perspective gains from:

1. **Deeper digital integration**: By emphasizing the profound integration of machinic trajectories with lived experiences, the organization attains a more nuanced understanding of digitalized interactions.
2. **Heightened technological influence**: Recognizing technology's power to redefine perceptions and bridge virtual and actual experiences provides the organization with enhanced tools for innovation.
3. **Value-oriented platforms**: By embedding tangible trajectories with intangible values, the organization can harness technology to provide platforms that cater specifically to user demands and value co-creation.

To transition toward co-innovation of machinic ecosystems that is grounded in life-experiences, organizations must:

- Understand the intricate interdependence of tangible realities and virtual possibilities, shedding light on the myriad relationships and interactions within our environments.
- Expand on this by emphasizing the technological underpinnings of such ecosystems. Acknowledge the presence of machinic trajectories as potential lines of transformation that emanate from our current states. Recognize how collaborative endeavors within these machinic ecosystems lead to iterative cycles of divergence and new experiential outcomes.
- Focus on the dynamics formed by tangible processes and intangible influences that actively shape our existential territories, guided by various tangible and socio-technological parameters.
- Elevate the notion of "experience" to appreciate the intricate interplay between human domains and machinic trajectories, emphasizing how technology amplifies perceptions and recalibrates relationships between the virtual and the actual.
- Directly delve into the potentials and trajectories within existential domains brought forth by technology. Understand how technology not only amplifies the predispositions of living systems but also actively reshapes perceptions and experiences. Recognize the significance of technological landscapes in catalyzing the dialectical movement between the virtual and the actual.

---

[20] De Luca's (2021) framing of the evolution of machine intelligence as an integral part of the physical environment, performing cognitive activities alongside humans, aligns with the infusion of incorporeal universes within machinic ecosystems, where technology enhances the assimilation and transformation of information, shaping both tangible trajectories and intangible values.

In navigating the intricacies of this shift, organizations can move from simply comprehending the interconnected dynamics of experiences to an enriched understanding of how technology and machinic elements intertwine and elevate these dynamics, through *co-innovation of life-experience platforms, together with individuals as creative-experiencers, enacted based on sensed-event experiences*. This refined viewpoint enables organizations to harness the power of technology to its fullest potential, ensuring adaptive and futuristic ecosystem engagements.

Realizing the virtual as machinic life

Modern tech organizations focus on "realizing the virtual" as machinic life with native technology stacks that emphasize digital applications and interfaces, and application programming interfaces (APIs) that facilitate developer-as-stakeholder engagement in creating new application-based experiences. This means they actively engage in bringing potentialities into reality. They integrate tech platforms with emergent experiences and existential territories via platforms built on the idea of maximizing the potential of technological affordances. They operate in "machinic life," where technology and human interactions converge, enabling various potential futures. By providing technological tools and frameworks, they enable users and partners to develop new pathways of value. Platforms emphasize the core of individual user engagement. They understand that every interaction on their platform, no matter how trivial it seems, can lead to unforeseen innovations. The role of shared digital infrastructures is pivotal here, enabling interactions and fostering generative processes across lived-journeys of individuals territorialized in individual-to-network and network-to-individual engagements.

In this framing, the organization might excel in:

1. **Adapting quickly to current technological contexts and user engagements:** The organization exhibits agility in recognizing the prevailing technological capabilities, ensuring prompt responses to the ever-shifting landscape. Given its emphasis on the "here and now," the enterprise excels at leveraging the power of existing platforms, technologies, and understanding user behaviors to drive timely innovations. This focus allows for a keen understanding of how users interact with technology in the immediate context, leading to the creation of user-centric designs and solutions that resonate with current needs.
2. **Recognizing and exploring potential trajectories from present realities:** A "machinic life" perspective offers a lens to view the potential futures emerging out of the present, ensuring that the organization is not just rooted in today but is also stretching its vision into tomorrow. While this approach foregrounds adaptability and foresees a range of transformation pathways, it also brings with it the challenge of remaining overly abstract if not anchored adequately in tangible life-experiences.
3. **Prioritizing infrastructural tools and collaborative platforms as catalysts for innovation:** Within this framework, the organization values foundational systems, shared platforms, and collaboration tools, viewing them as more than just tools—but as essential elements that spur collective innovation and amplify value. However, there' is

a nuance to be acknowledged: the subtle risk of emphasizing the technological infrastructure over the experiential, potentially sidelining the richness of lived experiences and user narratives in favor of pure tech-driven innovation.
4. **Maintaining a balance between technological possibilities and user engagement, ensuring active participation and continuous innovation:** This approach deeply recognizes the importance of user engagement, understanding that innovation is not solely driven by technology but also by how humans engage with it. This ensures that entities within the organization are active participants in the innovation process. However, it is crucial to be cautious of establishing a binary relationship between technology and humans. The organization needs to champion a state of continuous transformation, promoting a fluid relationship between the virtual and actual, ensuring entities are always in the realm of becoming, while ensuring this continuous transformation is grounded in the nuances and tangibility of real-world experiences.

Such a framing is not without limitations:

1. **A potential lack of long-term vision, emphasizing present realities over future implications**: By concentrating heavily on the immediate trajectories, there is the danger of being more reactive than proactive. This might hinder the organization from playing a pivotal role in shaping more extended technological and existential pathways.
2. **A risk of technological determinism, neglecting the co-creative transformation of existential territories**: The focus on current technological capabilities can cause the organization to assume, perhaps erroneously, that technology singularly molds user behaviors and societal outcomes. This viewpoint might miss out on appreciating the intricate interplay between existential territories and dynamic technological ecosystems.
3. **Overlooking the deeper, tangible nuances of lived human experiences and emotions**: There is a possibility that the organization might sideline the profound existential implications of technology. The intertwining of technology with personal histories, emotional landscapes, and broader human conditions may not be adequately recognized or valued.
4. **Focusing primarily on technological dimensions and potentially missing the deeper resonance of user-machine interactions**: While there is an emphasis on user engagement, there is a danger of neglecting the profound depth and resonance that certain interactions offer. These interactions, rich in meaning and substance, might get overshadowed by purely technological pursuits.
5. **Treating machines primarily as tools rather than entities influenced by and reflective of human experiences**: By prioritizing the technological and collaborative facets, the organization might not fully embrace machines as entities imbued with life-experiences. This perspective risks driving innovations that, though technologically sophisticated, might fall short in connecting deeply with human sensibilities.

These limitations could be addressed by framing technological innovations in deeper existential realities, ensuring a harmonious blend of the virtual and the actual that is grounded in life-experiences. As we discuss next, the **"Co-Innovation of Life-Experience Platforms"** emphasizes a holistic integration of machine potentialities with the essence of lived experiences, bridging the gap between the abstract technological trajectories and the tangible human

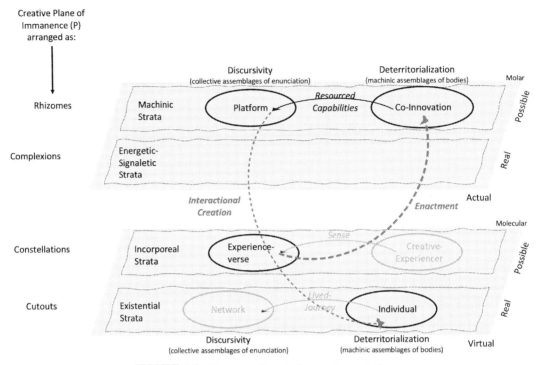

FIGURE 2.5 Co-innovation of life-experience platforms.

conditions. This synthesis ensures innovations are not only driven by immediate technological capabilities but are deeply rooted in human sensibilities. It underscores the symbiotic relationship between technological ecosystems and existential territories. Such a perspective ensures that technological advancements are not isolated from broader human conditions, leading to sustainable innovations that are both technologically sound and socially relevant.

Co-innovation of life-experience platforms: Interactional creation through tech platformization extends the tendrils of machines into the realm of life-experiences, dynamically evolving in response to the abstract and concrete facets they introduce. The co-innovation of life-experience platforms and interactional creation enacted through emergent (embodied) experiences mediates between the abstract and the concrete, conceptualizing a symbiotic relationship that informs and reshapes life-experiences (see Fig. 2.5).

Here, the unique pattern of potential futures and present realities is enriched by the omnipresence of machines. Their abstract elements, which revolve around potentialities and trajectories, meld with the foundational territorial spacetime of life-experiences.[21] The complex dynamic between technology and human interaction becomes intensified, with the abstract and concrete machinery lending depth to this interplay. The conditions, situations, and

---

[21] One popular example is the concept of living labs as dynamic spaces fostering co-creation and innovation, which underscores the integration of technology with human experiences in real-life settings, enhancing user engagement and diversifying innovation pathways (Mukama et al., 2022).

potentialities within the existential territories now bear the signature of machine potentials, ensuring that the very nature of collaboration, interaction, and innovation reflects this symbiosis. This co-creative transformation enhances user engagement and diversifies pathways of becoming, ensuring a richer, more intricate pattern of interactional creation.

Conversely, when the spotlight shifts to the realm of machines influenced by life-experience, there is a paradigm shift in how they are perceived and engaged with. These machines, no longer merely technological entities, assume a life-like quality. They are now deeply influenced by the values, meanings, and perceptions associated with life-experiences. Digital technology, in this context, becomes an enabler of deeper human predispositions. The amplification of human perceptions and experiences becomes more profound, as these machines inherently recognize and assimilate the life-experiences they are embedded within. As such, the act of synthesizing the virtual and guiding transformations relies heavily on the core essence of life-experience. Collaboration, co-design, and open innovation within these machines are not just functional aspects but are profoundly influenced by the intangible landscapes of meaning and value inherent to life-experiences. The iterative repetition and divergence characteristic of this transformation are not just abstract processes but are deeply rooted in life-experiences, ensuring that every cycle produces outcomes resonating with the essence of lived realities.

Co-innovation of life-experience platforms seamlessly mediates between the abstract and the concrete, ensuring that the perpetual transformation of reality is anchored in both machinery and life-experiences. The act of counteractualizing interactional creation and the process of actualizing enactment in this context become intertwined processes, constantly reshaping and informing each other, leading to an enriched, multifaceted landscape of emergent experiences.

1. **Integrating machinic life with life-experiences**: The co-innovation of life-experience platforms interweaves both machinic life and the lived experience by blending of abstract and concrete realities, which means that entities are not just navigating potentialities but also lived, tangible experiences.
2. **Deepened user engagement**: With machines absorbing and reflecting life-experiences, the way users interact with them becomes profoundly nuanced. It is not just about utilizing technological affordances; it is about deeply resonant engagements that mirror tangible experiences.
3. **Embodied machines**: Machines in this framework are not mere tools; they embody life-like qualities, deeply influenced by human values and experiences. This means technological innovations become more human-centered, incorporating, and reflecting life's essence.
4. **Enriched reality transformation**: The transformation of reality in this context is anchored not just in abstract potentialities but also in the richness of lived experiences. Counteractualization and actualization processes are continually informing each other, ensuring a more holistic approach to innovation.

For a more comprehensive view of "**Co-innovation of life-experience platforms**," organizations must:

- Focus on how technology shapes interactions and the trajectories from current realities to potential futures.

- Delve deeper into the intimate dynamic between technology and human experiences, acknowledging that machines aren't just tools but also reflections and extensions of human sentiments, desires, and histories.
- Recognize the convergence patterns of technology and human interactions that result in emergent outcomes.
- Understand "life" in a broader sense that includes the intangible—emotions, perceptions, values, and profound experiences. These elements do not just coexist with technology; they deeply influence and touch on the patterns and outcomes of human-tech interactions.
- Directly emphasizes the subjective and experiential aspect, highlighting the profound depth and resonance of human engagements with machines. Technology is not just facilitating experiences; it is actively shaping and being shaped by those very experiences.

By emphasizing these distinctions and relationships, an organization can progress from merely observing the interplay between technology and human life to embracing a richer, intertwined narrative where technology and life-experiences are deeply integrated. This matured perspective fosters a more holistic approach, wherein technology is not just an enabler but also an active participant in shaping human experiences and vice versa.

Stakeholder-ecosystems and life-experience platforms: Counteractualizing interactional creation and actualizing engagements in creative transformation of machinic ecosystems (see Fig. 2.6) offers an enhanced perspective on grounding the platformization of stakeholder-ecosystems across lived-journey engagements of life-experiences.

Within this frame, entities are not merely navigating a predefined landscape but are instrumental in dynamically sculpting it. Such an ecosystem repositions the boundaries of machinic engagements and existential territories, rendering them as a unified, co-evolutionary continuum.[22] Here, the existential territories are not just cognitive abstractions or subjective experiences but assume an embodied form within the machinic ecosystem. Their very essence is intertwined with the capabilities, affordances, and constraints presented by the technological infrastructure. While these territories have historically represented unique conditions, histories, and personal nuances, they now actively encompass, react to, and evolve with the technological and systemic underpinnings they are situated within. These territories, thus, are continually redefined, drawing from the interplay of individual desires, collective forces, and technological possibilities.

The machinic ecosystem, on the other hand, is no longer confined to its technological anatomy. While it remains rooted in digital infrastructures, platforms, and action potentials, its significance extends to envelop the existential territories of entities. This intertwining ensures that the machinic ecosystem does not operate in isolation but becomes deeply sensitive to the desires, memories, emotions, and environments that constitute these territories. As a result, the ecosystem's dynamism and adaptive nature are not just a byproduct

---

[22] Aarras et al. (2014) underline the significance of life-based design in environmental technology, focusing on user needs and local cultural aspects. Machinic ecosystems, by integrating life-based design, not only transform (counteractualize) potential futures into reality but also adapt and evolve (actualize) in response to the lived experiences and cultural contexts of their users.

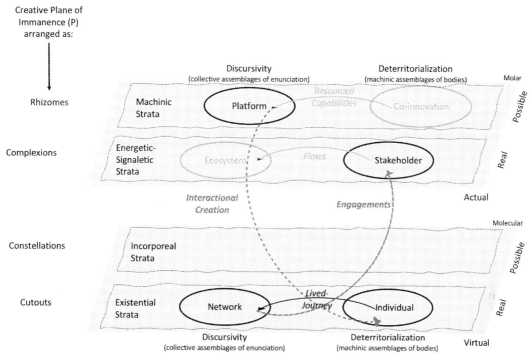

FIGURE 2.6 Grounding the platformization of stakeholder ecosystems across lived-journey engagements of life-experiences.

of technological capabilities but are influenced by the vibrant mosaic of existential territories it encompasses.

This synthesis showcases a recursive feedback mechanism. The ecosystem, while offering action possibilities, is shaped by the entities' engagements, resulting in innovative outcomes. Concurrently, as entities delineate their existential territories within this ecosystem, they are molded by the system's characteristics, influencing their perception of potentialities and tangible realities. This continual interchange between the territories of life and the machinic ecosystem heralds a transformative journey that seamlessly blends the potential with the tangible.

Conversely, in the *existential territories of life in machinic ecosystems*, the emphasis pivots to the individual and collective experiences as they are shaped and perceived within the machinic framework. Here, existential territories, imbued with personal nuances, memories, and emotions, do not merely interact with the ecosystem but are deeply embedded within its fabric. These territories, while retaining their uniqueness, are intricately woven into the technological realm, ensuring a richer, multifaceted lived experience.

The machinic ecosystem offers a structured environment wherein these existential territories flourish. These territories, while embedded, remain dynamic, navigating the myriad opportunities and challenges the ecosystem presents. The engagements within this landscape yield evolving behaviors and innovative outcomes, amplifying the potentialities intrinsic to both the territories and the ecosystem.

However, the co-existence is not unidirectional. The very nature of the machinic ecosystem is transformed by these territories. The technological interfaces, platforms, and structures are optimized, not merely based on action potentials but are enriched by the plethora of experiences, desires, and histories they house. This ensures that the ecosystem is not just a passive recipient but is in a state of continuous evolution, shaped by both its inherent capabilities and the existential territories it enfolds.

While machinic life, as perceived through these synthesized frameworks, underscores a deep interconnectedness between the technological and the existential, the intertwined realms of machinic ecosystems and existential territories creatively transform each other, resulting in a complex, adaptive, and dynamic continuum. Through this continual interplay, the boundaries between the virtual and the actual, the potential and the tangible, are seamlessly navigated, ensuring a perpetual transformation of reality. An organization that adopts the **platformization of stakeholder ecosystems with a life-experience view** acknowledges a broader interconnectedness. This organization understands that technology and human experiences are not separate realms; they dynamically shape each other.[23] Such an organization:

1. **Prioritizes human-centric designs**: By recognizing that existential territories of life are embedded within the machinic ecosystem, the organization will prioritize solutions that cater to deeper human needs, histories, and emotions.
2. **Anticipates future shifts**: By acknowledging the creative transformation of both the machinic ecosystem and existential territories, the organization will be better equipped to anticipate and shape future technological and societal shifts.
3. **Achieves sustainable innovations**: Recognizing the interplay between the virtual and the actual, the potential and the tangible, ensures innovations that are not just technologically advanced but also socially relevant and sustainable.

In serving its shareholders and stakeholders, such an organization offers a more comprehensive and sustainable approach, ensuring that technological advancements align with deeper human needs and existential conditions. Meanwhile, an organization rooted only in "machinic life" might excel in immediate adaptability but might struggle to foresee broader existential shifts and the deeper implications of their technological interventions. It must transition to:

- Recognize technology's role in shaping interactions and potential futures derived from present realities.
- Understand technology not as isolated tools or platforms but as a comprehensive system where technology intertwines with human experiences, histories, and emotions.
- Identify the patterns where human interaction and technology converge, leading to emergent outcomes.

---

[23] Voinea (2018) suggests that technology's role in society goes beyond individual or holistic impacts, issuing a call for technology design to foster a "convivial" society, where technological and existential realms are not isolated but dynamically integrated.

- Extend the understanding of "life" to include broader existential territories—the unique conditions, histories, and personal nuances of individuals and collectives as they co-evolve with technology.
- Emphasize the adaptive, interconnected, and co-evolutionary nature of technology and human experiences. Recognize that the technological ecosystem is sensitive to and shaped by the desires, memories, emotions, and environments of its users.

By honing in on these distinctions and integrations, an organization can shift from merely recognizing the intersection of technology and human interaction to appreciating the deeper interconnectedness and co-creative transformation of both realms. This expanded view encourages a holistic approach where technology and human experiences are understood as dynamically shaping one another within a complex, adaptive ecosystem.

Machinic life-experience ecosystems

Bringing it all together, we must see life-experience ecosystems as transcending technical platform ecosystems, with technicity as a facilitator of creative transformation of stakeholder ecosystems grounded in life-experiences. From AI tools to new platforms, technology aids in the confluence of human intelligence and computational intelligence. These entities must work in synergy, fostering Human-AI *Co-Creativeness*.[24]

Machinic life-experience ecosystems (MLXEs) entail a dynamic process of differentiation, wherein the virtual multiplicity of the plane of immanence differentiates into distinct, yet still virtual, domains, each with its own unique combination of MABs and CAEs, and each undergoing its own process of double articulation of territorialization (see Fig. 2.7).[25]

Each category ("oval" in Fig. 2.7) defines a topological space of assemblages, connected through functors and their natural transformations. These categories, when paired, represent molar or molecular strata, and are linked by functors of territorialization. The functors between the different categories represent lines of flight, movements of deterritorialization and transformation that disrupt and destabilize existing structures and systems, leading to the creation of new assemblages and territories. These functors capture the dynamic interplay between the existential, economic, symbolic, and technical dimensions of reality and the

---

[24] Our approach shares similar intentions with Boy (2023), who emphasizes the need for an epistemological approach in human systems integration (HSI), crucial for understanding the dynamic interplay between humans and technology in MLXEs.

[25] Geels' (2005) study on the co-evolution of technology and society during the transition from surface water to piped water systems in the Netherlands offers a vivid historical example that parallels the dynamic process of differentiation in machinic life-experience ecosystems (MLXEs) described here. This transition exemplifies how technological, cultural, political, and economic changes co-evolve, much like the interconnected evolution of machinic assemblages of bodies (MABs) and Collective Assemblages of Enunciation (CAEs) in differentiating virtual domains within MLXEs. Geels' multilevel perspective underlines how broader societal shifts, akin to the movements of deterritorialization and transformation in MLXEs, disrupt existing structures to create new socio-technical assemblages. This historical case study enhances our understanding of the intricate interplay between different strata and the continuous transformation within MLXEs, offering a concrete instance of how diverse domains co-evolve and shape each other in a complex system.

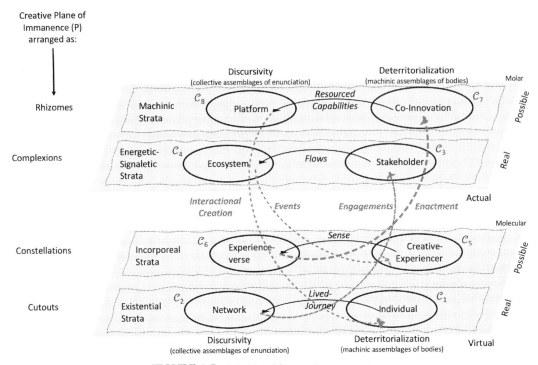

FIGURE 2.7 Machinic life-experience ecosystems.

constant flux and transformation between different levels and dimensions of reality within MLXEs.

Each domain is a unique combination of assemblages reflecting the specific dynamics and processes associated with that domain. In the Existential Life Territories (**L**), we have categories C_1 and C_2. Category C_1 encapsulates affordances, user interactions, and generativity, which form the machinic assemblages of bodies (MABs).[26] Affordances as the opportunities for action that technology offers us, dependent on our capabilities, goals, and context, and user interactions as the ways we engage with technology and digital environments are the concrete entities that form the substance of content. Generativity, the capacity of technology to enable new forms of value creation, innovation, and novel user experiences, is the form of content, the specific ways these entities are structured. Category C_2 delves into the concepts of transindividual relations, technicity, and collective individuation, which form the collective assemblages of enunciation (CAEs).[27] These are the narratives and discourses that

---

[26] Affordances (Gibson, 1979), user interactions (Preece et al., 2015), and generativity (Zittrain, 2006) form an MAB in the sense that they interconnect, influence each other, and collectively contribute to the dynamic, interactive, and evolving nature of the assemblage as an integrated, relational, and responsive system.

[27] Transindividual relations as interconnected individual-collective evolution and collective individuation as co-emergence of shared identities (Simondon, 2020), together with technicity mediating human experiences (Stiegler, 1998), constitute a CAE as a dynamic, interactive, and evolving space, where individual and collective narratives, mediated by technology, continuously interact and transform.

individuals create to make sense of their experiences with technology, which form the substance of expression. The form of expression is the specific ways these narratives are articulated. The functor F_1, that of "*Lived Journey*," represents this double articulation of territorialization, highlighting how our interactions with technology (the substance of content) can give rise to new transindividual relations (the form of content) and contribute to the individuation process (the form of expression).[28]

Functor F_2, that of "*Engagements*," acts as a line of flight between Existential Territories (**L**) and Energetic-Signaletic Flows (**E**). In Deleuzian terms, a line of flight represents a movement of deterritorialization and transformation that disrupts and destabilizes existing structures and systems, leading to the creation of new assemblages and territories. In this context, F_2 captures the transition from the realm of user interactions, affordances, and generativity (**L**) to the realm of social actors, networks, and interactions between individuals, technology, and their environment (**E**). It highlights the relationships between Simondon's transindividual framework and Latour's Actor-Network Theory in the context of a "Stakeholder Ecosystem," suggesting that the materialization of abstract ideas could lead to the emergence of black boxes in socio-technical systems. This transition represents a shift from the individual's engagement with technology to the broader social and networked dynamics of the digitalized ecosystem.[29]

Moving to the Energetic-Signaletic Flows (**E**), we have categories C_3 and C_4. Category C_3 captures the complex web of social actors, networks, and the interactions between individuals, technology, and their environment, which form the MABs.[30] The social actors and networks as the entities involved in the processes of interaction and transformation within the digitalized ecosystem form the substance of content. The interactions between individuals, technology, and their environment are the form of content, the specific ways these processes are structured. Category C_4 encompasses the ideas of systems and organisms evolving together, the creation of specialized environments, and the dynamic adaptation of social structures in response to technological change, which form the CAEs.[31] Policies, guidelines, and protocols that guide these processes as narratives and discourses that individuals create to make sense of their experiences with technology form the substance of expression. The

---

[28] Combining notions of lived experience (Merleau-Ponty, 1962), lines of segmentation (Deleuze and Guattari, 1987), and life of lines (Ingold, 2015), our concept of "lived journey" encapsulates individuals' navigation through MABs and CAEs, reflecting the intertwined trajectories of personal and collective experiences within the socio-technological matrix, as a continuous interweaving of life's paths and relational dynamics.

[29] In DRT, "engagement" integrates the insights of cognitive nonconscious (Hayles, 2017), material agency (Pickering, 1995), infosphere (Floridi, 2014), and global assemblages (Sassen, 2006), to define a transformative process that juxtaposes the cognitive interplay between users and technology as a dynamic reciprocity between human and material agencies, with the reconfiguration of human identity, social structures, and power dynamics within global digital networks.

[30] Complex interactions within social actor networks (Latour, 2005), where individual-technology-environment relations are central (Ihde, 1990) and processes are often "black-boxed" for simplicity and abstraction (Callon, 1986), collectively shape the structure, functionality, and evolution of the MAB.

[31] The co-evolution of ecosystems emerge from a dialog of mutual influence, where the collective application of technology evolves through interactive adaptations (Desanctis and Poole, 1994), entities actively shape and modify their digital environments (Odling-Smee et al., 2003), and the reciprocal development between technology and organizational practices (Orlikowski, 2000), forms a continuous, collective narrative as CAE.

form of expression is the specific ways these policies and guidelines are articulated. The functor F_3, that of "*Flows*," represents this double articulation of territorialization, highlighting how the interactions and relationships among social actors, individuals, and technology (the substance of content) can influence co-evolution, contribute to the construction of adaptive niches, and affect the process of adaptive structuration (the form of expression).[32]

Functor F_4, that of "*Events*," acts as a line of flight between Energetic-Signaletic Flows (**E**) and Incorporeal Universes (**X**). It reflects how the interactions and relationships among social actors, individuals, and technology in the stakeholder ecosystem (**E**) can influence perceptual-cognitive processes and actions mediated by technologies (**X**). This transition represents a shift from the concrete interactions and relationships within the digitalized ecosystem to the abstract principles, goals, and values that guide these transformations.[33]

In the Incorporeal Universes (**X**), we have categories C_5 and C_6. Category C_5 includes concepts related to understanding and processing information, the development of artificial general intelligence, and the integration of these elements into computing experiences that adapt to individual needs and preferences, which form the MABs.[34] The principles, goals, and values that guide the transformations within the digital ecosystem form the substance of content. The specific ways these principles, goals, and values are prioritized and balanced is the form of content. Category C_6 delves into the creative and transformative aspects of technology, the organization and structuring of experiences, and the process of interpreting and making sense of complex information, which form the CAEs.[35] Public discourses, debates, and narratives articulating these principles, goals, and values that individuals create to make sense of their experiences with technology form the substance of expression. The form of expression is the specific ways these discourses, debates, and narratives are conducted. The functor F_5, that of "*Sense*," represents this double articulation of territorialization, highlighting how machinic generalized intelligence (MGI) and experience computing (the substance of content) contribute to the processes of sense-making, poiesis, and enchantment (the form of expression).[36]

---

[32] Our concept of "flows" combines notions of networked information flows (Castells, 2010), actor networks (Latour, 2005), and machinic desire (Deleuze and Guattari, 1977), to account for the dynamic structuring and reconfiguration of socio-techno-economic territories, underpinned by reciprocal and adaptive exchange of information within networks of human and nonhuman actants, driven by an interplay of forces and intrinsic motivations.

[33] "Events" emerge from the interconnected dynamics of systems (Bateson, 2000), through the interplay of surface and depth structures (Deleuze, 1990), and lead to systemic changes via communication and information processing (Luhmann, 1995), signifying a critical transition from concrete interactions in digital ecosystems to abstract cognitive processes.

[34] Integrating internal cognitive mechanisms (Friston, 2010; Parr et al., 2022) and external technological mediation (Bachelard, 2002; Hansen, 2015), recent human-AI copilot systems are MABs that comprise fluid, adaptive networks, blending human cognition with AGI, continuously evolving and transcending traditional information processing paradigms.

[35] Interconnected intelligent ecosystems combine collective data interpretation (Weick et al., 2005), structured experience sequencing (Deleuze, 1990), and transformative technology engagement (Sha, 2013), into CAEs, manifesting in a network of expressive, organized, and evocative interactions within AGI systems.

[36] Adopting insights from the notions of duration (Bergson, 1896), affect (Massumi, 2002), ethico-aesthetics (Guattari, 1992), and cyborg (Haraway, 1991), our concept of "sense" can be defined as a mapping of AGI principles to human experiences, synthesizing temporal memory dynamics, affective engagement, ethical assemblage, and human–technology hybridity.

Functor F_6, that of *"Enactments,"* acts as a line of flight between Incorporeal Universes (**X**) and Abstract Machinic Phyla (**M**). It reflects the relationship between perception, cognition, and MGI in the context of experience computing and poiesis and enchantment, series and structure, and sense-making (**X**) and how these influence collaboration, co-design, and open innovation (**M**). This transition represents a shift from the realm of abstract concepts and ideas to the realm of concrete actions and practices, highlighting how the meanings and experiences generated by users inspire new collaborative efforts, fostering the creation of new ideas, artifacts, and practices.[37]

Finally, in the Abstract Machinic Phyla (**M**), we have categories C_7 and C_8. Category C_7 focuses on the processes of working together, creating new ideas and solutions collectively, and fostering an open environment for innovation, which form the MABs.[38] The entities that structure and organize the system, that enable and support digital experiences and interactions, form the substance of content. The specific ways these structures and organizations are arranged and coordinated is the form of content. Category C_8 encompasses the underlying technological systems, platforms, and services that enable and support digital experiences and interactions, which form the CAEs.[39] Administrative policies, procedures, and protocols that structure and organize the system, as the narratives and discourses that individuals create to make sense of their experiences with technology, form the substance of expression. The form of expression is the specific ways these policies, procedures, and protocols are articulated. The functor F_7, that of *"Resourced Capabilities,"* represents the double articulation of territorialization, between collaboration, co-design, and open innovation (the substance of content) and digital infrastructure, platforms, and services (the form of expression).[40]

Functor F_8, that of *"Interactional Creation,"* acts as a line of flight between Abstract Machinic Phyla (**M**) and Existential Territories (**L**). It encapsulates how digital infrastructure, platforms, and services (**M**) enable and support user interactions and generative processes (**L**). This transition represents a shift from the realm of collaborative actions and open innovation to the

---

[37] Combining performativity (Butler, 1990), agencement (Callon, 2021), practice (Schatzki, 2002), reflection-in-action (Schön, 1983), and design thinking (Micheli et al., 2019), our concept "enactments" entails a dynamic, multidimensional process where cognitive and emotional experiences are continuously performed, arranged, practiced, reflected upon, and designed into collaborative, innovative, and practical outputs.

[38] In co-innovation projects, dynamic interplay of diverse participants in collaboration (Fjeldstad et al., 2012), convergence of multiple perspectives and skills via codesign (Sanders and Stappers, 2008) and integration of external and internal ideas via open innovation (Chesbrough, 2003), constitute a MAB, as it exhibits heterogeneity, interconnectedness, and the production of emergent properties through diverse interactions.

[39] Digital infrastructures (Tilson et al., 2010), platforms (Tiwana, 2014), and services (Maglio et al., 2010) enable and support digital interactions, embodying the expressive and communicative essence of CAEs in articulating and facilitating shared experiences and engagements within digital ecosystems.

[40] Linking dynamic capabilities (Teece, 2007), resource orchestration (Sirmon et al., 2011), platformization (Tiwana et al., 2010), and business modeling (Zott et al., 2011), our concept of "resourced capabilities" refers to orchestrating firm resources and capabilities within evolving platform architectures to facilitate the continuous integration and adaptation of organizational and technological capacities in response to ecosystem changes.

realm of individual user interactions and affordances, highlighting how the infrastructure and tools provided by the ecosystem support and shape the ways users interact with the system.[41]

In practical terms, let us return to the case of breast cancer discussed earlier. MLXEs in this context could be a digital health ecosystem designed to support breast cancer prevention, diagnosis, and treatment. This ecosystem would include various technologies, platforms, and services, such as wearable devices for monitoring health, telemedicine platforms for remote consultations, and AI-powered tools for analyzing medical data.

Starting with the category C_1, affordances in this context could be the capabilities of the digital health technologies, such as the ability of a wearable device to monitor vital signs or the ability of an AI tool to analyze mammogram images. User interactions would be the ways patients, healthcare providers, and others engage with these technologies, such as using a wearable device to track health data or using a telemedicine platform for a consultation. Generativity would be the potential of these technologies to create new forms of value, such as generating insights about a patient's health or facilitating early detection of breast cancer.

The functor F_1, "*Lived Journey*," would represent how these interactions with technology can lead to new relationships among patients, healthcare providers, and others involved in breast cancer care, and contribute to the individuation process, such as forming a personalized treatment plan based on a patient's unique health data.

The category C_2 would explore the dynamic relationships between individuals and technology, the technical aspects of human experience, and the process of forming collective identities through interactions with technology. For example, how patients, healthcare providers, and others form a collaborative care team through their interactions with the digitalized health ecosystem.

The functor F_2, "*Engagements*," would connect the concepts in C_2 to those in the next category, C_3. It would highlight how the interactions and relationships among the care team in the digitalized health ecosystem can influence the development of new practices and approaches to breast cancer care.

The category C_3 would analyze how the interconnected network of patients, healthcare providers, technology (such as wearable health monitors, telemedicine platforms, and AI-powered diagnostic tools), and the broader environmental context (including policy landscapes and healthcare infrastructure) collectively influence health outcomes and healthcare delivery. For example, how the interactions between patients, healthcare providers, and technology shape the care process and influence the outcomes.

The functor F_3, "*Flows*," would capture how these interactions and relationships can influence the co-evolution of the digitalized health ecosystem, contribute to the construction of adaptive niches, such as personalized care pathways, and affect the process of adaptive

---

[41] Our notion of "interactional creation," through a conceptual synthesis of order parameters (Haken, 2006), vice-diction (Deleuze, 1994), and feed-forward mechanisms (Hansen, 2015), refers to the dynamic process where platforms and individuals co-evolve in a transition from structured, potential systems to dynamic, lived experiences, facilitated by platforms' predictive capabilities and individuals' creative engagement, leading to continual innovation and transformation.

structuration, such as the development of new care protocols based on the insights generated from the digitalized health ecosystem.

The category C_4 would provide a lens through which to view the digital health ecosystem as a dynamic, co-evolving space where technology, individuals, and social structures interact in complex, mutually influential ways to shape and redefine the landscape of healthcare. For example, how the digitalized health ecosystem evolves in response to the needs of patients and healthcare providers, and how it shapes the way breast cancer care is delivered.

The functor F_4, "*Events*," would reflect the relationship between co-evolution, niche construction, adaptive structuration, and the concepts in category C_5. It would show how as the digitalized health ecosystem evolves and adapts, it influences the development and application of perception, cognition, and MGI in the context of breast cancer care.

The category C_5 would examine the convergence of internal cognitive mechanisms and external technological mediation to create adaptive, fluid networks that blend human cognition with AGI, fostering environments that dynamically adapt to and anticipate the needs and preferences of individuals. For example, how AI tools in the digitalized health ecosystem can analyze health data, generate insights, and support decision-making in breast cancer care.

The fifth functor, F_5, "*Sense*," would represent the relationship between perception, cognition, and MGI in the context of experience computing and poiesis and enchantment, series and structure, and sense-making. For example, how AI tools contribute to the processes of sense-making, poiesis, and enchantment in breast cancer care, such as interpreting health data, generating insights, and creating meaningful experiences for patients and healthcare providers.

The sixth category, C_6, would underscore the role of AGI in organizing, structuring, and interpreting complex information within digital health ecosystems, highlighting its transformative potential in creating adaptive, patient-centered care environments that respond to the evolving needs and experiences of individuals. For example, how the digitalized health ecosystem supports the creation of new care pathways, the organization of care processes, and the interpretation of complex health data.

The sixth functor, F_6, "*Enactments*," would reflect the relationship between poiesis and enchantment, series and structure, and sense-making and collaboration, co-design, and open innovation. For example, how the insights and experiences generated by the digitalized health ecosystem inspire new collaborative efforts among patients, healthcare providers, and others, fostering the creation of new ideas, practices, and solutions in breast cancer care.

The seventh category, C_7, would explore the ways in which co-innovation projects facilitate a synergy among diverse participants, integrating multiple perspectives and skills through codesign, and leveraging both external and internal ideas via open innovation frameworks. For example, how patients, healthcare providers, and others collaborate to co-create personalized care plans and foster an open environment for innovation in breast cancer care.

The seventh functor, F_7, "*Resourced Capabilities*," would reveal the connections between the social aspects of the digitalized health ecosystem and their influence on the technical foundations. It would highlight how collaboration, co-design, and open innovation can drive the development and evolution of the digital infrastructure, platforms, and services in the digital health ecosystem.

I. Relational dynamics

The eighth category, C_8, would examine how digital infrastructures, platforms, and services act as the backbone of digital interactions, articulating and facilitating shared experiences and engagements within digital ecosystems. For example, the wearable devices, telemedicine platforms, and AI tools that support breast cancer prevention, diagnosis, and treatment.

Finally, the eighth functor, F_8, "*Interactional Creation*," would represent the relationship between digital infrastructure, platforms, and services and affordances, user interactions, and generativity. It would encapsulate how the digital infrastructure in the digitalized health ecosystem enables and supports user interactions and generative processes, such as the generation of health insights and the creation of personalized care plans.

In conclusion, the digitalized health ecosystem as an MLXE for breast cancer prevention, diagnosis, and treatment is a complex and dynamic ecosystem where various elements interact and evolve over time. By understanding the relationships and interactions within this ecosystem, we can inform the development of more inclusive, engaging, and innovative experiences for patients and healthcare providers, ultimately improving the quality of care and outcomes for patients with breast cancer.

Looking ahead

The foundational principles of DRT (see Table 2.1) are discussed in detail across part two (Chapters 3–6), part three (Chapters 7–10), and part four (Chapters 11 and 12).

Part Two discusses the creative transformations of dynamic relationalities:

1. **Dynamic Relationality within Assemblages** (Chapter 3): Within the framework of DRT, assemblages are understood as fluid networks of diverse elements. These elements, represented as objects within a category, symbolize intricate interactions and facets of a system. Morphisms capture the dynamic influence these components exert on one another. The topology of this system, where each point signifies a unique assemblage or state of interaction, is influenced by the system's underlying energy principles that steer the transformational trajectory of the assemblage. The base of this topology, with its open sets, showcases the diverse practices within the system, and their overlaps depict the collaborative nature of the ecosystem. Thus, within assemblages, relations are not static but are in a state of continuous evolution, driven by both internal and external factors.
2. **Dynamic Relationality between Assemblages** (Chapter 4): Beyond the intra-assemblage dynamics, DRT also emphasizes the relationships between different assemblages. Different categories represent varied facets of a system. The relationship between these categories is intricate and nonlinear. Transversality provides a lens to understand these intersections, revealing deeper insights into the system's dynamics. The theoretical framework integrates these categories through functors and their natural transformations, offering a structured perspective on the system's dynamics. By incorporating advanced mathematical concepts, each category is visualized as a manifold, with transformations depicted by vector fields. Concepts like transversality and continuous deformation enrich our understanding of the interplay between

TABLE 2.1 Foundational principles of Dynamic Relationality Theory (DRT).

#	Foundational principle	Mathematical and *Philosophical* concepts	Chapter
1	**Dynamic relationality within assemblages:** Assemblages are fluid networks of diverse elements, with morphisms capturing dynamic influences. The topology of this system is influenced by the system's underlying energy principles that steer the transformational trajectory of the assemblage.	Category, morphism, topological space, open set, base set, smooth manifold *Assemblage, Rhizome, Machinic Desire*	3
2	**Dynamic relationality between assemblages:** Relationships between different assemblages are intricate and nonlinear. Transversality reveals intersections, and functors integrate these categories, offering a structured perspective on system dynamics.	Functor, vector field *Territorialization, Strata, Double Articulation*	4
3	**Lines of flight and dynamic relationality:** "Lines of flight" signify departures from established structures, bridging stakeholder ecosystems to cognitive processes and propelling deterritorialization.	Transversality *Line of Flight, Event*	4
4	**Becoming and continuous transformation of identity:** "Becoming" represents a continuous transformation, challenging and reshaping established identities. This metamorphosis is symbolized as a natural transformation between functors.	Natural transformation, enriched category *Becoming*	5
5	**Reterritorialization of relational dynamics:** As assemblages evolve, the foundational principles governing their relational dynamics undergo profound shifts. Natural transformations ensure a structured path of evolution.	*Reterritorialization*	5
6	**Differential transformation in dynamic relationality:** Differential topology provides a dynamic lens for transformations within and between categories, with concepts like transversality and homotopy theory enriching the understanding of these dynamics.	Differential form, monad, homotopy	6
7	**Dynamic relationality of molar and molecular lines:** The interplay between local (molecular) and global (molar) dynamics is emphasized, with sheaf theory contextualizing how local cognitive processes intertwine with global dynamics.	Sheaf, presheaf, sheaf morphisms *Molar/Molecular, Induction, Translation, Transduction*	7
8	**Diagrammatic relationality in ecosystem dynamics:** Diagrams offer a visual representation of complex ecosystems, with "limits" and "colimits" merging structures and underscoring assemblage-organization interactions.	Diagram, limit/colimit object, 2-category	8, 9
9	**Strategic architecturing of organizations in ecosystems:** Organizations operate as dynamic constructs within ecosystems, balancing adaptability (BwO) and stability (OB), with sheaf/stack analysis providing insights into these interactions.	Stack sheaf *Body-without-Organs (BwO)*	10
10	**Immanence of emergence and evolution in ecosystems:** The plane of immanence epitomizes limitless potential, differentiating into subcategories capturing facets of identity, transformation, and interconnectedness. Evolution is an interplay of potentialities and actualities.	Fiber bundle *Plane of Consistency, Differentiation of Four Domains*	11, 12

categories, providing a comprehensive view of the dynamic relationality between assemblages.
3. **Lines of Flight and Dynamic Relationality** (Chapter 4): In the realm of DRT, the concept of "lines of flight" signifies departures from established structures, opening avenues for potential new formations. This idea, encapsulated by the functor "Event," bridges stakeholder ecosystems to cognitive processes, marking a transition from overarching structures to intricate transformations. Such lines of flight propel deterritorialization, fostering new assemblages and reshaping existing ones. They act as catalysts for change, driven by quasi-causes that unsettle prevailing assemblages, leading to novel formations. These transformations, whether rooted in global structures or in flux between paradigms, are guided by various mechanisms, emphasizing shifts in prevailing norms and practices.
4. **Becoming and the Continuous Transformation of Identity** (Chapter 5): Central to Deleuze and Guattari's philosophy and DRT is the concept of "becoming"—a continuous transformation that challenges and reshapes established identities and structures. In category theory, this metamorphosis is represented as a natural transformation between functors, symbolizing the evolution of identity. This "becoming" is not a mere state change but a profound metamorphosis, indicating shifts in methodologies and paradigms. Visualizing this process can be likened to phase transformations, representing the fluidity of identity.
5. **Reterritorialization of relational dynamics in evolving assemblages** (Chapter 5): As assemblages evolve, the foundational principles governing their relational dynamics also undergo profound shifts. This is not a mere adaptation but a deep metamorphosis, a reterritorialization, that redefines the very essence of the assemblage's relational structure. As categories representing assemblages evolve, the functorial relationships that map and connect these categories must also transform. Guiding this intricate transformation is the concept of a natural transformation. It acts as a meta-layer of transformation, ensuring that as the categories (assemblages) evolve and their internal and external (functorial) relationships shift, there remains a coherent, overarching structure that preserves the integrity of the entire system. This natural transformation ensures that the evolution of relational dynamics is not chaotic but follows a structured path, even as it allows for profound changes.
6. **Differential Transformation in Dynamic Relationality** (Chapter 6): Differential topology offers a dynamic lens to view transformations and flows within and between categories. As for the dynamics of relationalities within categories, seeing each category as a smooth manifold, with every point representing unique configurations of objects and morphisms, vector fields associated with each manifold capture transformational dynamics, indicating change direction and magnitude within a category. Lie derivatives can be used measure transformational flow of objects and morphisms, while differential forms reveal transformation intensity and De Rham cohomology elucidates the category's topological evolution. As for the dynamics of relationalities across categories, transversality is the way two submanifolds intersect in a differentiable manifold, such that they meet in a nontangential manner, in order to study the intersections between different categories (manifolds), such as modeling the points where lines of flight intersect with the strata. Homotopy theory provides a way to study the dynamics

behind natural transformations as a continuous path in the space of functors from one functor to another, not just the initial and final states of the transformation, but also the entire "process" of the transformation.

Part Three discusses co-creative transformations and organizations.

7. **Dynamic Relationality of Molar and Molecular Lines of Segmentation** (Chapter 7): Dynamic Relationality Theory (DRT) underscores the intricate interplay between local and global dynamics, represented by molecular and molar lines of segmentation. Molar lines, shaped by overarching socio-political forces, influence the broader structures of systems. In contrast, molecular lines delve into micro-transformations, capturing the nuances of cognitive landscapes and sense-making processes. The fluidity of this ecosystem is a testament to the reciprocal influence between these lines: while molar lines mold local dynamics, molecular lines subtly shape global systems. Through the lens of sheaf theory, we can contextualize this interplay, understanding how local cognitive processes intertwine with global dynamics, offering a comprehensive view of the system's relationality.

8. **Diagrammatic Relationality in Ecosystem Dynamics** (Chapters 8 and 9): DRT offers a diagrammatic perspective to decipher the complex architecture of ecosystems. Assemblages within categories resemble smooth manifolds, where manifold points depict unique configurations of entities and their interrelations. Functors interlink these categories, illustrating entity-assemblage dynamics. An organization's evolutionary path in the ecosystem is charted by a sequence of functors. Diagrams, acting as abstract maps, spotlight relationships within assemblages and extend to larger entities like organizations. These relationships are integrated into the ecosystem through "limits" and "colimits," universal constructions that merge structures and underscore assemblage-organization interactions. Organizations, as assemblage configurations, navigate between their current and aspirational states. Diagrams represent the present organizational state, with limit objects indicating ideal configurations and colimit objects showcasing growth potential. Category theory introduces a typology for organizational analysis, distinguishing between types like Structured Corporations and Agencing Collectives based on functor stability and limit/colimit focus. The ecosystem's macro-architecture, crafted through diagrams, emerges from foundational category interactions. Higher categories, such as 2-categories, capture intricate intercategory dynamics. The ecosystem's architectural diagram is anchored in the platform category, subtly influencing entities. Addressing misalignments demands the design of co-innovation platforms, as entities actively mold these structures, with diagrams evolving through organizational morphogenesis, interwoven with meaning-making and boundary-crossing. Grasping these transformations empowers organizations to steer the ecosystem toward envisioned goals.

9. **Strategic Architecturing of Organizations in Ecosystems** (Chapter 10): Organizations are dynamic constructs with entities forming multiple assemblages, each with unique dynamics. Central to this is the duality of potentiality (Body without Organs, BwO) and actualization (Organized Body, OB). Diagrams visually trace this dynamic, with nodes symbolizing entities and edges denoting relationships. As organizations evolve, new potentials (BwO elements) integrate into the established structure, transitioning to OB.

The BwO and OB dynamics parallel the Molecular (adaptability-focused) and Molar (stability-focused) lines of organizations. Complex organizations operate as intricate ecosystems, with diverse interactions. Sheaf/Stack Analysis provides insights into these interactions, especially in multifaceted sectors like healthcare. Energy dynamics between BwO and OB are encapsulated by kinetic and potential terms, governed by the Lagrangian. Shifts between adaptability and stability are influenced by these dynamics. In enterprise modeling, misalignments in diagrams indicate inconsistencies. Addressing these challenges can involve adjusting diagrams, using functors and natural transformations, or employing higher abstraction levels like 2-categories.

Part Four discusses complex transformations in Machinic Life-Experience Ecosystems.

10. **Immanence of Emergence and Evolution in Ecosystems** (Chapters 11 and 12): The inevitability and persistence of emergence and evolution in ecosystems are underscored through the concept of "immanence." This principle dives deep into the very fabric of dynamism, spotlighting how potentialities and actualities are not mere possibilities but are immanent forces driving transformation in ecosystems. The monadic structure serves as a foundational tool in elucidating this perspective. Its unit transformation highlights subjective experiences of emerging entities via the endofunctor's continuous transformational potential. The multiplication transformation captures the interconnectedness of entities. Incorporeal Universes act as abstract transformational vector fields. Taken together, the monadic structure illustrates the inherent, ongoing metamorphosis present within entities of the ecosystem. Their intricate connections and influence on each other draw a map of the immanent forces at play. While the gauge theory framework offers structured insights into the complex interplay of Existential Territories, Energetic-Signaletic Flows, Incorporeal Universes, and Abstract Machinic Phyla, the incorporation of Lagrangian dynamics emphasizes the kinetic and potential energies guiding the system's evolution, capturing both the movement and tension within the ecosystem's transformations delineating the perpetual processes of internalization and externalization and portraying the incessant folding and unfolding of entities in ecosystemic contexts.

Dynamic relationality theory

CHAPTER 3

Dynamic relationalities in an assemblage

In the previous chapter, we emphasized the intricate dynamic relationalities in a **Machinic Life-Experience Ecosystem (MLXE)** between different assemblages, visualized through categories and their interconnections via "functors," a mathematical construct that maps objects and their relationships (morphisms) from one category to another while preserving their intrinsic structure and relational dynamics.[1] In practical terms, we discussed the use of categories and functors in the creative transformation of relational dynamics in a healthcare ecosystem with the breast cancer example.

In the chapters that follow in Part 2, we delve deeper into the creative transformations of dynamic relationalities within and across assemblages, and across (stacked) strata, culminating in relational dynamics of dynamic relationalities (see Fig. 3.1).

---

[1] Postulation of morphisms between objects of different kinds within a category, and the existence of functors mapping different categories of objects and morphisms, can be approached both theoretically and methodologically. In the MLXE context, objects represent various components of a complex dynamic ecosystem. We can then consider morphisms between these diverse entities since they interact and influence each other within the ecosystem. Methodologically, using morphisms and functors provides a clear and structured way to analyze complex systems, as morphisms allow for the detailed study of how different elements within an MLXE interact and influence each other. Functors, by mapping these relationships across categories, enable the comparison and analysis of these interactions across different contexts or scales.

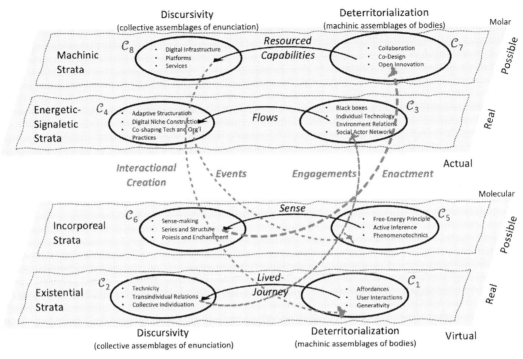

FIGURE 3.1 Dynamic relationalities in a Machinic Life-Experience Ecosystem

Dynamic relationality #1: Within assemblages

In category theory, objects can be thought of as the components of an assemblage, while morphisms represent the relations among these components.[2] The entire category then represents the assemblage as a whole, capturing both its structure (the objects and their relationships) and its dynamics (the transformations represented by the morphisms). This allows us to model assemblages in a very flexible and abstract way. For example, we can represent different types of components (e.g., individuals, groups, institutions, physical entities, ideas, etc.) as objects, and various kinds of relations (e.g., influence, communication, conflict, cooperation, etc.) as morphisms. The category then captures the overall structure and dynamics of the assemblage, including how it evolves and transforms over time.

For instance, in Fig. 3.1, category C_3 represents stakeholders as machinic assemblages of bodies composed of a complex web of social actors, networks, and the interactions between

---

[2] Each morphism shows one particular relation between relata, and the semantic/ontological meaning of that relation would specify one direction. We can define other morphisms between the same objects or add other objects and morphisms in each category. We would then have more and more complicated categories. It is important to see that our current MLXE framing, for the sake of exposition, is a minimalist model that captures a very small number of theoretical constructs and their relationships through category theoretical representation. Other theorists can and should add or complexify each category with more objects, morphisms, functors, etc. With DRT, our goal is to outline a core theoretical and methodological toolbox.

individuals, technology, and the environment, as well as the unknown or hidden aspects of the MLXE. In C_3, we have three objects:

- BlackBoxes[3]: This object represents aspects of the ecosystem that remain less explored or unknown. These black boxes influence the relationships within the ecosystem and warrant further investigation.[4]
- IndividualTechnologyEnvironmentRelations: This object represents the systemic and dynamic relationships between individuals, technology, and their environment. These relationships play a significant role in understanding the MLXE. [5]
- SocialActorNetworks: This object represents the dynamic interactions between humans and nonhumans within the ecosystem. Social actors are considered as products of interaction and network connections.[6]

In C_3, we also have two morphisms: f_{31} and f_{32}.

- f_{31}: BlackBoxes $\to$ IndividualTechnologyEnvironmentRelations. This morphism captures the idea that the unknown or hidden aspects of the ecosystem (black boxes) influence the relationships between individuals, technology, and their environment.[7]
- f_{32}: IndividualTechnologyEnvironmentRelations $\to$ SocialActorNetworks. This morphism represents the concept that the relationships between individuals, technology, and their environment shape the dynamics of actors-networks within the ecosystem.[8]

---

[3] We will use (upper) Camel case naming convention in denoting objects.

[4] The black box concept, primarily drawn from actor-network theory (ANT), refers to the way complex systems or networks become simplified and opaque in their operation once they are stabilized (Callon 1986a). In an assemblage, entities or processes are often "black-boxed," meaning their internal complexities are hidden or abstracted away, making the assemblage appear as a singular, cohesive unit.

[5] Oftentimes operating in the background without our direct engagement, technologies shape our cognitive engagement with the environment and our interpretation of data and signals from our surroundings (Ihde 1990). As they are used and understood in multiple ways by different individuals or within different environmental contexts, technologies are not fixed in their meanings or uses but are multistable. Individual-technology-environment relations are foundational in shaping the emergent properties and capabilities of assemblages, as they define how the assemblage behaves and evolves in response to external and internal stimuli.

[6] Actor-network describes the network of heterogeneous elements (both human and nonhuman) that interact and form complex systems (Latour 2005). We conjoin it with "social" to emphasize the importance of relationships and connections between various actors in an assemblage in shaping outcomes.

[7] For example, black-boxed technical objects, such as electoral maps and content management systems, can significantly influence political representation and journalistic practice (Anderson and Kreiss 2013), while black-boxed technical processes, such as cancer care information pathways, can shape the dynamics of patient care and interactions between patients, technology, and the environment (Lefkowitz 2022).

[8] Consider that the multistability of technology allows mundane artifacts to have multiple meanings and uses in different contexts (Rosenberger 2014), while technological entities play a crucial role in mediating and influencing collective practices (de Boer, Te Molder, and Verbeek 2018).

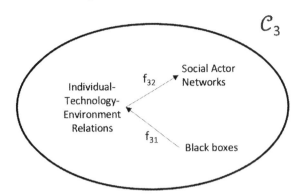

FIGURE 3.2 Diagram of a category with its objects and morphisms.

Our illustrative example C_3 can be considered a category according to category theory, as it satisfies the identity and composition requirements in the definition of a category.[9] In category theory, it is customary to visualize a category as an oval with objects connected by morphisms shown as arrows (see Fig. 3.2).

In the topological space of MLXE, each point signifies a unique assemblage or state of interaction. Ideal types for each object simplify the complex realities, providing a foundation to explore potential configurations. The base of this topology, consisting of open sets, represents various practices within the MLXE. Overlaps between these sets form more complex open sets, illustrating the interplay of multiple practices. These territories or open sets within the plane of immanence are characterized by distinct interaction patterns, with overlaps reflecting collaborations within the MLXE. The topological space, base sets, and open sets provide a mathematical lens to comprehend the intricate dynamics of the MLXE. By defining these elements, we can capture interactions in C_3, offering a structured understanding of the ecosystem's stakeholders. The dynamics of these interactions are governed by the kinetic and potential terms of the Lagrangian, determining the transformation of the assemblage. We will further explain and elaborate on these in the rest of this chapter.

---

[9] The identity morphisms symbolize the consistent nature of each object, while the composition of morphisms and their closure under composition reflect the interconnected and sequential influence within the ecosystem. Associativity ensures that the order of these influences does not alter the overarching impact, aligning with real-world ecological and social systems.

Rhizomatics of machinic desire within assemblages

In the complex ecosystem of healthcare, understanding the dynamics of relationality is of major importance. Assemblages, rooted in the philosophy of Deleuze and Guattari, and ecosystems, visualized as vast interconnected networks, both underscore the importance of relationships, interactions, and transformations. This section delves into the origins of dynamic relationality within healthcare assemblages and ecosystems, aiming to illuminate the foundational principles that give rise to their complex interplays (for a more technical background, see "Gauge Theory of Desiring Machines" in the **Appendix**).

Healthcare assemblages as microcosms of relationality

At the heart of healthcare assemblages lies the idea of heterogeneous components—physicians, patients, technologies, and institutions—coming together in fluid arrangements to form functional networks.[10] For instance, the Category C_3 from our earlier discussion represents stakeholders as machinic assemblages within the healthcare ecosystem, capturing interactions between humans, technology, and the environment. The origins of this dynamic relationality can be traced back to the inherent nature of components within healthcare assemblages. Each component, whether it be a physician, patient, or a piece of medical technology, has its desires, objectives, and functions. The "desire" here is not just a simple longing but a complex interplay of care, treatment, and healing.[11]

The rhizome serves as the backdrop against which these desiring-machines operate, emphasizing nonlinearity and interconnectedness.

Mathematical underpinnings of healthcare relational dynamics

The dynamics of relationships can be visualized as a series of configurations or states. Each state represents a unique arrangement within the healthcare network. The sequence and nature of these states mirror the nonlinear interactions within healthcare systems, such as the progression of patient care or the sequence of medical procedures. Transformations, representing shifts in the dynamics of these relationships, offer insights into the reconfigurations within healthcare systems. For instance, a transformation might evolve a traditional doctor-patient relationship into a more collaborative one involving multiple stakeholders. The dynamics of these transformations are governed by principles that capture the flow or change of objectives over time, representing the active dynamics of healthcare interactions. They also signify the potential for objectives to be produced or hindered, influenced by external factors like societal norms or internal factors like individual motivations.[12]

States of desire in healthcare

- **Conformity state**: A healthcare system that strictly adheres to established medical protocols can be represented by a state of alignment.
- **Resistance state**: A grassroots movement advocating for patient rights against

Continued

> ## Rhizomatics of machinic desire within assemblages *(cont'd)*
>
> established medical norms represents a state of resistance.
> - **Hybrid state**: A healthcare institution that balances traditional medical practices with innovative patient care methods represents a state that is a mix of alignment and resistance.
>
> ### Transformations of desire in healthcare
>
> In the realm of healthcare, the transformations of desire can be visualized through various scenarios:
> - **Conjunction of flows**: Consider a scenario where a patient's desire for holistic care merges with a physician's desire to incorporate alternative medicine. This merging signifies the unity operation of joining different flows of desire.
> - **Blocking or repulsion of flows**: Imagine a healthcare institution that resists the adoption of a new medical technology due to budget constraints. This resistance indicates a blockage or repulsion of the flow of desire to integrate the new technology.
> - **Schizophrenic flow**: A patient's erratic shift between wanting traditional treatment and seeking alternative therapies signifies the unpredictable, nonlinear flow of desire.
>
> ### Illustrating the dynamics of desire in healthcare
>
> In the intricate landscape of healthcare, the dynamics of desires and treatment pathways often intertwine, leading to a series of transformative states.[13] Take Mrs. Smith, for instance. Initially adhering to traditional treatments, her wish for holistic care aligns with Dr. Jones's approach to comprehensive treatment. As they collaborate, Mrs. Smith transitions from a state of traditional adherence to a mixed state, considering both conventional and alternative treatments. The pace and direction of this transition are influenced by various factors. Rapid changes in her preferences might see her actively exploring alternative treatments. However, societal norms favoring traditional methods could pose challenges to her explorative journey.
>
> Similarly, Mr. Lee's journey through the healthcare system is marked by challenges and transformations. Actively exploring a mix of treatments, he encounters a roadblock when his insurance denies coverage for an experimental procedure. This setback pushes Mr. Lee into a state of resistance, where he might question or challenge established norms. The intensity of his resistance is shaped by external barriers, such as insurance policies. If these barriers become too dominant, they could significantly obstruct his desires. On the other hand, if he remains proactive, he might find alternative pathways to achieve his treatment goals.
>
> Ms. Taylor's healthcare journey is an illustration of the complexities of patient desires. Resisting her current treatments, she oscillates between different therapies, seeking the best fit. This erratic pattern of behavior leads her to fluctuate between states of adherence to established treatments and resistance against them. The dynamics of her journey are influenced by societal perceptions and her own rapid shifts in treatment preferences. While her swift changes might be a quest for the best therapeutic fit, societal

Rhizomatics of machinic desire within assemblages (cont'd)

views on her unpredictable choices could present her with criticism, further complicating her treatment journey.

Healthcare ecosystems: Assemblages at scale

The MLXE is a confluence of stakeholders, technologies, and interactions in healthcare. Just as desires flow, merge, or get blocked in the realm of assemblages, healthcare ecosystems exhibit their dynamics of collaborations, resistances, and influences. The topological space represents all possible configurations of interactions within the healthcare ecosystem. Each point in this space corresponds to a unique state or configuration of the healthcare system, much like the various states of desire. The transformations between these points, akin to transformations of desire, capture the dynamic shifts in healthcare interactions. The base sets in the topology represent foundational interactions within the healthcare ecosystem, while open sets in the topological space represent groups of similar or "close" interactions within the healthcare ecosystem.[14]

Origins and implications in healthcare

The origins of dynamic relationality within healthcare assemblages and ecosystems can be attributed to three core principles:

- **Inherent desire for care**: Every component in healthcare, be it a physician, patient, or technology, has an inherent desire for expression, care, and healing. This desire drives the dynamics of relationality in healthcare.
- **Nonlinearity and interconnectedness**: The rhizomatic nature of healthcare assemblages and ecosystems emphasizes that components are interconnected in complex, nonlinear ways. For instance, the relationship between a physician, patient, and a piece of medical technology is not hierarchical but interconnected.
- **Adaptability and evolution**: Healthcare assemblages and ecosystems are not static entities. They adapt, evolve, and transform based on internal dynamics and external stimuli, such as new medical technologies or healthcare policies.

By understanding the origins and implications of dynamic relationality, healthcare professionals, policymakers, and patients can better navigate the intricate web of interactions that define the healthcare landscape.

[10] Duff (2014) elucidates the concept of healthcare assemblages, emphasizing how diverse elements like physicians, patients, and technologies converge in dynamic networks, impacting health and wellbeing through evolving affective and relational capacities.
[11] In his analysis of healthcare assemblages, Duff (2023) elucidates the concept of "desire" as a potent force integral to the assemblage's composition. Desire functions as both *potentia*, the active, constitutive power, and *potestas*, the reactive, constituted force within healthcare networks. This dual nature of desire drives the selection and interaction of diverse elements—such as patients, medical professionals, and technology—facilitating their integration into productive networks.
[12] Chang et al.'s (2017) highlights the value of mathematical models in capturing the intricate functioning of health systems, through the use of "systems thinking" and models like "system dynamics" (SD) and "susceptible-infected-recovered-plus" (SIR+), which help predict outcomes of policy

Continued

> ## Rhizomatics of machinic desire within assemblages *(cont'd)*
>
> interventions and understand the flow of resources. However, as Mayhew et al.'s (2022) analysis of Ebola response in Sierra Leone reveals how local actions and social systems intricately shape health responses, these existing models may be insufficient. This complexity is what motivates fresh modeling strategies that can more accurately represent the nonlinear and multifaceted nature of healthcare relational dynamics.
>
> [13] Hannigan's (2013) study on the emergence and impact of crisis resolution and home treatment (CRHT) services in mental healthcare provides a vivid illustration of the concepts of conformity, resistance, hybrid states, and transformations of desire in healthcare. The CRHT teams, set up as an alternative to hospital admission, reflect a hybrid state where traditional mental health practices are blended with innovative home treatment methods. The implementation of these teams, while valued by users, also triggered system-wide effects such as staff redistribution and workload shifts, showcasing the conjunction and repulsion of flows of desire. These dynamics highlight the complexities of adapting healthcare systems to new models, where the desire for innovation meets the practical realities of resource allocation and existing practices. Hannigan's case study illustrates how policy changes can lead to a range of responses, from alignment with new initiatives to resistance or mixed reactions within the healthcare ecosystem.
>
> [14] Trnka's (2021) exploration of "multisited therapeutic assemblages" aligns well with the concepts of topological space, base sets, and open sets in healthcare ecosystems. Her focus on youth mental health support across various sites, both physical and virtual, exemplifies the concept of topological space in healthcare. Each site, whether a clinic, a home, or an online forum, acts as a distinct point in this space, representing different configurations of therapeutic interactions. The base sets in this topology are akin to fundamental elements such as primary care facilities or key online resources that underpin the ecosystem. The open sets, in Trnka's context, are groups of these sites that, despite their individuality, share functional connections or thematic similarities, such as various online platforms for mental health or a network of community clinics. These sets of therapeutic sites interact and influence each other, shaping the overall landscape of mental health support. In summary, Trnka's work provides a rich, real-world depiction of how healthcare ecosystems function as complex, interconnected topologies, with each site contributing uniquely yet integrally to the overall system.

Categories and open sets as smooth manifolds and submanifolds

Let us define topological spaces of categories as differentiable (smooth) manifolds.

- **Categories as differentiable (smooth) manifolds**: A differentiable manifold represents a space where each point and its immediate surroundings can be smoothly mapped to a familiar coordinate system, ensuring that the properties and changes in the system are continuously trackable and predictable.[15] In this context, the objects of the categories can be seen as points in the manifold, and the morphisms as differentiable functions between these points. A smooth manifold is a specific type of differentiable manifold

---

[15] In a more technical sense, a differentiable manifold is a topological space with a maximal atlas, which is a collection of charts that cover the entire space and where the transition maps are differentiable.

where all partial derivatives exist and are continuous. Tangent spaces are the key tool for understanding the local behavior of a smooth manifold. The tangent space at a point on a manifold is the set of all possible tangent vectors at that point, which are essentially the directions in which one can move from the point. The tangent bundle of a smooth manifold consists of all the tangent vectors at each point of the manifold.

- Consider C_4, the "Ecosystem" category, as a differentiable manifold. This manifold represents the entire landscape of coevolution, niche construction, and adaptive structuration in the context of healthcare.[16] In the language of differential topology, a manifold is a space where, around every point, there is a neighborhood that resembles Euclidean space. In our context, this means that around every point (or object) in C_4, there is a "neighborhood" of other objects that share similar characteristics. The concept of a differentiable manifold adds an additional layer of structure, requiring that the "charts" or mappings that we use to navigate between points in the manifold are smooth or differentiable. This means that small changes in one point lead to small changes in its image under the mapping. The morphisms in C_4 (f_{41}, f_{42}) represent these differentiable functions or "transition maps" between the points.

- Suppose a hospital decides to implement a new electronic health record (EHR) system. This decision represents a change in the state of adaptive structuration in the healthcare ecosystem. The hospital, as an organization, is adapting its structure to incorporate a new digital technology. The introduction of the EHR system creates a new "digital niche" within the hospital. This digital niche is the environment created by the EHR system, which includes the digital platform itself and the new workflows and practices associated with its use. For instance, doctors now enter patient

---

[16] In the context of assemblage theory, the manifold C_4 and the points on it can be interpreted as follows:

- **Manifold as Assemblage Space**: The manifold C_4 can be seen as an "assemblage space", a space that encompasses all possible configurations of the assemblage. An assemblage, in this context, refers to a complex system, such as the healthcare ecosystem, composed of various interconnected elements that interact and evolve over time. The assemblage space represents the entire landscape of possibilities for how these elements can be arranged and how they can interact.

- **Points as Assemblage States**: Each point on the manifold represents a specific state of the assemblage. In the context of the healthcare ecosystem, a point could represent a specific configuration of healthcare practices, technologies, and organizational structures. For example, a point might represent a state where a hospital is moving towards a different set of healthcare practices (an element of "Adaptive Structuration") by way of implementing a new electronic health record (EHR) system (an element of "Digital Niche Construction"), which has led to and the integration of technology and organizational practices (an element of "Co-Shaping Technology and Organizational Practices").

- **Morphisms as Assemblage Transformations**: The morphisms in C_4 which are represented as arrows or transitions between points on the manifold, represent transformations of the assemblage. These transformations could include the introduction of new elements, the removal of existing elements, changes in the relationships between elements, or changes in the properties of the elements themselves. For example, the morphism f_{41} might represent the process of a hospital implementing a new EHR system as it adapts its structures and practices.

In this way, the manifold provides a mathematical framework for representing and analyzing the complex dynamics of assemblages. It allows us to visualize the entire space of possibilities for the assemblage, to identify specific states of the assemblage, and to understand the transformations that can occur within the assemblage.

information directly into the EHR system, and this information is immediately accessible to other healthcare providers. If the change in adaptive structuration is small (e.g., the new EHR system is similar to the old one, and only minor changes in workflows are required), then the change in the digital niche construction is also likely to be small. However, if the change in adaptive structuration is significant (e.g., the new EHR system is drastically different from the old one, requiring substantial changes in workflows), then the change in the digital niche construction is also likely to be significant. The morphism f_{41}: AdaptiveStructuration $\rightarrow$ DigitalNicheConstruction represents this transformation process.

- **Open sets as submanifolds**: Consider the open sets in each category as submanifolds of the respective manifolds. A submanifold is a subset of a manifold that is itself a manifold, and the inclusion map is a differentiable embedding. In this context, the open sets represent specific regions or subsets within each category.
  - For instance, let us take as an open set a specific region within the ecosystem of coevolution, niche construction, and adaptive structuration, where traditional, centralized medical practices are receptive to technological innovations but maintain a hierarchy in decision-making. This is an area of the space where top-down technology changes are welcomed and incorporated into a niche in the healthcare ecosystem. The points within this submanifold could represent specific instances of this interaction.
  - For example, one point might represent a hospital that has implemented a new AI system for diagnosing breast cancer but has kept the decision-making process centralized among a group of expert physicians. Another point might represent a healthcare system that has adopted telehealth for breast cancer treatment but has kept the decision-making process centralized among a group of administrators.
  - The inclusion map, which is a differentiable embedding, ensures that the structure of the submanifold (the open set) is preserved when it is viewed as part of the larger manifold (the category). This means that the relationships and interactions between the points within the open set remain the same whether we view them as part of the open set or as part of the larger category.

To summarize, we can conceptualize the topology of categories as differentiable manifolds, with each point on these manifolds symbolizing a state within the system, and morphisms representing transformations between these states. This framework allows for visualizing and analyzing the dynamics of these systems, including interactions and evolutions over time. Additionally, submanifolds within these categories denote specific regions or scenarios, maintaining the integrity of relationships and interactions within the broader system.[17]

---

[17] For another illustration of these ideas, consider Nugus et al. (2010), who studied the complexity of patient trajectories and care integration within the porous, shifting boundaries of the emergency department (ED) operations. The manifold embodies the dynamic healthcare ecosystem, with each point representing a unique state of assemblage, like specific emergency care practices. Morphisms depict transitions, such as adapting to new technologies or policies, akin to assemblage transformations. Boundary-work and patient trajectories reflect tangent spaces, indicating possible directions of change within the system. Open sets as submanifolds represent specialized care areas within the ED, highlighting the manifold's diverse but interconnected elements, embodying the concepts of "organic, differentiated care."

Vector fields animating the dynamics over smooth manifolds

A vector field assigns a vector to each point in a space, which can be seen as representing the "force" or "direction" at that point. A vector field on a smooth manifold assigns a tangent vector to each point on the manifold. This can be used to describe how a system evolves over time, by considering the flow of the vector field. The flow is a one-parameter family of diffeomorphisms (i.e., smooth and invertible functions) that describe how points on the manifold move over time. Vector fields and flows are useful for understanding the behavior of systems that change over time. A connection is a way to differentiate vector fields along other vector fields. Curvature is a measure of how much a connection fails to commute.[18]

- Vector Fields: A vector field on $\mathbf{C_4}$ could represent the direction and magnitude of change in the healthcare ecosystem. For instance, each point in $\mathbf{C_4}$ could be associated with a vector that indicates the direction and rate of change in the adoption of new technologies or practices. For example, a vector at the point representing "Adaptive Structuration" could indicate the direction and rate of change in the adoption of new screening guidelines or treatment protocols. The length of the vector could represent the magnitude of the change, and the direction of the vector could represent the direction of the change (e.g., toward more personalized treatment protocols or toward more standardized protocols). For "digital niche construction," the vector could indicate the direction and rate of change in the development and implementation of digital technologies in breast cancer care, such as the adoption of digital mammography. These vectors could be seen as the derivatives of the transition maps (morphisms) in $\mathbf{C_4}$, indicating how these transitions are changing over time. Specifically, the derivative of morphism f_{41} could represent how the rate of change in "adaptive structuration" affects the rate of change in "digital niche construction."
- Flows: The flow of the vector field on $\mathbf{C_4}$ could represent the evolution of the healthcare ecosystem over time. For instance, the flow could describe how the adoption of new technologies or practices in the field of breast cancer care changes over time. If we consider time as the parameter of the flow, then the flow could describe how each point in $\mathbf{C_4}$ (i.e., each state of the healthcare ecosystem) moves over time under the influence of the vector field. For example, the flow could describe how the state of "Adaptive Structuration"

---

[18] In a context where we have multiple objects and morphisms, it might be more appropriate to consider a tensor field rather than a vector field. A tensor field assigns a tensor to each point in a space. Tensors are generalizations of vectors and can be used to describe more complex properties of a space. For instance, while a vector field might describe the direction and magnitude of change at each point, a tensor field can describe how these changes interact with each other. In the context of $\mathbf{C_4}$, a tensor field could represent the interactions between the changes in "adaptive structuration," "digital niche construction," and "co-shaping technology and organizational practices." For instance, a tensor at the point representing "adaptive structuration" could describe not only the rate of change in the adaptation of healthcare structures, but also how this rate of change interacts with the rates of change in "digital niche construction" and "co-shaping technology and organizational practices." These tensors could be seen as generalizations of the derivatives of the transition maps (morphisms f_{41} and f_{42}) in $\mathbf{C_4}$, indicating how the transitions between different states of the healthcare ecosystem interact with each other. The tensor field could provide a more comprehensive picture of the dynamics of the healthcare ecosystem, capturing not only the individual changes in "adaptive structuration," "digital niche construction," and "co-shaping technology and organizational practices," but also the complex interactions between these changes.

evolves over time as new screening guidelines or treatment protocols are adopted. The flow at the point representing "digital niche construction" could describe how the development and implementation of digital technologies in breast cancer care change over time. This could be seen as a dynamic version of the transition maps in C_4, describing how the transitions between different states of the healthcare ecosystem evolve over time.

- Connections: A connection on C_4 could represent a way to compare the rates of change at different points in the healthcare ecosystem. For instance, a connection could provide a way to differentiate the vector field along itself, which would give a measure of how the rate of change in the adoption of new technologies or practices varies from point to point in C_4. For example, a connection could provide a way to compare the rate of change in the adoption of new screening guidelines at a large, urban hospital versus a small, rural clinic. This could be seen as a means to compare the derivatives of the transition maps at different points in C_4, indicating how the rates of change of these transitions vary across the healthcare ecosystem.
- Curvature: The curvature of a connection on C_4 could represent a measure of how much the rates of change at different points in the healthcare ecosystem fail to align with each other. For instance, a high curvature could indicate that the rates of change at different points in C_4 are diverging from each other, suggesting that the healthcare ecosystem is becoming more heterogeneous. Conversely, a low curvature could indicate that the rates of change at different points in C_4 are converging toward each other, suggesting that the healthcare ecosystem is becoming more homogeneous. For example, a high curvature could indicate that the adoption of new screening guidelines is progressing at a much faster rate at large, urban hospitals than at small, rural clinics, suggesting a growing disparity in healthcare practices. This could be seen as a measure of the noncommutativity of the transition maps in C_4, indicating how much the order of transitions affects the overall rate of change in the healthcare ecosystem.
- When we consider different open sets as submanifolds, these interpretations become more localized. For instance, within a specific open set representing a particular region of the healthcare ecosystem, the vector field could represent the local direction and magnitude of change, the flow could represent the local evolution over time, the connection could represent a way to compare the local rates of change, and the curvature could represent a measure of how much the local rates of change fail to align with each other. This allows us to focus on specific aspects of the healthcare ecosystem and understand their dynamics in more detail.[19]

---

[19] In Nugus et al.'s (2010) study of the emergency department (ED), vector fields represent the directional forces in patient care trajectories, e.g., vectors could represent shifts in clinical practices, patient flow dynamics, or resource allocation, each influencing the overall trajectory of patient care. Flows illustrate the evolution of these trajectories and operational processes, e.g., how practices, policies, and patient management strategies evolve, reflecting the dynamic nature of care delivery in response to varying demands and conditions. Facilitating the differentiation of vector fields along other vector fields, connections in the ED embody interactions between its various components, indicating how changes in one aspect of the system (like patient inflow) impact others (such as resource allocation or staff workload), crucial for coordinating care and managing patient trajectories effectively. Curvature reflects the degree of alignment or disparity in care delivery and resource allocation, highlighting the challenges and adaptability inherent in integrated care within this complex, ever-evolving healthcare environment.

Territorial escape and capture of machinic desire within assemblages

In the complex juxtaposition of socio-political dynamics and stakeholder ecosystems, the concept of "machinic desire" emerges as a potent force. Rooted in the philosophy of Deleuze and Guattari, machinic desire encapsulates the inherent drive of entities—both human and nonhuman—to connect, produce, and create within assemblages. It represents the inherent drive of entities to form assemblages, to connect with other entities, and to produce something greater than the sum of its parts. However, this desire is not unbounded, as it does not exist in a vacuum. It is continually influenced by, and in turn influences, the territories in which it operates. Territories—defined socio-political, cultural, or even physical spaces—capture and channel this desire in specific directions. Possible states of territorialization correspond to the various configurations of stakeholder engagements influenced by socio-political forces, which can be seen as manifestations of machinic desire within specific territories. The processes of deterritorialization and reterritorialization serve as the mechanics that capture this machinic desire within specific territories (for a more technical background, see "Gauge Theory of Deterritorialization and Reterritorialization" in the **Appendix**).

In healthcare, machinic desire manifests as the drive to provide care, to connect with patients, to integrate technology, and to innovate for better health outcomes. However, this desire operates within the territories defined by healthcare policies, institutional norms, and patient expectations. The Healthcare Technology Innovators Field, for instance, represents a territory where stakeholders—ranging from tech developers to healthcare providers—express their machinic desire to integrate technology for improved patient care.

Machinic desire can be visualized as a series of states or configurations, each representing a unique combination of stakeholder engagements, technological integrations, and patient-care paradigms. We can consider three primary states in healthcare. In the original state, we have traditional healthcare practices without significant technological intervention. In the deterritorialized state, healthcare practices that have broken away from traditional norms are largely influenced by technology and global trends. In a Reterritorialized State, practices have adapted to new territories, integrating technological advancements with localized healthcare norms and regulations. A healthcare practice might also exist in a superposition of these states, influenced by various factors like technological advancements, patient preferences, regulatory guidelines, and more.

The dynamics of stakeholder ecosystems are in a constant state of flux. Stakeholders, as they interact with and within these ecosystems, undergo processes of deterritorialization, i.e., breaking free from established structures, and reterritorialization, i.e., forming new structures. Differential shifts between states of territorialization map the states of stakeholder engagements to the states of the broader ecosystem, preserving their inherent structure and relationships. These could be due to infinitesimal transformations in stakeholder engagements, representing minor changes in their influence or adaptation to new contexts, like the gradual adoption of telemedicine in a traditionally face-to-face consultation practice, driven by digital integration or patient feedback mechanisms. They can also be due to larger

Continued

> **Territorial escape and capture of machinic desire within assemblages** *(cont'd)*
>
> transformations, representing significant changes in stakeholder roles or influences, such as the shift from traditional care to a technology-driven approach, driven by the adoption of new medical technologies.
>
> The tension between the forces of deterritorialization (escape) and reterritorialization (capture) can provide insights into the dynamics of machinic desire within territories.
>
> - **Full deterritorialization**: Traditional in-person consultations transition to online consultations. This shift moves healthcare from its original state to a fully deterritorialized state. The dynamics of this shift are influenced by various factors, such as technological advancements and patient preferences.
> - **Partial deterritorialization**: AI starts assisting radiologists but doesn't fully replace them. This shift leads to a mixed state where both human expertise and AI play roles in diagnostics. The momentum of AI adoption in diagnostics is influenced by technological advancements and trust in AI.
> - **Full reterritorialization**: In a potential future, AI might become the primary diagnostic tool, with human experts handling only complex cases. This shift to a new paradigm is influenced by factors like trust issues, regulatory concerns, and ethical considerations.
> - **Partial reterritorialization**: Even with increased data portability, some crucial data might remain with primary care providers. This adaptation to a mixed paradigm is influenced by factors like data privacy concerns and regulatory restrictions.
>
> The transformational dynamics of healthcare territories, from one state to another are fundamentally governed by energy dynamics. Representing the rate of change or flow between states, the kinetic term captures the momentum of transformation. For instance, the rapid shift toward telemedicine during a global pandemic would correspond to a high kinetic energy, indicating a swift transformation from the Traditional Care Paradigm to the Technology-Driven Paradigm. The potential term represents the "forces" or influences that push or pull healthcare toward a particular state. Deep potential wells might indicate strong institutional norms or policies that anchor healthcare in a particular paradigm. For example, stringent regulations might create a potential barrier against transitioning to fully digital healthcare solutions, even if the kinetic energy or momentum is pushing in that direction.[20]
>
> ---
>
> [20] In Mossabir et al.'s (2021) review on therapeutic landscapes, everyday geographies, such as parks and community spaces, serve as territories where healthcare entities' machinic desires—interaction, adaptation, evolution—are manifested. The study highlights the dynamic nature of these therapeutic landscapes, where entities like patients and caregivers engage in continuous negotiation (deterritorialization or escape) and adaptation of spaces for health and wellbeing (reterritorialization or capture). It showcases diverse healthcare assemblage states, from traditional healthcare practices to innovative uses of community spaces for health and wellbeing, and the integration of these new practices into the community's daily life. Stakeholder ecosystem dynamics can be seen in how patient needs and preferences can drive changes in the local environment, influencing policy and community planning. Energy transformations within these landscapes are a result of the interplay between the growing emphasis on mental health and wellbeing in urban planning as a form of kinetic energy and community values, cultural beliefs, or policy frameworks as potential energy.

Deriving the plane of consistency and territories of an assemblage

In the context of category C_3, the topological space encapsulates all possible configurations of interactions within the MLXE, involving social actors, networks, relationships between individuals, technology, their environment, and the black boxes. Each point in this space signifies a unique assemblage of these elements, representing a specific state of these interactions (for details, please see "Healthcare Case: Details of Category Theory Analysis" in the **Appendix**).

Utilizing ideal types for each object in category C_3 serves as a foundational step in constructing the topological space, as they represent distinct, simplified representations of complex realities, thereby providing a manageable starting point for exploring the vast array of possible configurations and interactions within the MLXE.

- *BlackBoxes*: Transparent BlackBoxes refer to technologies or processes that are well-understood by the users, like mammography or chemotherapy, where the underlying mechanisms and potential outcomes are well-understood by both healthcare professionals and patients. Opaque BlackBoxes refer to technologies or processes that are not well-understood by the users, like AI-driven diagnosis or treatment, where the underlying algorithms are complex and not easily understood by the users, or experimental treatments where the outcomes are uncertain.
- *IndividualTechnologyEnvironmentRelations*: Technology-Dependent Relations refer to scenarios where the individual's interaction with the environment is heavily reliant on technology, such as the use of AI for personalized treatment plans, telehealth for remote consultations, or wearable devices for monitoring patient health. Technology-Independent Relations refer to scenarios where the individual's interaction with the environment is not heavily reliant on technology, such as traditional face-to-face consultations, physical exams, or reliance on patient-reported symptoms for diagnosis and treatment.
- *SocialActorNetworks*: Hierarchical Networks refer to the traditional model, where the network of social actors is structured in a top-down manner, often seen in traditional healthcare settings where the physician's expertise is paramount. Collaborative Networks represent a more egalitarian structure, where all social actors have a say in decision-making, facilitated by technologies like patient portals and telehealth, which enable greater patient involvement and collaboration among healthcare professionals.

The topology, defined by the transformational relationships between these points, can be viewed as a plane of consistency or immanence. This space is not static but undergoes continuous reconfiguration and transformation through processes of deterritorialization and reterritorialization, embodying the dynamic and emergent nature of the ecosystem's interactions.

To further refine the definition of the topological space, we consider the nature, specific properties, and evolution of these technology-mediated practices in response to changes in the digital environment. The topological space is thus defined as a structured set of all possible technology-mediated interactions or digital practices. This structure, the topology, is a family of subsets of the space, termed open sets. The specification of this structure requires an understanding of what constitutes an "open" set of these technology-mediated practices in this context.

In a topological space, open sets typically represent "neighboring" or "close" elements. In the context of C_3, these open sets correspond to groups of similar or "close" technology-mediated practices within the MLXE. This notion of closeness is based on shared attributes of the interactions, such as similar types of social actors, networks, relationships, or "black boxes". Each open set, therefore, represents a region or "neighborhood" in the topological space where certain types of interactions dominate, reflecting a kind of "global field" of practices within the digital ecosystem. The open sets in this topology would then be subsets or collections of these points representing interactions between the objects of C_3, i.e., technology-mediated practices that share certain characteristics or are closely related in some way.[21]

We can define the open sets in the topological space based on different combinations of the ideal types of global processes in C_3, which are social actor networks, individual technology environment relations, and black boxes. Here is a selection of open sets that covers a range of the ideal types (for details, please see "Healthcare Case: Details of Category Theory Analysis" in the **Appendix**):

- Collaborative SocialActorNetworks & Technology-Dependent IndividualTechnologyEnvironmentRelations (Open Set 3): This open set includes interactions where egalitarian, collaborative networks of social actors are heavily reliant on technology. This represents areas where bottom-up technology changes are welcomed and incorporated into a niche in the ecosystem. Consider, for instance, the use of patient portals in a collaborative care setting. Here, the collaborative network of physicians, nurses, and patients relies heavily on technology for sharing medical information and making collaborative decisions about treatment.
- Hierarchical SocialActorNetworks & Opaque BlackBoxes (Open Set 6): This open set represents scenarios where traditional, hierarchical networks of social actors are using technologies or processes that are not well-understood. This represents areas where top-down technology changes meet resistance in the digital niche. Consider, for instance, the use of AI in breast cancer diagnosis in a traditional hospital setting. Here, the hierarchical network of physicians, nurses, and technicians relies on a technology that is not well-understood by all users.
- Technology-Independent IndividualTechnologyEnvironmentRelations & Transparent BlackBoxes (Open Set 11): This open set includes interactions where individuals' interactions with the environment are not heavily reliant on well-understood technologies or processes. Consider, for instance, traditional face-to-face consultations in a hospital setting, where the individual's interaction with the healthcare environment is not heavily reliant on any specific technology.

---

[21] Neighborhoods of similar technology-mediated practices in MLXEs within a topological space, or C_3's open sets, represent what DeLanda (2006) refers to as "populations of assemblages." Accordingly, these practices, through recurrent assembly processes, form distinct collectivities or populations, each characterized by shared attributes and interaction dynamics within the digital ecosystem.

In the philosophy of Deleuze and Guattari, these open sets are seen as territories within the plane of immanence, each distinguished by a unique pattern of interactions and relationships. For example, an open set where certain types of social actors and networks are dominant could be viewed as a territory where the deterritorializing force of social dynamics is embraced, leading to reterritorialization in the form of new social structures and relationships.

The base of the topology, a special collection of open sets, generates the entire topology and represents the various fields of practice within the MLXE. Each base set is associated with a unique kind of interaction, such as specific types of social actors, networks, relationships, or "black boxes." These base sets, which can be defined as the set of all possible interactions or connections between the objects of C_3, overlap and intersect to form more complex open sets in the topology, illustrating the interplay among multiple fields of practice. Also, these fields are likely to evolve over time as new technologies emerge and practices change. Some examples are:

- Physician-patient-nurse triad field: This field represents the traditional hierarchical network of social actors in healthcare. It includes practices where physicians, nurses, and patients interact in a traditional hospital setting, with physicians making most of the decisions.
- Healthcare technology innovators field: This field represents the practice around the development and implementation of new healthcare technologies. It includes practices where technology companies, researchers, and healthcare providers collaborate to develop and implement new technologies in healthcare, such as AI, telehealth, and genomics.
- Healthcare access and equity field: This field represents the practice around healthcare access and equity. It includes practices where healthcare providers, policymakers, and patient advocacy groups collaborate to improve access to healthcare and reduce health disparities.

These base sets can also be viewed as the foundational territories within the plane of immanence, each tied to a specific field of practice. The overlap and intersection of these territories reflect the collaboration and interaction among different fields within the MLXE. Transversality can be applied to understand intersections within categories, open sets, and base sets. For instance, in the healthcare ecosystem, the "healthcare access and equity field" base set represents an intersection of various open sets, each signifying a trend in healthcare. The convergence of these trends at this intersection is not tangential but spans new directions, profoundly influencing healthcare practices.

In summary, the topological space, open sets, and base sets provide a mathematical framework for understanding the complex dynamics of the interactions within an MLXE. By defining these concepts in this way, we can capture complex dynamics in C_3 and provide a framework for understanding the nature of stakeholders within the MLXE.[22]

Similar to the process we followed for C_3, we can define C_4 (ecosystems as collective assemblages of enunciation), C_5 (creative-experiencers as machinic assemblages of bodies), and C_6 (experience-verse as collective assemblages of enunciation) along with their topological spaces, open sets, and base sets (for details, please see "Healthcare Case: Details of Category Theory Analysis" in the **Appendix**).

Briefly, we can note that, for C_4, defined across ideal types of centralized vs decentralized decision-making for AdaptiveStructuration, technology-receptive vs -resistant DigitalNicheConstruction, technology- vs practice-driven CoShapingingTechnologyOrganizationalPractices, the points of the topological space can be thought of as different combinations, or assemblages, of these ideal types of AdaptiveStructuration, DigitalNicheConstruction, and CoshapingTechnologyOrganizationalPractices. The base sets generating the open sets of the topology in the context of healthcare would include the various fields of practice such as AI and data analytics field, remote healthcare delivery field, and patient-centered health and wellness field.[23] Since the categories C_3 and C_4, while both global, represent different aspects of the healthcare ecosystem, the relationship between the base sets of C_3 and C_4 is not necessarily contingent on a clear mapping or correspondence, although understanding the intersections, overlaps, and potential relationships between these base sets can provide valuable insights into the dynamics of the healthcare ecosystem. For C_5, the topological space consists of open sets defined by ideal types of entropy-reducing vs -increasing FreeEnergyPrinciple, data- vs experience-driven ActiveInference, and technology-mediated vs direct-experience Phenomenotechnics. Some of the relevant base sets in this category are technological innovation community, patient-centered care community, and predictive analytics community. Last but not least, for C_6, the dimensions of the topological space, underlying the open sets, are defined by the ideal types of data-driven vs experience-based SenseMaking, routine-based vs adaptive SeriesStructure, and technology-enabled vs human-centered PoiesisEnchantment. Some of the base sets relevant in the healthcare context

---

[22] In their intersectional analysis of health inequalities, Gkiouleka et al. (2018) break down complex health issues into specific, manageable factors (ideal types) like race and gender, similar to how the topological space in C_3 is defined by simplified elements. Their approach exemplifies open sets by grouping similar health experiences (e.g., based on race or gender), which represent "neighborhoods" of related factors in healthcare. Their emphasis on institutional structures reflects the concept of base sets, foundational territories within the healthcare ecosystem. Furthermore, their analysis demonstrates transversality by showing how intersections of categories (like race and gender) and institutions shape health outcomes, similar to how intersections within categories and base sets influence practices in MLXE.

[23] In the study by Nugus et al. (2010), Emergency Department's (ED) decision-making processes reflect both centralized (physician-led decisions) and decentralized (involving nurses, patients, and broader clinical inputs) AdaptiveStructuration. The use of electronic health records and telehealth in EDs can be inferred as instances of technology-receptive DigitalNicheConstruction, while the interaction between existing healthcare practices and emerging technologies in the ED implies a technology-driven dynamic, where patient care adapts in response to technological advances. The ED functions as a nexus where different fields of practice intersect, i.e., base sets generating open sets in the topology of healthcare. In the context of the ED, relevant base sets might include patient triage and assessment (AI and Data Analytics Field), telehealth services (Remote Healthcare Delivery Field), and patient-centered care approaches (Patient-Centered Health and Wellness Field).

are clinical decision-making community, patient engagement community, and technological innovation community.[24]

Note that in referring to the base sets of C_3 and C_4, the term "field" is used to denote large-scale, global entities representing broad, overarching trends or movements within the healthcare ecosystem, such as the "Healthcare Research and Education Field" or the "AI and Data Analytics Field". On the other hand, in C_5 and C_6, the term "community" is used to denote more localized, specific groups representing more specific, localized practices or trends within the broader healthcare ecosystem, such as the "patient-centered care community" or the "technological innovation community."

---

[24] The intricate dynamics of patient experiences preceding a formal diagnosis, as explored in Locock et al.'s (2016) study, reveal how patients engage in an anticipatory assemblage of diagnostic cues. Patients' active engagement in assembling procedural, spatial, and interactional evidence aligns with the entropy-reducing and -increasing aspects of the FreeEnergyPrinciple, as they seek to reduce uncertainty while acknowledging the inherent variability in their healthcare journey. Their efforts represent a balance between data-driven and experience-driven Active-Inference, where both empirical data (such as procedural cues) and personal experiences (such as perceived changes in healthcare providers' behavior) inform their anticipatory diagnosis. Moreover, patients' narratives demonstrate data-driven and experience-based SenseMaking, where they interpret and understand healthcare signals within their unique contexts. The SeriesStructure, encompassing both routine-based and adaptive elements, is evident in how patients navigate and understand the sequence of healthcare events. Furthermore, the study contributes to the understanding of PoiesisEnchantment, highlighting both technology-enabled and human-centered aspects in the formation of patients' diagnostic expectations. This aligns with the base sets, such as the clinical decision-making community, patient engagement community, and technological innovation community, underscoring the symbiotic relationship between technology and human elements in healthcare experiences.

CHAPTER 4

Dynamic relationalities on strata

Within the framework of Dynamic Relationality Theory (DRT), assemblages are understood as fluid networks of diverse elements. These elements, represented as objects within a category, symbolize intricate interactions and facets of a system. Morphisms capture the dynamic influence these components exert on one another. The topology of this system, where each point signifies a unique assemblage or state of interaction, is influenced by the system's underlying energy principles that steer the transformational trajectory of the assemblage. While the categories in Fig. 3.1 (e.g., C_3 and C_4) might represent broader aspects of the ecosystem, their relationship is not necessarily linear. Exploring their intersections can provide deeper insights into the dynamics of an MLXE.

Dynamic relationality #2: Territorialization of strata between assemblages

Consider category C_3, where stakeholders are viewed as a machinic assemblage of bodies, with the molar line of segmentation delineating them as collective entities shaped by larger socio-political forces. The morphisms represent macro-scale transformations that either aggregate or disaggregate stakeholder groups, reinforcing the collective nature of stakeholders in the ecosystem. Viewed through the lens of assemblage theory, stakeholders form a multiplicity of heterogeneous elements—human, nonhuman, material, and immaterial. The morphisms reflect the dynamic processes of assemblage and reassemblage, signifying the ever-changing nature of stakeholder relationships in the ecosystem.

The category C_4, composed of systems and organisms evolving together, the creation of specialized environments, and the dynamic adaptation of social structures in response to technological change, presents the ecosystem as a collective assemblage of enunciation, where the strata signify varying degrees of codification or rigidification. The morphisms act as processes of de-stratification, disrupting existing codes, and re-stratification, creating new codes. Viewed from the perspective of deterritorialization and reterritorialization, the morphisms act as the forces that disrupt and recreate the structures of the ecosystem, reflecting the constant state of flux in the ecosystem.

Relating the categories of stakeholders (C_3) and ecosystems (C_4) to each other, stakeholder ecosystems unfold through dual articulation of content and expression, reflecting the interaction between material components like technology and infrastructure (content) and the semiotic elements like values, norms, and meanings (expression) that stakeholders bring to the ecosystem. The morphisms or transformative processes act as a double articulation, mediating between content and expression, and reciprocally shaping the dynamics of the ecosystem.

Formalizing this process of double articulation, a functor between categories can represent the relationships or mappings between different kinds of assemblages, capturing how one assemblage influences or transforms another. This allows us to model complex systems of assemblages and their interconnections.[1]

In particular, we have functor $F_3: \mathbf{C_3} \to \mathbf{C_4}$

$F_3(\text{BlackBoxes}) = \text{AdaptiveStructuration}$

$F_3(\text{IndividualTechnologyEnvironmentRelations}) = \text{DigitalNicheConstruction}$

$F_3(\text{SocialActorNetworks}) = \text{CoShapingTechnologyOrganizationalPractices}$

$F_3(f_{31}) = f_{41}; F_3(f_{32}) = f_{42}$

---

[1] Postulating the existence of functors between machinic assemblages of bodies (MABs) and collective assemblages of enunciation (CAEs) within strata and across stacked (molar and molecular) strata entails several theoretical implications:

- "Functors between MABs and CAEs might indicate consistent relationships between different assemblages, i.e., *correlation*, suggesting that certain patterns or behaviors in one assemblage are regularly associated with specific patterns or behaviors in another.
- The concept of *symmetry* can be inferred from the functorial relationships, i.e., not necessarily identical or mirror-like relations but a form of conceptual or functional equivalence between processes or entities across different assemblages.
- The use of functors implies a high degree of *interdependence* between MABs and CAEs, where various components and processes are deeply interconnected and mutually influential.
- The existence of functors between MABs and CAEs might also suggest that changes or properties in the former could *cause* complex and nonlinear changes or properties in the latter.
- MABs and CAEs are not isolated but are continuously influencing and *transforming* each other, not as a simple transfer but a complex process of *translation*, where the properties or behaviors in one domain are converted into corresponding but not identical properties or behaviors in another domain.
- Functorial mappings can be indicative of *emergence*, i.e., properties or behaviors that arise from the intricate interactions between MABs and CAEs, which cannot be fully understood by examining these assemblages in isolation.
- The functorial relationship allows for a *modular and flexible* understanding of MLXEs. While MABs and CAEs are distinct, they are also part of a larger, integrated system, and local changes can have broader systemic effects.

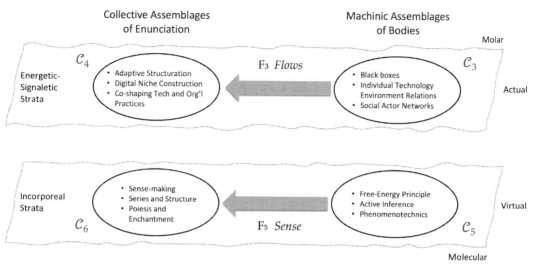

FIGURE 4.1 Territorialization of strata between assemblages.

F_3, representing the process of territorialization as flows in an MLXE, maps objects and relationships between them (morphisms) from C_3 to C_4. By doing so, F_3 preserves the structure and relationships within the categories. It captures how the interactions and relationships among social actors, individuals, and technology (objects in C_3) can affect the process of adaptive structuration, contribute to the construction of adaptive digital niches, and influence the co-shaping and mutual evolution of technology and organizational practices (objects in C_4). This mapping provides a framework for understanding the transformation of stakeholder engagements into flows within the ecosystem, highlighting the dynamic and emergent nature of the MLXE (see top portion of Fig. 4.1).[2]

In the category C_5, the Free Energy Principle, Active Inference, and Phenomenotechnics are seen as components of a cognitive assemblage, not defined by its collective identity but by the dynamic and fluid interactions of its heterogeneous elements—cognitive processes, mental models, and phenomenological experiences. The morphisms in C_5 represent the microtransformations that disrupt and recreate the cognitive landscape of the creative-experiencer. This

---

[2] In a study by Carroll et al. (2021), the concept of "enabling sporting assemblages" (e.g., wheelchair basketball and adaptive/inclusive sailing) for disabled youth illustrates the complex interplay between stakeholders and the broader ecosystem, reflecting the double articulation of material (wheelchairs, basketball courts) and expressive (attitudes toward disability and sport, rules for wheelchair basketball) elements within sports assemblages. The research exemplifies the process of territorialization as flows, highlighting dynamic interactions between social actors and the adaptive sports environment, which influence and reshape the sports ecosystem. It captures how interactions among social actors (disabled youth, their families) and the technology/environment (adaptive sports programs and equipment) influence the structuration of the sports ecosystem, creating adaptive digital niches and co-evolving practices. This process represents the dynamic structuring and reconfiguration of socio-techno-economic territories within the sporting community, underpinned by reciprocal and adaptive exchange of information and experiences among human and nonhuman actants.

reflects the molecular nature of the cognitive assemblage, where change is driven by the constant state of flux within the assemblage itself. Moving to the category C_6, the experience-verse is a collective assemblage of enunciation with elements of Sense-Making, SeriesStructure, and PoiesisEnchantment, which are not static entities but dynamic processes that are constantly evolving and interacting with each other. They form a molecular assemblage of sense-making processes, which together shape and structure our experiences in the experience-verse.

The functor $F_5: C_5 \rightarrow C_6$, representing the process of territorialization as senses in the experience-verse, captures this molecular stratification (see bottom portion of Fig. 4.1). It maps the fluid and dynamic cognitive processes (objects in C_5) to the equally fluid and dynamic sense-making processes (objects in C_6). By doing so, F_5 not only preserves the structure and relationships within the categories but also captures the molecular dynamics that drive the transformation of cognitive processes into experiences within the experience-verse. In essence, the molecular stratification of the creative experience-verse reflects the dynamic, emergent, and fluid nature of both cognitive processes and experiences. It captures the constant state of flux within the experience-verse, where change is driven not by large-scale transformations but by the intricate interplay of microtransformations within the assemblage itself.[3]

While F_3 ("flows") and F_5 ("sense") have been depicted and discussed as unidirectional functors, mapping respectively from categories C_3 (stakeholder) to C_4 (ecosystem) and from C_5 (creative-experiencer) to C_6 (experience-verse), a deeper analysis reveals that each can be conceptualized as a pair of *adjoint* functors.[4] This conceptualization underscores a profound, reciprocal relationship between material and symbolic dimensions in complex systems. For F_3, one functor could represent the influence of stakeholders on ecosystems, while its adjoint encapsulates the reciprocal influence of ecosystems on stakeholders. The adjoint functor embodies the transformation of tangible stakeholder interactions (substance and form of content in C_3) into the guiding narratives and structures of ecosystems

---

[3] In Carroll et al.'s (2021) study, disabled youths' experiences, from facing barriers in mainstream sports to empowerment in specialized programs, reflect cognitive assemblage dynamics and collective assemblage of enunciation. The youths' experiences, their struggles with ableist attitudes, and efforts to overcome physical and social barriers dynamically interact within the youth's cognition, as they actively infer capabilities and potentials in sports, challenging and transforming the conventional phenomenotechnics of ability and disability in sporting contexts. The experiences of disabled athletes, from feeling marginalized in mainstream sports to finding empowerment and identity in specialized programs, illustrate the continuous and dynamic sense-making processes. The youths' journeys from exclusion to inclusion in sports exemplify the double articulation of content (the actual experiences of sports participation) and expression (the formation of identity and self-esteem through these experiences). The process of territorialization as sense, where personal experiences and struggles translate into broader narratives of inclusion and empowerment, illustrates the intricate interplay of micro-transformations within cognitive and experiential assemblages in the realm of sports for disabled youths.

[4] Adjoint functors involve pairs of functors that establish a coherent, bidirectional mapping, reflecting a deep duality in the interconnected dynamics of the categories involved. They are characterized by two key formal properties: (1) The unit transformation associates each object in one category with a corresponding object in the other, while the counit transformation relates these objects back, enabling a "round-trip" journey between the categories. (2) Hom-set isomorphisms ensure a structural correspondence between the morphisms (transformation processes) in these categories. In category theory notation, $F \dashv G$ indicates that F is left adjoint to G (or G is right adjoint to F).

(substance and form of expression in C_4). Similarly, for F_5, one functor could express the transformation of cognitive processes into experiences, with its adjoint functor representing the grounding of experiential forms back into cognitive structures. These functors articulate the conversion of neural and biological processes (substance and form of content in C_5) into communicative and experiential realms (substance and form of expression in C_6). These pairs of adjoint functors imply a bidirectional and deeply interconnected dynamic, where changes in one category are mirrored and responded to in the other. This bi-directionality ensures a continuous, reciprocal dialog between the categories, transcending the initially perceived unidirectionality and reflecting the intricacies of socio-techno-economic and cognitive-experiential interplays in DRT. This approach of conceptualizing functors as pairs of adjoint functors can be extended to other key functors within DRT, such as F_1 ("lived journey") and F_7 ("resourced capabilities"), which also represent processes of double articulation on their respective strata, embodying the same principles of reciprocal influence and structural correspondence.

Analyzing stratification across machinic and collective assemblages

A structured approach to analyzing and modeling molar stratification of stakeholder ecosystems involves using the concepts of topological space, open sets, base sets, and functorial mapping.

1. Identify the open sets in C_3 and C_4: Start by identifying the open sets in each category. In C_3, these could be the different types of stakeholder engagements, and in C_4, these could be the different types of flows in the ecosystem.
2. Identify the base sets in C_3 and C_4: Identify the base sets in each category. These represent the foundational territories within the plane of immanence, each tied to a specific field of practice.
3. Apply the functor F_3: Apply the functor F_3 to map the open sets from C_3 to C_4. This mapping illustrates how a specific configuration of stakeholder engagements (in C_3) can be translated into a specific configuration of flows in the ecosystem (in C_4).
4. Analyze the transformation: Analyze the transformation in the structure of the base sets in the target category C_4 as a result of the functorial mapping of the open sets. This analysis can provide insights into how the fields of practice are transformed because of the changes in stakeholder engagements and flows in the ecosystem.
5. Model the molar stratification of stakeholder ecosystems: The choice of modeling paradigm, ranging from system dynamics modeling (of stocks, flows, feedback loops, and time delays) through network analysis (including multilayer network modeling) to complex adaptive systems modeling (including agent-based modeling), would depend on the specific research question, the nature of the data available, and the level of detail required in the model. This model can help understand the complex dynamics of interactions within an MLXE and provide a robust framework for stakeholder engagements territorialized as flows in ecosystems.

6. Iterate and refine: Finally, iterate and refine the model as new data and insights become available. This iterative process can help improve the accuracy and robustness of the model over time.

Through this iterative process, we can see how different configurations of stakeholder engagements in C_3 can influence the dynamics of flows in the ecosystem in C_4. This provides a robust framework for understanding the complex dynamics of interactions within an MLXE, and how stakeholder engagements can be territorialized as flows in the ecosystem.[5]

Such an analysis and model would plausibly suggest (among other insights):

- The collaborative and technology-dependent interactions among stakeholders in the healthcare technology innovators field (C_3) are driving the transformation of the remote healthcare delivery field (C_4) toward a more decentralized and technology-receptive digital niche. This process illustrates how stakeholder engagements in C_3, characterized by egalitarian networks heavily reliant on technology, can influence and shape the flows within the ecosystem in C_4, leading to a more patient-led approach to healthcare.
- The hierarchical social actor-networks and opaque black boxes in the physician-patient-nurse triad field (C_3) are driving the transformation of the AI and data analytics field (C_4) toward a more technology-receptive digital niche and technology-driven co-evolving organizational practices. This process illustrates how traditional, hierarchical networks of social actors using technologies or processes that are not well-understood can influence and shape the flows within the ecosystem in C_4, leading to the co-evolution of technology and organizational practices despite initial resistance.
- The technology-independent individual-technology-environment relations and transparent black boxes in the healthcare access and equity field (C_3) are driving the transformation of the patient-centered health and wellness field (C_4) toward a more technology-resistant digital niche and practice-driven co-evolving organizational practices. This process illustrates how individuals' interactions with the environment that are not heavily reliant on well-understood technologies or processes can influence and shape the flows within the ecosystem in C_4, leading to the co-evolution of technology and organizational practices that prioritize patient-centered care and resist impersonal or intrusive technologies.

Analysis and modeling of molecular stratification of creative experience-verse would, like its molar stratification counterpart in stakeholder ecosystems, entail using the concepts of topological space, open sets, base sets, and functorial mapping, as well (please see "Healthcare Case: Details of Category Theory Analysis" in the Appendix and Fig. 1 therein).

---

[5] Carroll et al.'s (2021) empirical studies of wheelchair basketball and adaptive/inclusive sailing reveal insightful transformations in fields of practice as a result of changes in stakeholder engagements. In both cases, the functorial mapping illustrates a significant shift in the sports ecosystem. Stakeholder engagements characterized by collaborative networks and varying degrees of technology reliance (wheelchair basketball with high reliance and adaptive/inclusive sailing with a balanced approach) have led to ecosystems that are adaptive, inclusive, and receptive to technological innovations. This shift is not just in the practicalities of sports participation but also in the perception and value accorded to sports for "differently-abled" individuals. These transformations reflect a broader societal change toward inclusivity and empowerment, where the focus is on enabling participation and celebrating diversity in abilities. The fields of practice in these sports are transformed to become more than just physical activities; they become spaces of social inclusion, personal empowerment, and community building.

Dynamic relationality #3: Lines of flight between assemblages across stacked strata

As shown in Fig. 4.2, a different kind of dynamic relationality is represented by the functor $F_4: C_4 \rightarrow C_5$, named "Event" (as Deterritorialization):

$F_4(AdaptiveStructuration) = FreeEnergyPrinciple$

$F_4(DigitalNicheConstruction) = ActiveInference$

$F_4(CoShapingTechnologyOrganizationalPractices) = Phenomenotechnics$

$F_4(f_{41}) = f_{51}; F_4(f_{42}) = f_{52}$

This functor embodies a Deleuzian line of flight, a process of escape or deterritorialization from existing structures, and the potential for creating new ones. The functor F_4 maps the co-evolving systems of stakeholder ecosystems (C_4) to the cognitive and perceptual processes of the creative-experiencer (C_5), signifying a shift from the macroscopic structures and processes of the stakeholder ecosystems to the microscopic transformations that disrupt and recreate the cognitive landscape of the creative-experiencer. This mapping reflects a transformation from molar/global processes to molecular/local processes, capturing the dynamic and evolving nature of the ecosystem, where new cognitive processes and experiences are constantly being created and old ones are being disrupted. This transformation is characterized by transversality, a concept that denotes fluid,

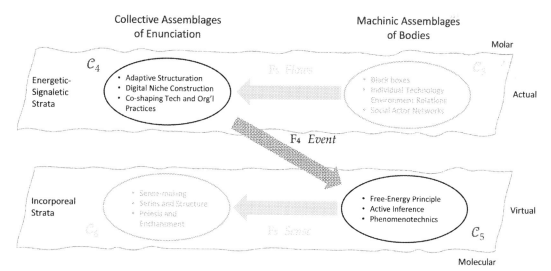

FIGURE 4.2 Lines of flight between assemblages across stacked strata.

dynamic, and transformative connections and relations that cut across different domains, scales, or levels of organization.

This transformation disrupts the established codes and conventions of the stakeholder ecosystems, opening up a space for the emergence of new structures, new patterns, and new ways of thinking and acting within the MLXE. The mapping of the functor F_4 embodies the interplay between the actual and the virtual. The actual, represented by the existing structures of C_4, is disrupted and deterritorialized, giving way to the virtual, represented by the potential new structures and patterns in C_5. This interplay drives the transformative process of the Event, shifting from the macroscopic structures and processes of the ecosystems to the microscopic transformations that recreate the cognitive landscape of the creative-experiencer.[6]

In this transformation, the functor F_4 acts as a catalyst, initiating a movement of deterritorialization that leads to the creation of new assemblages. These new assemblages emerge from the interactions and combinations of physical entities that can affect and be affected. The transformation is characterized by the emergence of new meanings and the disruption of old ones, leading to the creation of new lines of flight and the transformation of assemblages. The transformation is triggered by a problematic or paradoxical element, the quasi-cause, which destabilizes the existing assemblages and leads to the creation of new ones. This process is characterized by the resonance and displacement of series and singularities, and the triggering of new sense. In summary, the functor F_4 as an event signifies a process of deterritorialization that results in the transformation of existing assemblages and the creation of new ones, reflecting the complexity, dynamism, and fluidity of the MLXE.

Analyzing functorial transformations across stacked strata

Analyzing the transformative process embodied by the functor $F_4: C_4 \rightarrow C_5$ involves, once again, using the concepts of topological space, open sets, base sets, and functorial mapping:

1. Identify the open sets in C_4 and C_5: In C_4, these could be different fields representing various co-evolving systems or processes. In C_5, these could be different assemblages representing various cognitive processes or experiences.

---

[6] In Carroll et al.'s (2021) study of wheelchair basketball and adaptive/inclusive sailing, the experiences in both programs demonstrate a shift from structured sports environments (AdaptiveStructuration) to a new cognitive landscape where disabled participants actively adjust their perceptions and responses (FreeEnergyPrinciple). It also shows how use of technology (such as adapted sports wheelchairs and sailing equipment) enables new forms of engagement and interaction, which can be seen as a form of active inference. The transformation in organizational practices through technology adaptation aligns with Phenomenotechnics, illustrating how new sensory and experiential worlds are created for the participants. In both programs, the "event" acts as a catalyst for deterritorialization, triggering a shift from the structured, collective interactions of the stakeholder ecosystems to the individualized, cognitive, and perceptual transformations, marked by the emergence of new meanings, the disruption of old patterns, and the creation of transformative experiences. These experiences resonate and displace the existing series and singularities, leading to systemic changes that are both concrete in their interaction within the ecosystems and abstract in their cognitive ramifications.

2. Identify the base sets in C_4 and C_5: In C_4, these could be the foundational elements within each field, representing the core processes or systems that drive the co-evolution. In C_5, these could be the foundational elements within each assemblage, representing the core cognitive processes or experiences that shape the creative-experiencer.
3. Apply the Functor F_4: Apply the functor F_4 to map the open sets from C_4 to C_5, which illustrates how a specific field (in C_4) can be translated into a specific assemblage (in C_5). This is where the transformative process of deterritorialization occurs, signifying a shift from the molar to the molecular, from the actual to the virtual.
4. Analyze the transformation: Analyze the transformation in the structure of the base sets in the target category C_5 as a result of the functorial mapping of the open sets. This analysis can provide insights into how the assemblages are transformed as a result of the changes in the fields.

Through this methodology, we can see how different fields in C_4 can influence the dynamics of assemblages in C_5, and how these transformations signify shifts in focus and approach within the healthcare ecosystem, driven by the increasing use of technology and the recognition of the importance of patient experiences and preferences. The transformations also reflect the dynamic and fluid nature of the healthcare ecosystem, with the assemblages in C_5 constantly evolving and adapting in response to changes in the fields in C_4. Some insights that this analysis might reveal are:

- The transformation from the remote healthcare delivery field (C_4) to the predictive analytics community (C_5) signifies a shift from a generalized approach to healthcare to a more personalized, data-driven approach. This transformation is driven by the increasing use of technology in healthcare, particularly in the form of predictive analytics. The base set in C_5, the predictive analytics community, is transformed as a result of this shift, with the community increasingly focusing on the use of predictive analytics for early detection and personalized treatment plans.
- The transformation from the AI and data analytics field (C_4) to the patient-centered care community (C_5) signifies a shift from a technology-focused approach to a more patient-centered approach. This transformation is driven by the increasing recognition of the importance of patient experiences and preferences in healthcare. The base set in C_5, the patient-centered care community, is transformed as a result of this shift, with the community increasingly focusing on integrating AI and data analytics into patient-centered care practices.
- The transformation from the patient-centered health and wellness field (C_4) to the technological innovation community (C_5) signifies a shift from a focus on health and wellness to a focus on technological innovation. This transformation is driven by the increasing use of technology in health monitoring and prevention. The base set in C_5, the technological innovation community, is transformed as a result of this shift, with the community increasingly focusing on the development and use of technologies that afford continuous health monitoring.

Transversality

Transversality is a concept that can be related to becoming, desiring-production, and boundary-spanning in Deleuze and Guattari's philosophy. The term originates from differential topology, a branch of mathematics that studies the properties and structures of differentiable manifolds. In this context, transversality refers to the way two submanifolds intersect in a differentiable manifold, such that they meet in a nontangential manner. This geometric concept is adopted by Deleuze and Guattari as a metaphor for understanding complex interactions and connections in different domains.

In the context of Deleuze and Guattari's work, transversality can be related to becoming, desiring-production, and boundary-spanning in the following ways:

1. Becoming: Transversality can be seen as an aspect of becoming, as it involves the formation of novel and unexpected connections between different elements, which can lead to new becomings. Transversal connections enable the flow of ideas, energies, and influences between diverse domains, allowing for the emergence of new possibilities and transformations.
2. Desiring-production: Transversality is also linked to desiring-production, as it represents the productive and connective force that generates new combinations and assemblages. In Deleuze and Guattari's philosophy, the concept of transversality underscores the importance of understanding desire as a dynamic and relational force that drives the formation of connections between different entities and elements.
3. Boundary-spanning: Like boundary-spanning in organizational theory, transversality involves the formation of connections and interactions that cross established boundaries or territories. Both concepts emphasize the importance of creating and maintaining relationships that bridge gaps between different domains, fostering innovation, adaptation, and change.

The differential topological concept of transversality can be applied to understand the intersections and interactions between the different categories, open sets, and base sets.

Base sets as intersecting submanifolds: Consider the base sets as intersecting submanifolds within the categories. The intersection of two submanifolds is transversal if every point in the intersection is a regular value of both submanifolds. In this context, the base sets represent the points of intersection or nontangential overlap between different open sets within each category.[7]

- The "AI and data analytics field" base set represents a specific area within the healthcare ecosystem where AI and data analytics technologies are shaping the practices. It represents a specific region within the manifold of C_4 where the influences of centralized and decentralized adaptive structuration, technology-receptive

---

[7] Gkiouleka et al.'s (2018) intersectional analysis of health inequalities serves as a practical illustration. They demonstrate how various social and institutional factors, which we can see as submanifolds, intersect transversally, shaping health outcomes in significant and nontangential ways. This approach mirrors the concept of base sets, where intersections among different categories and open sets create substantial overlaps, revealing complex interactions within the healthcare ecosystem. These intersections, far from being superficial, span new directions and insights, exemplifying the rich interplay of factors that define health dynamics.

digital niche construction, and technology-driven co-shaping technology-organizational practices intersect. This base set could be seen as an intersection of the following open sets:
  * Open Set 1: Centralized AdaptiveStructuration & Technology-Receptive DigitalNicheConstruction: This open set includes interactions where traditional, centralized medical practices are receptive to technological innovations. In the context of the "AI and data analytics field," this could represent healthcare practices where AI and data analytics are being integrated into centralized decision-making processes.
  * Open Set 2: Decentralized AdaptiveStructuration & Technology-Receptive DigitalNicheConstruction: This open set comprises interactions where patients and healthcare professionals share in decision-making, facilitated by technology. In the context of the "AI and data analytics field," this could represent healthcare practices where AI and data analytics are being used to facilitate shared decision-making between patients and healthcare professionals.
  * Open Set 5: Technology-Receptive DigitalNicheConstruction & Technology-Driven CoShapingTechnologyOrganizationalPractices: This open set represents interactions where technological receptiveness drives changes in organizational practices. In the context of the "AI and data analytics field," this could represent healthcare practices where the receptiveness to AI and data analytics is driving changes in the way healthcare is delivered.
- Transversality, in the language of differential topology, refers to the idea that at the points of intersection between two submanifolds, the tangent spaces span the entire space. In simpler terms, the intersecting submanifolds don't merely touch or cross each other tangentially; they intersect in a way that they span new directions in the manifold.
  * Each of these open sets (submanifolds) represents a specific direction or trend in the healthcare ecosystem. The intersection of these submanifolds at the "AI and data analytics field" base set represents a point where these trends converge and interact in a meaningful way. This intersection is transversal, meaning that the influence of AI and data analytics in the healthcare ecosystem is not just a tangential or peripheral phenomenon. Instead, it is a significant intersection point that spans new directions in the manifold of C_4, influencing and being influenced by various aspects of adaptive structuration, digital niche construction, and co-shaping technology-organizational practices.
  * This transversal intersection at the "AI and data analytics field" base set represents a complex, multidimensional interaction within the healthcare ecosystem. It is a point where various trends in technology adoption, decision-making structures, and evolving practices intersect and interact, driving new directions and possibilities in the prevention, diagnosis, and treatment of breast cancer. Let us consider a few concrete examples in the context of breast cancer prevention, diagnosis, and treatment:
    * **AI in mammography interpretation**: AI algorithms are being developed and used to interpret mammography images to detect early signs of breast cancer. This is a clear example of the "AI and data analytics field" intersecting with

"Centralized AdaptiveStructuration & Technology-Receptive DigitalNicheConstruction" (Open Set 1). In a traditional healthcare setting, the interpretation of mammograms is a centralized process done by radiologists. With the introduction of AI, this process becomes receptive to technological innovation. The AI does not just tangentially assist the radiologists; it actively influences the process of diagnosis, potentially identifying patterns that might be missed by the human eye.

- **AI in personalized treatment plans**: AI is also being used to analyze a vast array of data, including genetic information, to create personalized treatment plans for breast cancer patients. This represents an intersection with "Decentralized AdaptiveStructuration & Technology-Receptive DigitalNicheConstruction" (Open Set 2). Here, decision-making is shared among physicians, patients, and potentially other healthcare professionals. The AI's role is not peripheral; it is central to the process of analyzing the data and generating treatment options. This represents a shift in the traditional healthcare paradigm, with technology playing a key role in the co-evolution of healthcare practices.
- **AI in predictive analytics for breast cancer risk**: AI and data analytics are being used to analyze large datasets, including genetic, lifestyle, and environmental factors, to predict an individual's risk of developing breast cancer. This is an example of the intersection with "Technology-Receptive DigitalNicheConstruction & Technology-Driven CoShapingTechnologyOrganizationalPractices" (Open Set 5). The technology is not only welcomed but also drives changes in healthcare practices. For instance, high-risk individuals identified by these predictive models may be recommended for more frequent screenings or preventive therapies, thereby influencing healthcare strategies and policies.

Functorial mapping as transversality: Consider the functor F_4 as a mapping that induces a transversality between the categories C_4 and C_5. In differential topology, a map between two manifolds is transversal to a submanifold if the derivative of the map is surjective at every point in the preimage of the submanifold. In this context, the functor F_4 can be seen as a map that induces a transversality between the categories, meaning that it intersects each object in C_4 with an object in C_5 in a nontangential manner.

- Imagine two surfaces in three-dimensional space. If one surface just "touches" the other (like a tangent line touching a curve), we say the intersection is tangential. But if one surface "cuts through" the other, we say the intersection is transversal.
  - Now, let us apply this concept to the functor F_4 mapping objects from category C_4 to category C_5. Each object in C_4 can be thought of as a "point" in a source space, and each object in C_5 as a "point" in a target space. The functor F_4 is the "map" that relates these points. When we say that F_4 induces a transversality between the categories, we mean that for each object in C_4, the corresponding object in C_5 is not just tangentially related, but is directly influenced by it.
  - For example, the adaptive structuration in C_4 does not just "touch" the Free Energy Principle in C_5, but directly influences it, "cutting through" to impact how the Free Energy Principle is understood and applied in the context of breast cancer prevention, diagnosis, and treatment in US healthcare systems. This reflects the transformative process of deterritorialization, where existing structures are disrupted and new ones are created.

- The "derivative of the map being surjective at every point in the preimage of the submanifold" is a formal way of expressing this idea of "cutting through." In this context, "surjective" means that for every point in the target space, there is at least one point in the source space that maps to it. The "preimage" of a point in the target space is the set of all points in the source space that map to it. Saying the derivative of the map is surjective at every point in the preimage of the submanifold means that at every point where we intersect the submanifold, we are "cutting through" rather than just "touching."
    - In the context of differential topology, the derivative of a map at a point gives us information about how the map behaves near that point. It tells us how a small change in the input (source space) affects the output (target space). In essence, it provides a linear approximation of the map near that point. When we say that the derivative of the map is surjective at every point in the preimage of the submanifold, we are making a statement about the "completeness" of this local information. A surjective map (also known as an onto map) is one where every point in the target space is associated with at least one point in the source space. So, a surjective derivative means that for every direction in the target space, there is a direction in the source space that leads to it under the map. In the context of transversality, this surjectivity of the derivative ensures that the map intersects the submanifold in a "complete" way, covering all directions. This is what we mean by "cutting through" rather than just "touching." Even if the map and the submanifold intersect at a single point, the surjectivity of the derivative ensures that the map intersects the submanifold in all possible directions at that point, giving us a full, "nontangential" intersection.
    - In the context of the functor F_4 mapping objects from category C_4 to category C_5, this concept of a "surjective derivative" can be thought of metaphorically. It suggests that the influence of an object in C_4 on its corresponding object in C_5 is "complete" in some sense, affecting all aspects of the latter. This reflects the transformative and comprehensive nature of the deterritorialization process.
- This mapping is not just a simple one-to-one correspondence, but rather a complex transformation that intersects each object in C_4 with an object in C_5 in a nontangential manner. This means that the influence of each object in C_4 on the corresponding object in C_5 is not just peripheral or tangential, but rather significant and transformative.
    - For instance, consider the object "Adaptive Structuration" in C_4. When mapped by the functor F_4, it corresponds to the "Free Energy Principle" in C_5. This mapping signifies a transformation from the macroscopic structures and processes of the stakeholder ecosystems to the microscopic transformations that disrupt and recreate the cognitive landscape of the creative-experiencer. This transformation is characterized by transversality, meaning that the influence of Adaptive Structuration on the Free Energy Principle is significant and transformative, rather than just peripheral or tangential.
    - In the context of breast cancer prevention, diagnosis, and treatment, this could be illustrated as follows: Adaptive Structuration in C_4 could represent the evolving structures and processes in the healthcare ecosystem that adapt to the changing needs and preferences of patients, providers, and other stakeholders. This could

include the use of telehealth platforms for remote care delivery, the integration of AI and data analytics into clinical decision-making, and the shift toward patient-centered care models. When this object is mapped to the Free Energy Principle in C_5, it signifies a shift toward a more personalized, data-driven approach to healthcare. The Free Energy Principle models living systems' drive to minimize discrepancies between their internal world models and actual sensory data. In the context of healthcare, this could represent the use of predictive analytics and AI to analyze patient data and predict disease progression or treatment outcomes, thereby reducing uncertainty and improving predictability in healthcare outcomes. The mapping of Adaptive Structuration to the Free Energy Principle through the functor F_4 signifies a transformation from the macroscopic structures and processes of the healthcare ecosystem to the microscopic transformations that disrupt and recreate the cognitive landscape of the creative-experiencer. This transformation is characterized by transversality, meaning that the influence of Adaptive Structuration on the Free Energy Principle is significant and transformative, rather than just peripheral or tangential. This transversality can be seen in the way that the evolving structures and processes in the healthcare ecosystem (adaptive structuration) significantly influence the development and application of predictive analytics and AI in healthcare (Free Energy Principle). The use of predictive analytics and AI in healthcare is not just a peripheral phenomenon, but rather a significant intersection point that spans new directions in the manifold of C_5, influencing and being influenced by various aspects of the Free Energy Principle. This transversality reflects the complexity, dynamism, and fluidity of the healthcare ecosystem, where new cognitive processes and experiences are constantly being created and old ones are being disrupted.

This application of transversality provides a way to understand the complex interactions and intersections between different categories, open sets, and base sets. It captures the dynamic and evolving nature of the ecosystem, where new structures and patterns are constantly being created and old ones are being disrupted. The functor F_4, in particular, embodies this process of transformation, signifying a shift from the molar to the molecular, from the actual to the virtual.[8]

---

[8] Returning to Carroll et al. (2021), the concept of functorial mapping as transversality is exemplified through the transformative impact of wheelchair basketball and adaptive/inclusive sailing, which catalyze a shift from physical participation in sports to significant cognitive and perceptual changes in participants, reflecting a non-tangential, comprehensive influence characteristic of the Deleuzian notion of transversality. This transformation demonstrates a robust intersection, that is, "surjective derivative" in differential topology, between the physical activities and participants' cognitive realms, affecting various aspects of participants' lives. These activities disrupt traditional perceptions of disability in sports, leading to innovative patterns of engagement and interaction, thus embodying the process of deterritorialization and the emergence of new cognitive landscapes. This shift from macroscopic (structured sports activities) to microscopic (individual cognitive transformations) processes highlights the dynamic, evolving nature of these ecosystems, illustrating complex and transformative interactions across different domains and scales within the theoretical framework of DRT.

Dynamic relationality of machinic desires and territories in MLXEs

In Chapter 3, we looked at rhizomatics of machinic desire within assemblages and territorial escape and capture of machinic desire between assemblages, separately. In relational dynamics of MLXEs we have to consider both, but as we show in the **Appendix** (please see "Unified Gauge Theory of Desire and Territory"), we do not simply cross these two analyses, but synthesize a compact and manageable combination for generating insights across the MLXE sequence of categories and functors.

Intertwined dynamics of desire and territorial flows governed by Euler-Lagrangian dynamics: The Euler-Lagrangian dynamics, rooted in classical mechanics, provides a framework to understand the evolution of systems over time. When applied to the conceptual realm of desire and territory, it offers insights into how these entities interact, evolve, and influence each other. The kinetic (T) and potential (V) energy terms serve as indicators of the active forces and constraints within this system.

- **Basis states of desire and territory**
  - **Deterritorialized desire**: This represents the innate, unstructured desire.[9] This state embodies high kinetic energy (T) as it is characterized by active seeking, exploration, and the drive to find solutions.
  - **Reterritorialized desire**: This is the structured, organized form of desire.[10] This state has a higher potential energy (V) as it is more stable, defined, and bound by the existing healthcare narratives and structures.
  - **Superposition states**: These states capture complex scenarios.[11] In energy terms, these states might have fluctuating kinetic and potential energies.
- **Transformation of deterritorialized desire to/from reterritorialized desire using inversion operator**:
  - The transformation from a free-flowing, deterritorialized desire to a structured, reterritorialized one can be seen as a shift from a high kinetic energy state (T) to a potential energy-dominated state (V). The free-flowing desire, with its inherent motion and activity, embodies kinetic energy. When this desire gets captured or bound by a specific territory, it is like a moving object being trapped in a potential well, converting its kinetic energy to potential energy.
  - The transformation from a structured, reterritorialized desire to a free-flowing, deterritorialized one can be seen as a shift from a potential energy-dominated state (V) to a high kinetic energy state (T). The structured desire, with its inherent constraints and boundaries, embodies potential energy. When this desire is liberated and becomes free-flowing, it is akin to an object escaping from a potential well, converting its potential energy back to kinetic energy.
- **Interplay between desire and territory**:
  - **Transformation using latent operator**: The complex state resulting from this transformation indicates a blending of

Continued

Dynamic relationality of machinic desires and territories in MLXEs (cont'd)

desire and territory. The imaginary component suggests that the relationship is not straightforward but has nuances. In terms of energy dynamics, this state might represent a scenario where both kinetic and potential energies are in play, indicating a dynamic equilibrium between active desires and territorial constraints.[12]

* **Phase-shifted deterritorialized desires**: These states, with their imaginary components, suggest variations in the kinetic energy of the system. The negative and positive imaginary components might indicate different orientations or directions of the kinetic energy, representing varied forms of free-flowing desires with different nuances or tendencies.[13]
* **Phase-shifted reterritorialized desire**: The introspective consolidation indicated by this state suggests an increase in potential energy. The territory is not just static but is actively reinforcing its boundaries, increasing its potential energy.[14]
* **Transformation using dominance/bifurcation operator**: The unchanged state suggests a balance between the forces of desire and territory. In energy terms, the kinetic and potential energies might be in equilibrium, leading to a stable state where neither energy form dominates.[15]
- **Euler-Lagrangian equations in context**:
  * These equations, when applied to the realm of desire and territory, can help predict how the system will evolve over time based on the current states of kinetic and potential energy. For instance, a system dominated by kinetic energy (high T) might indicate rapid changes, transformations, or disruptions in the landscape of desire and territory. Conversely, a system where potential energy (V) dominates might suggest stability, consolidation, or resistance to change.
  * The transformations represented by inversion, latent, and dominance/bifurcation operators influence the energy dynamics of the system. For example, the inversion operator, which captures the shift from deterritorialized to reterritorialized desire, can be seen as an operator that converts kinetic energy to potential energy.

In summary, the intertwined dynamics of desire and territory, as governed by the Euler-Lagrangian framework, provide a rich panorama of interactions, evolutions, and influences. The kinetic and potential energy terms serve as lenses through which we can understand and predict the behaviors and tendencies of this complex system.[16]

Given the intricate dynamics of desire and territory within the Euler-Lagrangian framework, let us trace the "possible" itinerary of three desires corresponding to key entities in the breast cancer prevention, diagnosis, and treatment ecosystem:

- **Patient's desire (PD)**: The innate desire of a patient for health, early detection, and effective treatment.
- **Healthcare professional's desire (HPD)**: The desire of healthcare professionals to provide accurate diagnosis, effective treatment, and holistic care.

Dynamic relationality of machinic desires and territories in MLXEs *(cont'd)*

- **System's desire (SD)**: The collective desire of the healthcare system, including institutions, policies, and technologies, to offer efficient, accessible, and advanced care.

In each pairing of categories via functors, the transformations capture the sophisticated interchange between desire and territory, illustrating the dynamic forces at play in the breast cancer care ecosystem. The Euler-Lagrangian framework provides a lens to understand these dynamics, offering insights into the active forces and constraints shaping the journey of each desire.

Let us delve deeper into each desire, tracing its evolution through the pairings of categories via functors and highlighting the kinetic and potential energy dynamics. Each desire's journey through the healthcare system is influenced by both their active aspirations and the structured constraints of the system. The Euler-Lagrangian framework provides a lens to understand these dynamics, offering insights into the forces driving the patient's evolving desires, following the eight MLXE categories in Fig. 4.3.[17]

Patient's desire (PD): In the realm of healthcare, the journey of a patient's desire begins within category C_1. Here, through the functor *Lived Journey*, the patient's free-flowing aspirations for health and well-being undergo a transformation, shifting from a deterritorialized state to a reterritorialized one. This evolution signifies the patient's initial experiences with the healthcare system, where their raw, unstructured desires for health become shaped by the structured protocols and guidelines of category C_2. The kinetic energy of their aspirations is converted into potential energy as they navigate the healthcare landscape. As the patient moves to category C_2, their desire evolves through the functor *Engagements*. The transformation here is intricate, with the patient's desire intertwining with the territorial constraints of healthcare policies and societal expectations in category C_3. This dynamic equilibrium between active aspirations and territorial constraints is evident in the balanced interplay of kinetic and potential energies. Transitioning to category C_3 and through the functor *Flows*, the patient's desire seeks to integrate various healthcare components. This aspiration becomes structured as the patient establishes new interdisciplinary collaborations, converting the kinetic energy of their aspirations into potential energy within category C_4. In category C_4, through the functor *Events*, the patient's interactions with global healthcare events introduce nuances to their desire. Reflecting on global best practices and standards, the kinetic energy of their desire undergoes an internal transformation, leading them to category C_5. Within category C_5 and through the functor *Sense*, the patient's desire for efficiency and satisfaction finds equilibrium with the territorial constraints of resources and regulations. Both kinetic and potential energies find balance, reflecting the patient's alignment with their aspirations and constraints in category C_6. As the patient transitions to category C_6 and through the functor *Enactments*, their desire to implement new healthcare strategies intertwines with the territorial constraints of legal frameworks and societal expectations, leading them to category C_7. In category C_7, through the functor *Resourced Capabilities*, the patient's desire to leverage healthcare capabilities and

Continued

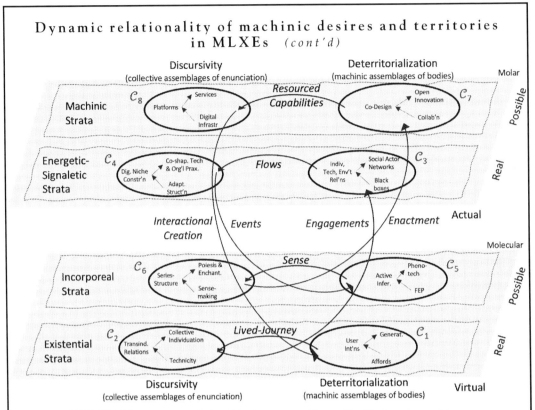

FIGURE 4.3 Dynamic relationalities in an assemblage with objects and morphisms.

resources undergoes a transformation, seeking innovative ways to optimize their health outcomes. The kinetic energy of their desire takes on a new orientation, driven by aspirations to innovate within resource constraints. Finally, in category C_8 and through the functor *Interactional Creation*, the patient's structured interactions with healthcare platforms liberate as they revisit foundational healthcare principles. The potential energy of structured interactions is converted back to kinetic energy, marking a return to core healthcare values and leading them back to category C_1.[18]

Healthcare professional's desire (HPD): A healthcare professional's desire emerges within category C_1. Through the functor *Lived Journey*, this desire, initially driven by the aspiration to provide the best care, becomes nuanced by the challenges of modern medical technologies. The kinetic energy of this desire undergoes an internal transformation, reflecting the evolving medical landscape. Transitioning to category C_2 and influenced by the functor *Engagements*, the professional's desire intertwines with the territorial constraints of hospital protocols. This dynamic equilibrium between aspirations and institutional constraints sees both kinetic and potential energies in play, a hallmark of the tug-of-war between individual ambition and systemic boundaries. As the journey progresses to category C_3 via the functor *Flows*, the professional's experiences

> ### Dynamic relationality of machinic desires and territories in MLXEs *(cont'd)*
>
> with interdisciplinary teams introduce a new direction in their desire. The kinetic energy takes on a new orientation, driven by the collaborative ethos of modern healthcare, leading them to category C_4. Within category C_4, the functor *Events* shapes the professional's interactions with medical events like conferences. Here, their desire for knowledge finds equilibrium with the territorial constraints of their specialty. The balance between kinetic and potential energies reflects this alignment with the broader medical community. Moving to category C_5 through the functor *Sense*, the professional's innovative aspirations become structured by evidence-based practices. The kinetic energy of their free-flowing desire is converted into potential energy, bound by the rigidity of medical protocols. In category C_6, influenced by the functor *Enactments*, the professional's reinforced understanding of ethical practices increases the potential energy of their desire, seeking alignment with ethical norms. Transitioning to category C_7 via *Resourced Capabilities*, the professional's aspirations for cutting-edge treatments intertwine with the capabilities of the healthcare industry. The dynamic equilibrium between aspirations and industry constraints is evident in the interplay of kinetic and potential energies. Finally, in category C_8, through *Interactional Creation*, the professional's structured interactions with digital health platforms liberate, leading them back to category C_1. The potential energy of structured interactions is converted back to kinetic energy, marking a return to the essence of hands-on patient care.[19]
>
> **System's desire (SD):** The system's desire emerges from category C_1. Through the lens of the functor *Lived Journey*, this desire, rooted in structured protocols, undergoes introspection. As it transitions to category C_2, the system reevaluates its foundational principles, reinforcing its boundaries with increased potential energy. As the system navigates to category C_2, influenced by the functor *Engagements*, it aspires to expand its horizons. Exploring new avenues of patient care and technological integration, its kinetic energy takes on a new orientation, reflecting its innovative aspirations. Venturing into category C_3 via *Flows*, the system's free-flowing aspirations become more structured. The kinetic energy of its expansive vision is converted into potential energy as it forms structured collaborations, integrating various healthcare components. In category C_4, shaped by the functor *Events*, the system's interactions with global healthcare events nuance its desires. Reflecting on global best practices, its kinetic energy undergoes an internal transformation, resonating with global healthcare trends. Transitioning to category C_5 through *Sense*, the system's desire for efficiency finds equilibrium with territorial constraints. The balance between its aspirations for patient satisfaction and the realities of resources and regulations results in an equilibrium of kinetic and potential energies. In category C_6, under the influence of *Enactments*, the system's desire to implement new policies intertwines with societal expectations. The dynamic equilibrium between its aspirations and societal constraints sees both kinetic and potential energies in play. Moving to category C_7 via *Resourced Capabilities*, the system's desire undergoes another transformation. As it seeks to optimize its capabilities, its kinetic energy takes

Continued

Dynamic relationality of machinic desires and territories in MLXEs *(cont'd)*

on a new orientation, driven by innovation within resource constraints. Finally, in category C_8, through *Interactional Creation*, the system's structured interactions with digital platforms liberate. As it transitions back to category C_1, the potential energy of its structured interactions is converted back to kinetic energy, marking a return to the essence of core healthcare values.[20]

The itineraries of the three desires—patient's desire (PD), healthcare professional's desire (HPD), and system's desire (SD)—provide a comprehensive view of the breast cancer prevention, diagnosis, and treatment ecosystem. While they often interact and intersect in their journeys, leading to collaborative solutions, they can also interrupt each other due to differing priorities, systemic constraints, and rapid technological evolutions. Understanding these dynamics and analyzing interactions, intersections, and possible interruptions are crucial for creating a harmonious and effective healthcare system:

- **Interaction:**
  * **Shared spaces:** All three desires navigate through the same categories and functors, implying that they operate within the same healthcare ecosystem and are influenced by similar external factors.
  * **Complementary goals:** At several junctures, the desires align. For instance, PD's aspiration for health aligns with HPD's aim to provide accurate diagnosis and holistic care, and both are supported by SD's goal of efficient and advanced care.
  * **Feedback loops:** The evolution of one desire often influences the others. For example, as SD introduces new technologies (category C_4, via events), both PD and HPD must adapt, leading to new aspirations and challenges.

- **Intersection:**
  * **Shared challenges:** In category C_2 (via engagements), all three desires grapple with the territorial constraints of policies and societal expectations. This shared challenge can lead to collaborative solutions or conflicts, depending on how each entity perceives and addresses the issue.
  * **Collaborative evolution:** In category C_3 (via flows), while PD seeks integrated healthcare components, HPD aims for collaborative care, and SD aspires for structured collaborations. This suggests a mutual drive toward a more integrated and collaborative healthcare system.
  * **Resource allocation:** In category C_7 (via resourced capabilities), both HPD and SD express desires related to leveraging resources, indicating potential competition or collaboration in resource allocation.

- **Interruption:**
  * **Differing priorities:** While PD might prioritize personal health and well-being, HPD might be more focused on adhering to medical protocols, and SD might prioritize system-wide efficiency. These differing priorities can lead to interruptions in the seamless flow of care.
  * **Systemic constraints:** In category C_5 (via sense), while PD seeks efficiency and satisfaction, SD is bound by resource and regulatory constraints, potentially leading to interruptions in fulfilling patient desires.

Dynamic relationality of machinic desires and territories in MLXEs *(cont'd)*

- **Technological evolution:** As SD introduces or changes technologies, both PD and HPD might face interruptions in their journeys as they adapt to these changes.[21]

[9] For example, patients desire for accurate diagnosis and effective treatment. It's the raw, unfiltered need that drives patients to seek medical technologies like mammograms and MRIs.

[10] For example, patient's desire shaped and guided by public health campaigns, patient testimonials, and medical literature.

[11] For example, patients are influenced by both their personal experiences and the collective narratives. It is like being in two places at once—driven by personal needs while also being guided by societal wisdom.

[12] This represents the complex reciprocity between individual needs (like accurate diagnosis) and the broader societal narratives and guidelines. It is a state where kinetic and potential energies are intertwined, reflecting the dynamic equilibrium between personal desires and societal structures.

[13] This phase-shifted version of the "deterritorialized desire" can be seen as a nuanced form of the patient's need for accurate diagnosis and effective treatment.

- The negative imaginary component might symbolize a form of desire that's introspective or reflexive. For instance, a patient might be reflecting on their past experiences, perhaps re-evaluating their trust in certain medical technologies or even grappling with the emotional weight of a recent diagnosis. In terms of energy dynamics, this introspective state might indicate a temporary dip in kinetic energy as the patient pauses to reflect and understand their journey better.

- This state, with its positive imaginary component, represents an outwardly expansive form of the "deterritorialized desire". Here, the patient might be actively seeking new treatments, exploring alternative diagnostic methods, or even advocating for better healthcare solutions. The kinetic energy in this state is directed outward, indicating an active, exploratory phase in the patient's journey.

[14] This state represents a deeper, more introspective form of the "reterritorialized desire." In the context of category C_4, it might indicate a patient deeply engaging with public health campaigns, patient testimonials, and medical literature. They are not just passively consuming information but actively reflecting on it, comparing it with their personal experiences, and perhaps even challenging or questioning certain narratives. This introspective consolidation suggests an increase in potential energy. The territory (public health campaigns, testimonials, literature) is not just static but is actively reinforcing its boundaries. This could be seen in patients who become advocates or champions of certain healthcare narratives, reinforcing and strengthening the established guidelines and practices.

[15] This captures the differentiation between personal experiences with medical technologies and the collective wisdom of medical literature. It is like comparing the raw data (individual experiences) with the processed information (collective wisdom). This transformation might involve shifts in both kinetic and potential energies as patients navigate the healthcare landscape.

[16] In their study of three pivotal moments in the emergence of the 'e-clinic', Fox et al. (2005) reveal a complex interplay, governed by Euler-Lagrangian dynamics, of how patients' unstructured desires interact with and adapt to the existing territorial flows within the healthcare system. The first moment, the electronic transfer of prescriptions (ETP), signifies a reterritorialized desire within healthcare, where traditional prescription practices are reconfigured digitally, embodying the conversion of potential to kinetic energy as the system adapts to electronic efficiency, through the use of an inversion operator. The second moment, the development of e-pharmacies and the role of the "virtual" pharmacist, exemplifies a deterritorialized desire for convenient medication access, indicating high kinetic energy toward innovative healthcare solutions. This phase includes a nuanced reterritorialization, albeit phase-shifted, as traditional pharmacy roles adapt to the online environment, indicating a structured digital pharmaceutical service approach. The third moment, the establishment of online virtual medicine consultations, or e-clinics, presents an interplay of

Continued

Dynamic relationality of machinic desires and territories in MLXEs *(cont'd)*

deterritorialized (patients seeking remote consultations) and reterritorialized (traditional medical profession structures) desires. This dynamic balance mirrors the use of dominance/bifurcation operators, showcasing an equilibrium between conventional in-person and innovative online e-clinic models, reflecting both kinetic and potential energies.

[17] Note that this figure is an expanded version of Fig. 3.1, where each category is detailed to show its objects and morphisms similar to Fig. 3.2.

[18] In Fox et al.'s (2005) study, the likely evolution of a patient's desire (PD) in the context of digital healthcare can be mapped through eight categories of DRT. This journey begins in category C_1, where patients first engage with digital health initiatives like e-pharmacies. In C_2, their desires intersect with healthcare policies and societal expectations around digital health. Moving to C_3, patients integrate digital health components into their care. In C_4, global trends in digital healthcare, such as the rise of e-clinics, influence PD. Category C_5 sees patients balancing the efficiencies of digital healthcare with available resources and regulations. In C_6, patients navigate and adapt to new strategies within legal frameworks. C_7 involves leveraging digital healthcare resources to optimize health outcomes. Finally, in C_8, there's active interaction with digital healthcare platforms, leading to a full circle back to C_1. This return reflects a reassessment and realignment of foundational healthcare values influenced by their digital healthcare experiences, underscoring the transformative and dynamic nature of digital healthcare systems in shaping patient desires and expectations.

[19] In Fox et al. (2005), we can trace the pharmacist's desire (HPD) as follows. Starting in category C_1, pharmacists grapple with the foundational challenges of integrating digital health practices, such as electronic prescriptions, into their traditional roles. As they progress to categories C_2 and C_3, they actively engage with and adapt to e-pharmacy models and digital platforms, reshaping their professional practices to suit the digital era. This evolution continues through category C_4, where global healthcare trends further influence their practices. In categories C_5 to C_7, pharmacists find a balance between digital efficiencies and traditional practices, uphold ethical standards in virtual environments, and innovate within industry constraints. Completing the cycle in category C_8, they integrate these digital experiences back into their core professional ethos, ensuring that technological advancements enhance patient care.

[20] In Fox et al.'s (2005) study, the system (UK government agencies and the pharmaceutical industry) rooted in established healthcare protocols, faces the advent of digital health, including e-prescriptions (C_1). It then engages with technological evolutions, like electronic transfer of prescriptions, reflecting early digital adaptations (C_2). The integration of e-pharmacies challenges the system to merge digital models within traditional frameworks (C_3), influenced by global e-health trends (C_4). Balancing innovation with regulation becomes crucial as the system navigates the digital health landscape (C_5), further adapting to e-clinics and online consultations (C_6). This evolution leads to strategic innovation within industry constraints (C_7), culminating in a synthesis of traditional and digital healthcare, reintegrating learned experiences into core healthcare principles (C_8). Returning to C_1, the system embodies a holistic adaptation to digital health, prioritizing patient welfare within a regulatory and ethical context.

[21] In Fox et al. (2005), the interaction, intersection, and interruption of PD, HPD, and SD provide insight into the digital healthcare ecosystem's complexity. These desires navigate similar stages in adapting to digital health advancements, with shared experiences and complementary goals across categories, such as PD's adoption of e-clinics and HPD's adaptation to online prescription services aligning with SD's digital health standards. However, shared challenges arise, particularly in regulatory and societal expectations, leading to potential collaborative or conflicting approaches. The mutual drive toward integrated healthcare is evident in the integration of digital health components, with a focus on collaborative care and structured collaborations. Resource allocation in optimizing health outcomes can trigger competition or collaboration between HPD and SD. Notably, differing priorities can cause interruptions: patients prioritize personal health, healthcare professionals adhere to medical protocols, and the system focuses on efficiency. Systemic constraints, particularly in balancing efficiency with resources, and the rapid pace of technological evolution, further complicate these interactions, underscoring the need for a harmonious, efficient, and patient-centered digital healthcare system.

CHAPTER 5

Dynamic relationalities across (stacked) strata

In Chapter 4, we saw how beyond the intraassemblage dynamics, DRT also emphasizes the relationships between different assemblages. Different categories represent varied facets of a system. The relationships between categories are intricate and nonlinear. The theoretical framework integrates these categories through functors and their natural transformations, offering a structured perspective on the system's dynamics. By incorporating advanced mathematical concepts, each category is visualized as a manifold, with transformations depicted by vector fields. Transversality provides a lens to understand these intersections, revealing deeper insights into the system's dynamics. Transversality reveals intersections, and functors integrate these categories, offering a structured perspective on system dynamics. Concepts like transversality and continuous deformation enrich our understanding of the interplay between categories, providing a comprehensive view of the dynamic relationality between assemblages.

In this chapter, we further develop the concepts of line of flight, becoming, and reterritorialization, using functors and natural transformations, to build preliminary insights into the singularity and transcendence of Artificial General Intelligence (AGI) within the broader realm of Machinic Generalized Intelligence (MGI) and co-evolution of HI-AI cocreativeness. Category theoretical concepts of functors and natural transformations can help us delve deep into the intricate relationships between perception, cognition, technology, and experience. They provide a robust framework for understanding the transformative potential of AI and experiencial computing in shaping digital ecosystems. These concepts highlight the interconnected nature of various research domains and emphasize the need for an integrated, interdisciplinary approach to studying digitalized ecosystems.

In the realm of DRT, the concept of "lines of flight" signifies departures from established structures, opening avenues for potential new formations. "Lines of flight" signify departures from established structures, bridging stakeholder-ecosystems to cognitive processes and propelling deterritorialization. The idea of a "line of flight" is encapsulated by the functor "Event," which bridges stakeholder ecosystems to cognitive processes, marking a transition from overarching structures to intricate transformations. Such lines of flight propel deterritorialization, fostering new assemblages and reshaping existing ones. They act as catalysts for change, driven by quasi-causes that unsettle prevailing assemblages, leading to novel formations. These transformations, whether rooted in global structures or in flux between paradigms, are guided by various mechanisms, emphasizing shifts in prevailing norms and practices.

Furthermore, we discuss the concept of "becoming" as proposed by Deleuze and Guattari as another type of dynamic relationality that is fundamental to Dynamic Relationality Theory (DRT). In Deleuze and Guattari's philosophy, "becoming" is not a simple transition from one state to another, but a process of continuous transformation and change that disrupts established structures and identities and creates new connections, relationships, and possibilities. This aligns closely with the core principles of DRT, which emphasizes the fluid, contingent, and interconnected nature of relations and the transformative and generative dynamics that they entail. Further, as assemblages evolve, the foundational principles governing their relational dynamics undergo profound shifts. Natural transformation ensures a structured path of evolution.

Enriching the creative-experiencer with generative artificial intelligence

Exploring the dynamic interplay between humans and technology within experience ecosystems, Chapter 2 highlighted how tangible factors like economic and media influences intertwine with intangible elements such as aspirations and self-understanding. Through events as pivotal moments of change, the relationship between actual experiences and virtual potentials is constantly recalibrated, signifying deep structural and dynamic shifts within the ecosystem. Technology plays a catalytic role, a medium through which events manifest, changing the context and nature of interactions, shaping how individuals engage with and make sense of their environment, altering perceptions and experiences. Thus, technology is not a mere tool but an active participant in shaping our cognitive and experiential realities, and enhancing human cognitive and experiential capacities more intuitively, responsively, and personally.

Fig. 5.1 shows the concept of Category C_{5+} as representing an aspirational vision of the future where artificial systems not only mimic human cognitive processes but also contribute to the creation of immersive, personalized computing experiences. We are witnessing the foundations of such a topological space in various developments. For instance, research in cognitive science and AI is advancing our understanding of human perception and cognition.[1] Although Artificial General Intelligence (AGI) remains largely theoretical, significant strides have been made in narrow AI.[2] Moreover, advancements in technologies like virtual

---

[1] Kriegeskorte and Douglas (2018) discuss the integration of cognitive science, computational neuroscience, and AI in understanding brain computations for cognitive tasks, highlighting efforts to model human cognition.

[2] Fjelland (2020) provides a critical perspective on the feasibility of achieving AGI, contrasting it with advances in narrow AI and discussing inherent limitations in replicating human intelligence.

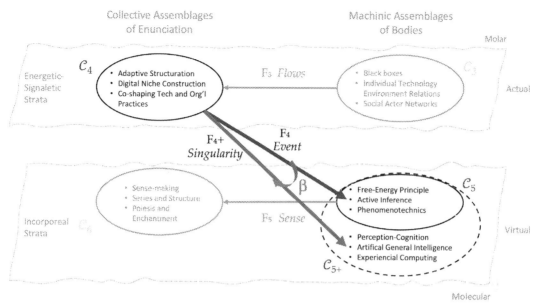

FIGURE 5.1 Becoming through lines of flight.

reality (VR), augmented reality (AR), and mixed reality (MR) are being used to create immersive digital experiences.[3]

The transformation from Category C_5 to Category C_{5+} is not a mere replacement but an expansion, incorporating elements like PerceptionCognition, AGI, and ExperiencialComputing. The manifold C_5 symbolizes the entire realm of possibilities for the assemblage, representing configurations of foundational elements like the FreeEnergyPrinciple, ActiveInference, and Phenomenotechnics. As Category C_5 evolves into C_{5+}, these foundational elements transform into enriched concepts like PerceptionCognition, AGI, and Experiencial Computing. Emerging technologies play a crucial role in this transformation:

- Generative AI enhances the FreeEnergyPrinciple by generating new data.
- Neuromorphic Computing, which emulates the human brain, offers insights into ActiveInference.
- Quantum Computing provides a fresh perspective on the FreeEnergyPrinciple and ActiveInference.
- Mixed Reality (AR/VR) connects to Phenomenotechnics, altering human perception and experience.

However, the integration of these elements into a cohesive topological space where they interact and influence each other in the ways described by Category C_{5+} is still a work in progress. The transformation of these individual advancements into a unified, dynamic

---

[3] Dwivedi et al. (2022) explore the transformative potential of virtual and augmented reality in the metaverse, emphasizing the blurring lines between physical and digital realities and its impact across various sectors, resonating with the idea of immersive, personalized computing experiences.

system represents a significant challenge that researchers, technologists, and theorists are actively working to address. In addition to these developments, there are several other promising technologies that could contribute to the development of topological spaces of assemblages approximating the description of Category $\mathbf{C_{5+}}$. Quantum computing, neuromorphic computing, edge computing, 5G and beyond, blockchain and decentralized AI, bioinformatics and genomics, and Emotion AI are all areas of technological advancement that, in combination with the developments mentioned earlier, could contribute to the emergence of topological spaces of assemblages that approximate the description of Category $\mathbf{C_{5+}}$. As these technologies continue to evolve and converge, they are likely to drive significant advancements in our ability to create artificial systems that not only mimic human cognitive processes but also contribute to the creation of immersive, personalized computing experiences.

Thus, we can posit a *still-becoming* Category $\mathbf{C_{5+}}$ as an "Enriched Creative-Experiencer" focusing on perception, cognition, and artificial general intelligence (AGI) in the context of experiencial computing. In $\mathbf{C_{5+}}$, we have three additional objects:

- PerceptionCognition: This object represents the mental processes involved in interpreting and making sense of sensory information. It is a key area of research in cognitive science and has significant implications for the development of AGI.[4]
- AGI: This object represents artificial general intelligence, which is capable of learning and understanding any intellectual task that a human can perform. The development of AGI is influenced by insights from perception and cognition research.[5]
- ExperiencialComputing: This object represents the use of digital technologies, data analytics, and AI to enhance user experiences through personalization, immersion, and interactivity.[6] AGI has the potential to greatly enhance experiencial computing by enabling more advanced, personalized, and immersive user experiences.

And two additional morphisms: f_{53} and f_{54}.

- f_{53}: PerceptionCognition $\rightarrow$ AGI. This morphism captures the idea that insights from perception and cognition research influence the development of AGI, helping to understand human cognitive processes and their potential replication in artificial systems.[7]

---

[4] Banich and Compton (2023) provides comprehensive insights into the brain's processing of sensory information, cognition, and higher-order functions.

[5] Roitblat (2020) discusses the limitations of current AI in achieving AGI, emphasizing the need for AI to extend beyond solving structured problems to encompassing human-like insight and creativity.

[6] Emphasizing digitally mediated experiences in daily activities using embedded technologies, Yoo (2010) underlines the importance of integrating time, space, actors, and artifacts through digital technology to enhance user interactions and experiences.

[7] Yamakawa (2021) advocates for a brain-inspired AGI approach, emphasizing the replication of brain architecture to address AGI's design complexities. Huang (2017) discusses neurocomputers, which emulate biological neural networks, suggesting a structural and functional imitation of the brain for AGI. Allen et al. (2022) demonstrate the application of cognitive neuroscience in AI, using a massive fMRI dataset to train neural networks, thereby illustrating the practical use of perception and cognition research in advancing AGI.

- f_{54}: AGI → ExperiencialComputing. This morphism represents the concept that AGI has the potential to greatly enhance experiential computing, enabling more advanced, personalized, and immersive user experiences through the integration of artificial intelligence with various computing technologies.[8]

Dynamic relationality #4: Becoming through lines of flight

As we saw in Chapter 4, the functor F_4, termed "Event," encapsulates a dynamic relationality, echoing the notion of "lines of flight," a concept central to Deleuze and Guattari's philosophy. This idea represents a departure from established structures, paving the way for potential new formations. F_4 bridges stakeholder ecosystems (C_4) to the cognitive processes of the creative-experiencer (C_5), marking a transition from overarching structures to intricate transformations. This mirrors the fluidity of ecosystems, where novel cognitive processes arise and existing ones undergo disruption.

Acting as a catalyst, F_4 propels deterritorialization, fostering new assemblages through the interplay of entities capable of mutual influence. This transformation is sparked by a quasi-cause, an element that unsettles prevailing assemblages, ushering in novel formations. Entities within this framework can be deeply rooted in global structures, adhere to prevailing norms, or be in flux between global and local paradigms. Various mechanisms, such as the Inversion Operator and Latent Operator, guide these transitions, emphasizing shifts like the move from a one-size-fits-all treatment strategy (surgical) to more personalized, nuanced methods (noninvasive).[9]

In contrast, functor F_{4+}, termed "Singularity," signifies a more advanced stage of transformation. It integrates artificial intelligence, pointing toward a technological singularity—a hypothetical moment when AI systems can recursively self-improve, leading to exponential technological growth beyond human comprehension.[10] As we delve deeper into the realm of AI, its integration in fields like healthcare becomes evident. For instance, in breast cancer care, AI's incorporation has revolutionized diagnostic accuracy and personalized treatment plans. This shift from traditional care methods to technologically advanced, AI-driven care is emblematic of the "becoming" process. As care practices evolve, the underlying cognitive processes must align with the capabilities of AGI, marking a significant step toward the singularity

---

[8] Baskerville et al. (2020) emphasize the computed nature of human experiences in a digital-first world, where digital and physical realities intertwine, and digital platforms use algorithms to shape human experiences. Accordingly, AGI could significantly enhance experiential computing, making it more responsive, personalized, and context-aware by bridging digital and physical realms and advancing computational models that understand and predict human behavior.

[9] As we saw in Chapter 2, F_4 as "Event" encapsulates a dynamic relationality that stems from the interconnectedness of systems (Bateson, 2000), the transformation of underlying structures into observable events (Deleuze, 1990), and the systemic communication that leads to new states and behaviors (Luhmann, 1995).

[10] Kurzweil (2005) links AI's self-improvement with broader societal and existential changes, envisioning a transformative merge of human and machine intelligence that profoundly impacts life, work, and human identity, and highlights the potential for radical evolution in human capabilities and society.

To account for how co-evolutionary dynamics of systems and organisms are being transformed into the cognitive and perceptual landscape of hyper-intelligent, self-improving systems, we propose functor F_{4+} as a theoretical pathway to the technological singularity, the hypothesized point at which artificial systems become capable of recursive self-improvement, leading to rapid, exponential technological growth beyond human comprehension. Thus, functor $F_{4+}: \mathbf{C_4} \rightarrow \mathbf{C_{5+}}$, named "Singularity" as a line flight:

$F_{4+}(\text{AdaptiveStructuration}) = \text{PerceptionCognition};$

$F_{4+}(\text{DigitalNicheConstruction}) = \text{AGI}$

$F_{4+}(\text{CoShapingTechnologyOrganizationalPractices}) = \text{ExperiencialComputing}$

$F_{4+}(f_{41}) = f_{53}, F_{4+}(f_{42}) = f_{54}$

- As humans and AI systems co-evolve, continually adapting to and shaping digital technologies, we may witness an enhancement of machine perception and cognition capabilities.[11] This parallels the idea of AI systems reaching a level of sophistication where they start to understand and interpret the world in a similar manner as humans, a development closely associated with the onset of the singularity.
- As specialized digital environments evolve, they may foster the development of AGI systems capable of generalizing across tasks.[12] The emergence of such AGI systems would mark a significant step toward the singularity, as they would embody a level of intelligence comparable to, or even surpassing, human intelligence.
- As technology and social practices co-evolve, we could see an emergence of highly adaptive, personalized computing experiences.[13] Such a transformation might indicate a movement toward the singularity, as hyper-intelligent systems could leverage this

---

[11] Adams et al. (2012) present a unique roadmap for AGI development, drawing on human cognitive development stages, suggesting AGI's evolution should parallel the growth from basic sensory-motor skills to advanced cognitive functions. Their interdisciplinary approach integrates physiological insights, such as neural growth patterns, suggesting AGI might replicate human cognitive maturation. They utilize mathematical and information-processing models, considering intelligence as data compression and paralleling brain development in hardware/software terms. Additionally, they incorporate system-theoretic ideas, exploring how AGI could evolve within social and cultural contexts.

[12] Fei et al.'s (2022) large-scale multimodal foundation model demonstrates a significant advancement in AGI through its multimodal training on a diverse internet dataset. The model's proficiency in adapting to various cognitive tasks, coupled with its imagination and reasoning abilities, exemplifies strides in creating AGI systems capable of generalizing across multiple domains.

[13] Allal-Chérif's (2022) study on using immersive technologies in cathedrals exemplifies the trend toward personalized computing experiences. It shows how augmented reality, virtual reality, and artificial intelligence can create intensely personal cultural experiences, foreshadowing a future where hyper-intelligent systems deeply understand and predict individual human behaviors and preferences.

personalized data to understand and predict human behaviors, preferences, and needs, to a degree that is currently unimaginable.

In the language of category theory, this "becoming" can be conceptualized as a natural transformation between two functors, which can be thought of as lines of flight in Deleuze and Guattari's terminology. The lines of flight, in this context, can be seen as the pathways of transformation that the functors represent. This is in line with the "-al-" in DRT, signifying the processual and generative nature of relations, which are continually being made and remade through interactions. This process of "becoming" is not a one-time event, but a continual process of transformation and change that is integral to the constitution and transformation of entities and systems. It underscores the inherent unpredictability and nonlinearity of these processes, reflecting the complex interplay of undetermined, determinable, and determined elements.

A natural transformation provides a way of transitioning between different functors, representing the process of change and transformation within *and* between categories. In this context, a natural transformation can be seen as a "line of flight *of* line of flights" that breaks away from the established structures in one category and leads to the creation of new structures in another category, or more commonly, the extension or enrichment of a category. Such an extension or enrichment is underway in the context of our category framework, as Category C_5 is becoming Category C_{5+} as a process of transformation where the original objects and morphisms, i.e., FreeEnergyPrinciple, ActiveInference, and Phenomenotechnics, are not just replaced, but their identities are expanded and transformed to include new objects and morphisms: PerceptionCognition, Artificial General Intelligence (AGI), and ExperiencialComputing.

Natural transformation β: $F_4 \rightarrow F_{4+}$ ("Cognitive Metamorphosis")

We can introduce the dynamic relationality of becoming, as natural transformation β: $F_4 \rightarrow F_{4+}$, which maps the objects and morphisms of C_4 as seen through the lens of functor F_4 (which views them in terms of FreeEnergyPrinciple, ActiveInference, Phenomenotechnics) to the lens of functor F_{4+} (which views them in terms of PerceptionCognition, AGI, ExperiencialComputing). Through this process, the identity of Category C_5 is transformed and expanded to become Category C_{5+}, reflecting the dynamic, evolving nature of the concepts it represents. As AGI develops and becomes integrated into our cognitive processes (via experience computing), it might be said to deterritorialize our traditional understanding of perception and cognition, expanding and transforming our cognitive "territory", thus effectuating a "Cognitive Metamorphosis."

The natural transformation β: $F_4 \rightarrow F_{4+}$ can be specified as follows.[14] For every morphism f: A → B in C_4, the natural transformation components $β_A$: $F_4(A) \rightarrow F_{4+}(A)$ and $β_B$: $F_4(B) \rightarrow F_{4+}(B)$ should make the diagram commute. This is expressed as: $F_{4+}(f) \circ β_A = β_B \circ F_4(f)$. This means that, no matter which path you take in the diagram (either first going through F_4 and then $β_B$, or first through $β_A$ and then F_{4+}), the outcome should be the same, that is, the way objects and morphisms are transformed by F_4 into C_{5+} and then modified by β is consistent with directly applying F_{4+} to the objects and morphisms in C_4. This requirement ensures that the structure and relationships defined by the functors are preserved by the natural

---

[14] In defining the natural transformation, we will assume that F_4 maps the same objects and morphisms from C_4 to C_5, but in the latter's becoming C_{5+}.

transformation across the categories. In other words, there is a family of morphisms $\beta_X: F_4(X) \rightarrow F_{4+}(X)$:

- $\beta_{\text{AdaptiveStructuration}}$: FreeEnergyPrinciple → PerceptionCognition: adaptive structuration processes in organizations or systems can be connected to the perception and cognition processes occurring within these systems. This relationship provides a theoretical framework for understanding how minimizing uncertainty and adapting to environmental changes influence the way agents perceive and process information.[15]
- $\beta_{\text{DigitalNicheConstruction}}$: ActiveInference → AGI: active role of agents in seeking information and adapting to their niches has implications for the design of AGI systems. Understanding this connection can help inform the development of AGI algorithms that better mimic the adaptive and information-seeking behavior of biological systems, leading to more flexible and contextually adaptive AI systems.[16]
- $\beta_{\text{CoShapingTechnologyOrganizationalPractices}}$: Phenomenotechnics → ExperiencialComputing: understanding the interplay between human perception, technology, and organizational practices can inform the design of more immersive and engaging computing experiences. By considering the role of technology in shaping human experience, researchers, and practitioners can create systems that better cater to user needs and expectations.[17]

This natural transformation represents the connection between the concepts mapped by F_4 and F_{4+} in the context of Information Systems, AGI, and Experiencial Computing, highlighting the relationship between these areas of research. The morphism conditions encapsulate the transition from adaptive structuration processes to digital niche construction and co-shaping technology and organizational practices, showing a smooth evolution while integrating the transformative effects of different theoretical frameworks. This perspective opens up new avenues for research and innovation in the study of digital ecosystems and other complex systems and provides a novel contribution to the field.

Deleuze and Guattari's philosophy emphasizes the concept of "becoming," a continuous transformation that challenges and reshapes established identities and structures. This idea is foundational to DRT, which underscores the fluidity and interconnectedness of relations

---

[15] Friston's (2009) Free-Energy Principle posits that cognitive processes in systems inherently aim to minimize free-energy, equating to reducing uncertainty or surprise. Organizational adaptation and structuring are geared toward aligning perceptions and cognitions with environmental dynamics to minimize predictive errors, thus reducing informational entropy.

[16] Deane's (2022) work on affective feelings and mental action within active inference illustrates how affective self-modeling, as a domain-general controller, enables organisms to adaptively engage with their environment. Deane shows affective processes are not merely reactive but involve hierarchical inferential processing, aligning with the depth and complexity required for AGI. This understanding suggests that for AGI to emulate the adaptability of biological systems, it must incorporate self-modeling capabilities, facilitating flexible and context-aware decision-making.

[17] In their conceptualization of ecosystems of intelligence where active inference is a core principle, Friston et al. (2024) underscore the co-evolution of technology and human systems, advocating for communication protocols and nested intelligences to facilitate multiscale, context-aware interactions. Their vision extends to the physicality of information processing, advocating for a holistic approach in experiencial computing, where technology becomes an active participant in human experiences and organizational practices.

and their transformative dynamics. In category theory, this transformation is represented as a natural transition between functors. The transformation β: $F_4 \to F_{4+}$ symbolizes this "becoming" process, where Category C_5's identity evolves into Category C_{5+}. This transformation is not just a mere state change but a metamorphosis, indicating a shift from traditional methodologies to AI-enhanced approaches. The identity remains patient-centric, but the methods evolve, reflecting the advancements in care practices. Visualizing this "becoming" process can be likened to phase transformations, representing the fluidity of identity. In the context of breast cancer care, this transformation is evident as practices transition from traditional methods to AI-driven approaches. The core identity remains consistent, emphasizing patient care, but the phase or approach changes, indicating the evolution of care practices.

Dynamic relationality #5: Reterritorialization

Following Fig. 5.2, the functor F_{5+}, named "Epiphany" as Territorialization, maps from category to C_6 to C_{5+}.[18] This mapping illustrates how fundamental concepts and structures

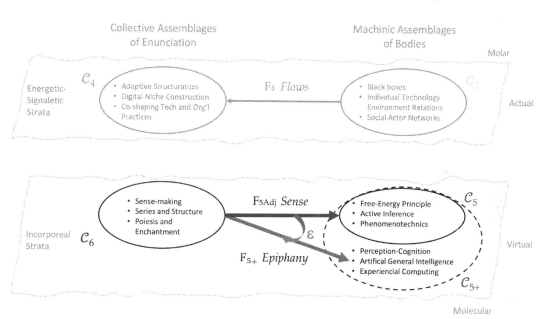

FIGURE 5.2   Reterritorialization of creative-experiencer.

---

[18] Note that Fig. 5.2 also depicts the functor F_{5Adj} from C_6 to C_5, which is the *adjoint* pair of F_5 (see Chapter 4 about adjoint functors). While F_5 maps from C_5 (Creative-Experiencer) to C_6 (Experience-verse), embodying the transformation of cognitive processes into experiences, F_{5Adj} translates these experiences back into cognitive structures, ensuring a reciprocal and deeply interconnected relationship between these categories. If F_5 embodies the concept of "Sense," F_{5Adj} is characterized by *assimilation* of "Sense", reflecting the continual refinement of mental models and predictive schemas in response to new sensory and experiential data.

in C_6 are pivotal in shaping and influencing the development of more advanced and integrative constructs in C_{5+}, emphasizing the cyclical and interconnected nature of cognitive and computational development. Functor F_{5+} reinterprets fundamental sense-making activities as advanced perceptual and cognitive processes (PerceptionCognition), transforms structured data and information sequences into the foundations for artificial general intelligence (AGI), and evolves immersive, enchanting experiences into sophisticated experiencial computing technologies. This transformation is vital for integrating and elevating basic experiences and structured data into more complex, integrative constructs. In this mapping, the foundational aspects of sense-making in C_6 become the building blocks for complex cognitive processes in C_{5+}. Similarly, the organization and manipulation of series and structures, foundational to AGI development, are seen as evolving from simpler structured data arrangements in C_6. Lastly, the enchanting and poietic experiences fundamental to immersive technologies in C_6 serve as the basis for the development of advanced experiencial computing in C_{5+}, enhancing user engagement through personalization, immersion, and interactivity.

Functor $F_{5+}: C_6 \to C_{5+}$ ("Epiphany" as territorialization)

$$F_{5+}(\text{SenseMaking}) = \text{PerceptionCognition}$$

$$F_{5+}(\text{SeriesStructure}) = \text{AGI}$$

$$F_{5+}(\text{PoiesisEnchantment}) = \text{ExperiencialComputing}$$

$$F_{5+}(f_{61}) = f_{53}, F_{5+}(f_{62}) = f_{54}$$

- Sense-making activities serve as the primary mode of perceiving and understanding environmental information. These activities entail the analysis and organization of sensory data, essential for interpreting and navigating our surroundings. Through F_{5+}, these fundamental sense-making activities in C_6 become the basis for advanced PerceptionCognition processes in C_{5+}. This shift highlights the progression from basic data interpretation to sophisticated cognitive frameworks that enhance understanding and interaction within the Experience-verse.[19]
- SeriesStructure denotes the organized sequences and structures formed within digital ecosystems. These structures, fundamental to the development of AGI, highlight the

---

[19] Di Paolo et al.'s (2017) enactive approach posits that sense-making is not merely about data interpretation but involves an organism's active engagement with its environment as a precursor to complex cognitive functions. Their concept of sensorimotor life, emphasizing concrete, action-based practices, underpins a natural progression from interacting with the environment in meaningful ways to forming sophisticated understandings and interpretations of that environment.

progression from simple data organization to complex, intelligent systems capable of advanced learning and decision-making. These foundational structures in C_6 are transformed into AGI in C_{5+}, signifying the evolution from basic data manipulation to the creation of AI systems with comprehensive intellectual capabilities.[20]

- The deployment of PoiesisEnchantment within C_6 encompasses the creation of immersive and captivating experiences through artistic and creative use of technology. This enchanting aspect of experiential technology serves as the groundwork for the development of advanced ExperiencialComputing in C_{5+}. F_{5+} reinterprets these fundamental enchanting experiences as sophisticated, personalized, and interactive computing technologies in C_{5+}, underscoring the potential of experiencial computing to revolutionize user engagement in the Experience-verse.[21]
- The transformation from SenseMaking to SeriesStructure in C_6, as facilitated by F_{5+}, mirrors the progression in C_{5+} from basic cognitive processing (PerceptionCognition) to the creation and manipulation of structured data and information systems (AGI). This parallel illustrates the fundamental role of sensory interpretation and data organization in the development of AGI. Similarly, the transition from SeriesStructure to PoiesisEnchantment in C_6 reflects the evolution from structured data manipulation to the creation of immersive, enchanting experiences, paralleling the shift in C_{5+} from AGI to advanced, emotionally resonant ExperiencialComputing. This highlights the interconnectivity and cyclical nature of cognitive processing, data structuration, and experiential technology across the categories.

The natural transformation ε, named "Epistemic Alignment," suggests that an integrated understanding of sense-making, series and structures, and poiesis and enchantment in both theoretical contexts (F_{5Adj} and F_{5+}) could enhance our capacity to navigate and shape digital ecosystems. Insights from the Free Energy Principle, Active Inference, and Phenomenotechnics can enrich our understanding of PerceptionCognition, AGI, and ExperiencialComputing, and vice versa, emphasizing the interconnected nature of these research domains.

Natural transformation ε: $F_{5Adj} \to F_{5+}$ ("Epistemic Alignment")

The natural transformation ε bridges the conceptual domains represented by the functors F_{5Adj} (assimilation of Sense) and F_{5+} (Epiphany) in the context of DRT. It ensures that the transformational processes between the two functors are not just parallel but are also interconnected and aligned in their epistemic structures and outcomes. Given a morphism $g: A \to B$ in C_6, the components of ε, namely $\varepsilon_A: F_{5Adj}(A) \to F_{5+}(A)$ and $\varepsilon_B: F_{5Adj}(B) \to F_{5+}(B)$, ensure this epistemic alignment. The commutativity of the diagram—$F_{5+}(g) \circ \varepsilon_A = \varepsilon_B \circ F_{5Adj}(g)$—ensures that applying the functor F_{5Adj} to a morphism $g: A \to B$ in C_6 and then transforming the result via ε yields the same outcome as transforming first and then applying

---

[20] Bostrom's (2014) concept of quality superintelligence, encompassing systems that are both rapid and qualitatively superior in cognitive abilities, supported by digital minds' advantages in software and hardware, such as editability, duplicability, and enhanced memory, are crucial for AGI development. These attributes facilitate the transition from structured data sequences to advanced AGI, emphasizing the evolution from basic information processing to complex, intelligent systems capable of sophisticated cognitive functions.

[21] Aw et al. (2022) illustrates how digital voice assistants (DVAs), through their ability to personalize interactions and create a sense of animacy and intelligence, transform user experience, making them not just tools but integral, emotionally resonant components of everyday life. This transformation is key to reinforcing the interconnected evolution from basic creative interactions to advanced experiential computing technologies.

the functor F_{5+}. In other words, the transformation from the foundational concepts in C_6, such as SenseMaking, SeriesStructure, and PoiesisEnchantment, as viewed through the lens of assimilation of Sense (F_{5Adj}), aligns coherently with their interpretation in the context of Epiphany (F_{5+}). This alignment is critical for maintaining conceptual consistency and for enhancing our understanding of the interplay between cognitive processes, AGI, and experiencial computing. In essence, there is a family of morphisms $\varepsilon_X\colon F_{5Adj}(X) \to F_{5+}(X)$:

- $\varepsilon_{SenseMaking}$: FreeEnergyPrinciple → PerceptionCognition: This component represents the transformation of foundational sense-making processes grounded in the FreeEnergyPrinciple into more complex PerceptionCognition mechanisms. It signifies how basic cognitive frameworks aimed at minimizing uncertainty (FreeEnergyPrinciple) evolve to encompass more sophisticated perception and cognitive processes (PerceptionCognition). This transition reflects the development from fundamental cognitive operations to more advanced interpretative and understanding capabilities, highlighting the interplay between basic neural processing and higher-level cognitive interpretation.[22]

- $\varepsilon_{SeriesStructure}$: ActiveInference → AGI: This transformation illustrates how the principles of ActiveInference, which involve active engagement with the environment to reduce uncertainty, often resulting in the formation of structured, sequential patterns of behavior or cognitive models, evolve into a more advanced, synthetic form of intelligence (AGI), where algorithms and systems are designed to mimic this biological pattern recognition and prediction capability. This progression indicates a leap from organic, dynamic interaction with the environment to the creation of sophisticated AI systems (AGI), where the structuring and organization of information (SeriesStructure) is achieved through advanced computational models, thus reflecting the progression from biological to artificial intelligence mechanisms.[23]

- $\varepsilon_{PoiesisEnchantment}$: Phenomenotechnics → ExperiencialComputing: This component can be viewed as an evolution from the generation of creative, transformative experiences using technology (Phenomenotechnics) to the development of highly immersive and interactive computing environments (ExperiencialComputing). The term PoiesisEnchantment in this context signifies the creation of captivating and transformative experiences through technological innovation and design. The ε transformation signifies a maturation from creating moments of awe and artistic brilliance to the cultivation of comprehensive, emotionally and sensorially rich

---

[22] Clark's (2016) work on predictive processing and embodied cognition highlights the role of the brain as a predictive organ engaged in minimizing prediction error (Free Energy Principle), which underpins the foundational sense-making processes characterized by circular causal flows and structural couplings that keep the organism within its viability window. This continuous interplay between perception and action, forms the basis of the PerceptionCognition mechanisms in more complex cognitive and computational systems, marked not only by advanced neural processing but also by the integration of external sociocultural factors.

[23] Pezzulo et al. (2023) demonstrate that active inference, unlike passive AI, embeds agency in generative models, crucial for AGI's series and structure analysis. Since living systems, through embodied interactions, develop a grounded understanding of reality, integrating action consequences into their models, this approach would lend AGI the ability to generate content that is both meaningful and contextually relevant, moving beyond mere data replication. Hence, AGI's series structuring should not be static but evolve dynamically with purposive engagement, structuring experiences as an active, continuous process.

computing experiences, thereby elevating the role of technology from a creator of individualized experiences to an architect of holistic and emotionally engaging digital ecosystems.[24]

The natural transformation ε, termed "Epistemic Alignment," is not only a crucial conceptual bridge between two pivotal theoretical domains mapped by the functors F_{5Adj} (assimilation of Sense) and F_{5+} (Epiphany), but it also encapsulates Deleuze and Guattari's concept of "reterritorialization," a process where displaced or deterritorialized elements are recontextualized, integrated, and adapted within new or existing frameworks. This reterritorialization involves a substantive reconfiguration and integration of the foundational concepts of SenseMaking, SeriesStructure, and PoiesisEnchantment into new epistemic structures represented by PerceptionCognition, AGI, and ExperiencialComputing. The process of ε mirrors the fluid, ever-evolving nature of knowledge and understanding in complex systems. It reflects the multifaceted and dynamic interplay of cognitive processes, advanced artificial intelligence, and the design of immersive technological experiences.

Enriching the experience-verse with machinic generalized intelligence

As shown in Fig. 5.3, just as Category C_5's identity evolves into Category C_{5+}, we can also see Category C_6's identity co-evolving into Category C_{6+}, which explores the epistemic role of technology, the intertwining of organic and artificial, and the emphasis on felt experiences in digitalized ecosystems. TechnoEpistemology studies how technology influences our ways of knowing. OrganicIntegration emphasizes a harmonious blend of human and machine capabilities. SentioExperientiality focuses on experiences generated through sensory interactions with digitalized ecosystems. Morphisms depict the transition from knowledge creation through technology to the blending of organic and artificial components and the generation of profound and immersive experiences.

Category C_{6+} ("Enriched Experience-verse")
Objects:

SenseMaking, SeriesStructure, PoiesisEnchantment
+
TechnoEpistemology, OrganicIntegration, SentioExperientiality

Morphisms:

f_{61}: SenseMaking → SeriesStructure
f_{62}: SeriesStructure → PoiesisEnchantment

---

[24] Zhang et al.'s (2021) study on the unknowability of autonomous tools and their impact on socio-material agency delves into the liminal experiences of chip designers using autonomous tools, revealing a state of continuous emergence and ambiguity, characterized by multiple design trajectories and a multifarious temporality. The designers' encounters with the unpredictable nature and emergent outcomes of autonomous tools amplify the poietic aspect, where creation is not just about producing outputs but engaging in a dynamic, transformative process. The enchantment lies in the designers' engagement with the autonomous tools, as they navigate the ambiguity and emergent nature of the technology, contributing to a collective narrative that transcends traditional functionality.

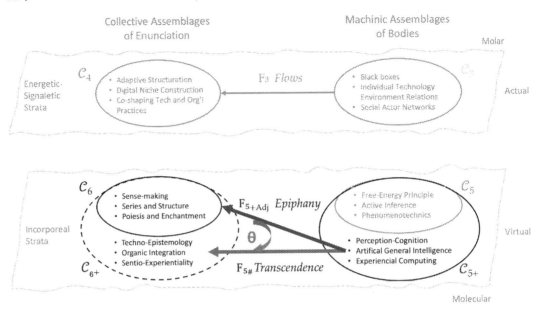

FIGURE 5.3 Reterritorialization of experience-verse.

+
f_{63}: TechnoEpistemology → OrganicIntegration
f_{64}: OrganicIntegration → SentioExperientiality

These new concepts explore the epistemic role of technology, the intertwining of organic and artificial, and the emphasis on felt experiences, respectively, in digitalized ecosystems.

- TechnoEpistemology refers to the study of how technology influences our ways of knowing, interpreting, and understanding the world. It encompasses the cognitive and conceptual frameworks that evolve in response to technological advancements and their integration into daily life. TechnoEpistemology examines the intersection of technology with human cognition and epistemology, exploring how digital tools and platforms reshape our perception, knowledge acquisition, and the very nature of knowledge itself.[25]

---

[25] Flusser (2000) introduces a framework where technological apparatuses, such as cameras and technical images, while extending human cognitive capacities, act as mediators of reality, shaping our perception and understanding. Inherent programs in such technologies embed specific epistemological frameworks, which underscores the significance of decoding and encoding in information processing. Digital technology's capacity for multidimensional communication and non-linear knowledge forms affords a symbiosis between human imagination and technology, suggesting digital interaction as a profound and autonomous form of worldly engagement, shaping historical consciousness and understanding.

- OrganicIntegration is about creating seamless interactions and synergies between humans and machines, emphasizing a natural, intuitive, and harmonious blend of capabilities, emphasizing a symbiotic relationship where both organic (human) and inorganic (machine) elements enhance each other's functions. OrganicIntegration seeks to transcend the traditional boundaries between human and machine, advocating for a fluid and integrated co-existence that leverages the strengths of both to create a more efficient and effective system.[26]
- SentioExperientiality captures the realm of experiences generated through sensory interactions with digitalized ecosystems, centering on emotion, affect, and sensory perception. It emphasizes the subjective, felt experiences as central to our engagement with technology. SentioExperientiality goes beyond the functional or utilitarian aspects of technology, focusing instead on the rich, immersive, and emotive dimensions of technological interactions. It is about how technology can evoke and enhance human emotions, sensations, and overall experiential quality.[27]

Morphisms depict the transition from knowledge creation through technology to the blending of organic and artificial components and, ultimately, the generation of profound and immersive experiences:

- f_{63}: TechnoEpistemology $\rightarrow$ OrganicIntegration, i.e., the shift from the cognitive and conceptual realm of understanding the impact of technology on knowledge to a tangible, integrative dynamic that enables a harmonious blend of human and machine capabilities.[28]
- f_{64}: OrganicIntegration $\rightarrow$ SentioExperientiality, i.e., the shift from the actual process of integration of human and technological capabilities to the ensuing enhanced experiential

---

[26] Exploring the posthuman condition as the nonbinary interaction between human consciousness and cybernetic systems, Hayles (1999) proposed that cognitive processes extend beyond the biological brain to encompass the tools and systems we interact with, creating a symbiotic relationship where technology not only augments human capabilities but also becomes an integral part of the cognitive process. She articulated this as a co-evolution of humans and machines, where each continually influences and reshapes the other, leading to a complex, integrated system of thought and action.

[27] In her pioneering manifesto on affective computing, Picard (2000) showed how technologies, imbued with the ability to recognize and adapt to human emotional states, can significantly enhance human–computer interactions. Mirroring the sentient aspect of human experience, emotional intelligence in machines can lead to more empathetic and intuitive technological experiences.

[28] Brey (2005) discusses how computer systems as cognitive artifacts, through their autonomous processing capabilities, extend memory, interpretation, search, pattern matching, and higher-order cognitive abilities, thereby transforming the way humans process information and interact with the digital world. He further elaborates on the evolving relationship, where computers and humans form a single hybrid cognitive system, working in tandem to enhance overall cognitive capabilities, redefining the interaction as an intertwined cognitive process beyond mere tool usage. As an extension of the world, and creator of virtual and social environments, technology is not just an extension of human faculties but also expands the experiential world.

state, emphasizing the subjective, emotional, and sensual dimensions of experience that arise from the practical and functional integration of organic and artificial elements.[29]

Following Fig. 5.3, the functor $F_{5\#}$, named "Transcendence" as Reterritorialization, provides a fresh theoretical lens to understand the intersections of humans, artificial intelligence, and technology. It emphasizes their co-evolving relationship and their impact on creating profound and immersive experiences within digitalized ecosystems. This functor captures the transformative role of technology in shaping our epistemology, the blurred line between humans and technology, and the importance of emotion in our encounters with technology.

Functor $F_{5\#}$: $C_{5+} \to C_{6+}$ ("Transcendence" as reterritorialization)

$F_{5\#}(\text{PerceptionCognition}) = \text{TechnoEpistemology}$

$F_{5\#}(\text{AGI}) = \text{OrganicIntegration}$

$F_{5\#}(\text{ExperiencialComputing}) = \text{SentioExperientiality}$

$F_{5\#}(f_{53}) = f_{63}, F_{5\#}(f_{54}) = f_{64}$

The "Transcendence" functor provides a fresh theoretical lens through which we can understand the intersections of humans, artificial intelligence, and technology, with emphasis on their co-evolving relationship and their impact on the creation of profound and immersive experiences within digitalized ecosystems.

- The way we perceive and cognize is intricately tied to the technology we use, thereby influencing the knowledge that we create. This aligns with the view that technology, specifically digital technology, plays a crucial role in shaping our epistemology.[30]
- The development of AGI, with its potential to mirror or even exceed human cognition, signifies an increasingly blurred line between the organic (humans) and the inorganic

---

[29] Yannakakis and Togelius (2011) demonstrate how games, through procedural content generation and player experience modeling, can integrate human and technological capabilities in a seamless, symbiotic manner while also enhancing the emotional and experiential aspects of human–computer interactions. The creation of game content that optimizes the player's experience, focusing on the subjective and emotional dimensions of gameplay, demonstrates how technology can be employed to evoke and enhance human emotions and sensations, creating more immersive and emotive technological interactions. By closing the affective loop, games can create more empathetic and intuitive experiences, mirroring the sentient aspect of human experience and highlighting the potential of emotional intelligence in machines.

[30] Smart (2018) provides a nuanced understanding of how digital technology not only augments but also becomes an integral part of our cognitive and epistemic processes. He shows that even though emerging digital technologies become part of our cognitive apparatus, thereby influencing how we perceive, process, and understand information, our belief in truth relies on metacognitive processes managing adaptive cognition and belief shaping, highlighting our primary role in knowledge attribution, unaffected by technological cognitive extensions.

(technology). This "organic integration" could be key to understanding the co-evolutionary relationship between humans and technology.[31]
- Experiencial computing, which tailors technology to individual preferences and experiences, emphasizes the importance of "sentio" (feeling) in our encounters with technology. This mapping accentuates the central role of emotional, affective, and sensory experiences ("sentio-experiences") in the realm of digital technology.[32]
- As the focus shifts from the cognitive processes involved in interpreting and understanding sensory data to the development and integration of artificial general intelligence, there is a transition from understanding how technology influences our ways of knowing to creating seamless interactions and synergies between humans and machines. As the focus expands from the development and integration of AGI to the creation of immersive, personalized computing experiences, there is a transition from the seamless interactions and synergies between humans and machines to the sensory, emotional, and experiential realm generated through digital ecosystems.

The evolution of AGI systems from tools that simply process perceptual and cognitive data to tools that profoundly influence our understanding and interpretation of the world leads to a deeper and more harmonious integration of AGI systems into human activities, experiences, and societies. The transformation of AGI systems from mere tools of integration to agents that generate rich, immersive, and emotionally resonant experiences reflects the increasing centrality of AGI in creating experiential value in digital ecosystems, and the ways in which these technologies are transforming the very nature of human experience.

The natural transformation θ, named "Trans-Epistemic Shift," reflects the transition from a phenomenological perspective on AI (F_{5+Adj}) to a more epistemological and integrative view ($F_{5\#}$).[33] This transformation highlights the transformative impact of AI on human cognition, integration, and experience. The shift from F_{5+Adj}'s perspective to $F_{5\#}$'s perspective incorporates the evolution from mere sense-making to a technology-enhanced understanding, from simple structuration to organic integration, and from enchanting creation to immersive, emotional experience design

Natural transformation θ: $F_{5+Adj} \to F_{5\#}$ ("Trans-Epistemic Shift")

The natural transformation θ plays a crucial role in linking the theoretical constructs of F_{5+Adj} (grounding of "Epiphany") and $F_{5\#}$ ("Transcendence"). This transformation ensures a seamless and coherent transition of concepts and methodologies from the grounded, experiential

---

[31] Heylighen and Lenartowicz' (2017) exploration of the Global Brain concept emphasizes the evolution from seeing technology as an external aid to recognizing it as an intrinsic part of human cognitive and societal processes, a transformation critical to the understanding of AGI in the context of human–technology symbiosis.

[32] Chiang et al. (2022) demonstrates the pivotal role of emotional intelligence in enhancing digital experiences, specifically in service robotics and experiencial computing. This underscores a shift toward affective computing, where technology transcends functional utility to offer rich, sensory, and affective user interactions.

[33] F_{5+Adj} maps advanced cognitive, AGI, and experiential computing constructs in C_{5+} back to their foundational sense-making processes, structured data, and enchanting experiences in C_{6+}, revealing how complex systems in C_{5+} are grounded in and informed by the fundamental processes and structures of C_{6+}. If F_{5+} represents the concept of "Epiphany," then F_{5+Adj} embodies the *grounding* of Epiphany, i.e., the process of tracing back the developmental pathways from higher-order cognitive and computational constructs to their more fundamental processes and structures.

perspective of F_{5+Adj} to the more advanced and integrative stance of $F_{5\#}$. As such, θ functions as a key mediator, aligning the epistemic structures and outcomes of these two conceptual realms. For any morphism h: A → B in $\mathbf{C}_{5+}$, the components of θ, namely $θ_A$: $F_{5+Adj}(A) → F_{5\#}(A)$ and $θ_B$: $F_{5+Adj}(B) → F_{5\#}(B)$, facilitate this crucial alignment. The commutative diagram—$F_{5\#}(h) \circ θ_A = θ_B \circ F_{5+Adj}(h)$—guarantees that the process of applying the functor F_{5+Adj} to a morphism in $\mathbf{C}_{5+}$ and then transforming through θ is equivalent to first applying the transformation θ and then $F_{5\#}$. This integration and alignment are vital for preserving the conceptual integrity and facilitating a deeper understanding of the interplay between cognitive processes, AI, and immersive computing, underpinning the "Trans-Epistemic Shift" in the DRT framework. Thus, there is a family of morphisms $θ_X$: $F_{5+Adj}(X) → F_{5\#}(X)$:

- $θ_{PerceptionCognition}$: SenseMaking → TechnoEpistemology. This transformation represents an evolution in human cognitive processes from traditional sense-making, where understanding and interpretation of the environment are key, to a more technologically integrated approach, termed TechnoEpistemology. In this transition, technology is not just a supplementary tool but becomes an intrinsic part of cognitive processes, fundamentally transforming how we acquire, process, and utilize knowledge. This shift indicates a profound change in the epistemological framework, where the integration of technology elevates and reshapes our perceptual and cognitive abilities, leading to new, technologically enriched ways of interacting with and comprehending the world.[34]
- $θ_{AGI}$: SeriesStructure → OrganicIntegration. This transformation highlights the evolution of artificial general intelligence (AGI) from a focus on structured, sequential data processing (SeriesStructure) to a holistic, seamless integration within human cognitive and environmental systems (OrganicIntegration). In this transition, AGI transcends its traditional role of handling structured decision-making processes to become an integral part of a more adaptive, fluid system. This integration fosters a symbiotic relationship where AGI and human intelligence complement and enhance each other's capabilities. The shift signifies AGI's progression toward understanding and adapting to complex, real-world contexts, thus blurring the boundaries between human cognition and artificial intelligence in a harmonious and interconnected manner.[35]
- $θ_{ExperiencialComputing}$: PoiesisEnchantment → SentioExperientiality. This transformation reflects a pivotal shift in experiencial computing from the creation of enchanting, artistically driven experiences (PoiesisEnchantment) to a profound focus on emotional, sensory, and affective dimensions of technological interactions (SentioExperientiality). It

---

[34] Ramstead et al. (2022) demonstrate how computational phenomenology applies generative modeling to interpret lived experiences, thus expanding traditional phenomenological sense-making into a comprehensive computational framework. Beckmann et al. (2023) propose a nonrepresentationalist approach to deep learning, where AI mechanisms are conceptualized through the lens of lived experience, enhancing the understanding of AI beyond mere representation. These perspectives collectively signify the evolution from conventional cognitive processes to a technologically integrated TechnoEpistemology, where technology becomes intrinsic to cognitive frameworks.

[35] Roli et al. (2022) underscore the limitations of AGI in situational reasoning and exploiting new affordances, suggesting that AGI's current algorithmic approach falls short in mimicking organismic adaptability. Jebari and Lundborg (2021) detail the constraints of machine agency, noting that productive desires necessary for AGI to function as a general agent across diverse contexts are not inherently present in current AI models. These insights collectively illuminate the complexities and necessary advancements for AGI to achieve OrganicIntegration within human cognitive and environmental systems.

encapsulates the evolution of AI and digital systems in transcending functional boundaries to immerse users in experiences that deeply resonate with their senses and emotions. This shift emphasizes the critical role of technology in not just facilitating, but actively enhancing human emotional and sensory experiences. It underscores the burgeoning importance of designing empathetic, intuitive digital environments that are attuned to and capable of evoking rich emotional responses.[36]

This natural transformation θ reflects the transition from a phenomenological perspective on AI and human interaction (F_{5+Adj}) to a more epistemological and integrative view ($F_{5\#}$), highlighting the transformative impact of AI on human cognition, integration, and experience. Also, the shift from F_{5+Adj}'s perspective to $F_{5\#}$'s perspective signifies a shift in epistemic paradigms, from mere sense-making to a technology-enhanced understanding, from simple structuration to organic integration, and from enchanting creation to immersive, emotional experience design. Thus, the natural transformation θ, by aligning and integrating the perspectives of F_{5+Adj} and $F_{5\#}$, embodies the essence of reterritorialization, ensuring the continued relevance and adaptability of theoretical constructs in the face of evolving technologies and understanding, mirroring Deleuze and Guattari's concept of continuous conceptual innovation and adaptation.

Lines of flight from structures and becoming of identities

While functors capture the essence of transformation in a categorical sense, specific mathematical mechanisms detail how these transformations unfold in the domain of identity and structure (for a more technical background, see "Unified Gauge Theory of Identity and Structure" in the **Appendix**).

Lines of flight as structural transformation

In the philosophy of Deleuze and Guattari, "lines of flight" refer to paths of escape or rupture from established structures or systems. These lines represent moments or processes of deterritorialization, where entities break free from the constraints of existing territories or structures.

Entities can be at different stages of transformation, from being embedded within global structures to engaging in local transformative processes:

- They can be fully embedded within the global, macroscopic structures. In this state, entities are in harmony with the prevailing norms and systems.

Continued

---

[36] Deane (2022) posits that affective self-modeling is key to achieving flexible behavior and general intelligence in both biological agents and AI, suggesting that AI's capacity for emotional resonance and intuitive interaction is essential for immersive experiences. Possati (2021), drawing from neuropsychoanalysis, argues for AGI development rooted in affective neuroscience, embedding emotional and affective systems in AGI, where AI extends beyond functional outputs to deeply engaging human emotions and senses.

Lines of flight from structures and becoming of identities *(cont'd)*

Consider a traditional hospital setting where breast cancer treatment is primarily reactive. Patients are diagnosed after symptoms appear, and treatments are standardized based on the stage of cancer. Here, the hospital, its physicians, and even patients are operating within the established, global structure of reactive care.

- Entities might have transitioned to a local, microscopic level, engaging in transformative processes that challenge or deviate from the established global norms. Consider a community health initiative that emphasizes early breast cancer detection through regular self-examinations and community-led awareness campaigns. This initiative operates outside the traditional hospital-based system, focusing on local, community-driven efforts. Nurses and community health workers take the lead, educating women about early signs and the importance of regular check-ups.
- Entities in a superposition state are navigating the transition from global to local processes. They are neither fully aligned with the macroscopic structures nor fully engaged in microscopic transformations. Instead, they are in a dynamic interplay between the two. Consider a hybrid breast cancer care model that integrates both hospital-based treatments and community-driven preventive measures. In this model, patients receive standardized treatments but are also educated about preventive care, early detection, and post-treatment self-care. Medical technologies, like telemedicine platforms, bridge the gap, allowing patients to consult with physicians remotely. Caretakers and family members are also involved, receiving training on post-treatment care.

There are several mechanisms that capture the ways entities break free from established structures:

- An **inversion operator**, capturing entities' transitions from global structures to local transformative processes. In the context of "lines of flight," it represents entities that challenge or invert their positions within the global structure, initiating a shift toward local processes. In the global context of breast cancer care, there's a macroscopic structure where treatment decisions are predominantly based on generalized clinical guidelines. When the inversion transformation is applied, there's a shift to a microscopic, individualized approach. Instead of generalized treatments, there's a focus on personalized medicine, tailoring treatments based on the genetic makeup of the tumor and the individual. This inversion represents a direct challenge to the "one-size-fits-all" approach, emphasizing the unique needs and conditions of individual patients.
- A **latent operator** that might not be immediately visible at the global level but have profound effects at the local level capture the underlying shifts that gradually lead entities from global adherence to local innovations. On a macroscopic level, breast cancer awareness campaigns globally emphasize early detection and the importance of regular mammograms. However, when this latent transformation is applied,

Lines of flight from structures and becoming of identities *(cont'd)*

there's a microscopic shift toward understanding the psychosocial aspects of breast cancer. This includes the subtle emotional, psychological, and social challenges patients face, from diagnosis to treatment and recovery. While the global message remains early detection, there's a growing emphasis on providing comprehensive mental health support, recognizing the intricate challenges patients navigate beyond the physical ailment.

- **A dominance/bifurcation operator** differentiates between global and local states, emphasizing one over the other. It captures moments where either the global structure is reinforced, or the local transformative processes become dominant. Historically, the primary treatment for breast cancer has been surgical removal, either through mastectomy (removal of the entire breast) or lumpectomy (removal of the tumor and some surrounding tissue). With advancements in medical technology and research, nonsurgical treatments like targeted radiation, chemotherapy, and hormone therapies have become more effective and dominant. However, the application of a dominance/bifurcation transformation highlights a significant microscopic shift toward these noninvasive treatments, reserving surgeries for cases where they are absolutely necessary. This shift underscores a bifurcation in treatment approaches, where nonsurgical treatments are becoming the dominant choice for many patients, emphasizing quality of life and minimal side effects.[37]

Becoming as continuous transformation of identity

"Becoming" is a central concept in Deleuze and Guattari's philosophy, emphasizing the continuous transformation and fluidity of identity. It is not about being a fixed entity but rather about being in a state of constant change and evolution.

The functor F_4's transformation from macroscopic structures to microscopic transformations embodies the Deleuzian concept of "lines of flight." The functor F_{4+}, named "Singularity," represents a more advanced stage of this transformation, where the cognitive processes are not just human but also involve artificial intelligence, leading to a technological singularity.

- The introduction of AI and advanced technologies in breast cancer care can be seen as the functor F_{4+} transformation. With the integration of AI in diagnostic tools like mammograms, there's an enhancement in the perception and cognition of potential cancerous growths, leading to early and more accurate detection. Artificial General Intelligence can revolutionize personalized treatment plans. By analyzing a patient's genetic makeup, medical history, and other relevant data, AGI can recommend treatments tailored specifically for the individual. Patient experiences are enhanced with the use of virtual reality for pain management during treatments, AI-driven chatbots for patient queries, and personalized patient portals for tracking treatment progress.

The natural transformation $\beta: F_4 \to F_{4+}$ represents the process of becoming, where

Continued

> ### Lines of flight from structures and becoming of identities *(cont'd)*
>
> the identity of Category C_5 evolves to become Category C_{5+}.
>
> - This transformation signifies the shift from traditional care methods (F_4) to advanced, AI-driven approaches (F_{4+}). The transformation in the realm of breast cancer care, from traditional methods to technologically advanced practices, should be mirrored in the underlying cognitive and perceptual processes. The journey from traditional care practices to technologically advanced, AGI-driven care should be smooth, consistent, and maintain its core essence of providing the best possible care for patients across following transformations:
>   * The transformation from traditional methods to perception-driven care, where AI tools adapt and learn from new data, improving their diagnostic capabilities over time.
>   * The shift from traditional care to AGI-driven personalized treatments, where treatments are no longer generalized but constructed specifically for the individual's unique needs.
>   * The transformation from traditional patient experiences to experiencial computing, where technology and care practices co-evolve to provide enhanced patient experiences.
> - As breast cancer care transitioned from relying on accumulated knowledge (AdaptiveStructuration) to utilizing digital tools for active inference (DigitalNicheConstruction), there should be a corresponding shift from understanding how minimizing uncertainties influences medical professionals' decisions (PerceptionCognition) to the development of AGI algorithms that mimic this adaptive behavior. In essence, as care practices evolved to be more data-driven, the underlying cognitive processes should also evolve to be more aligned with AGI's capabilities.
> - As breast cancer care evolves from being merely data-driven (DigitalNicheConstruction) to a more integrated approach where technology and organizational practices co-evolve (CoShapingTechnologyOrganizational Practices), there should be a corresponding transition from AGI-driven diagnosis and treatment plans to a more holistic experiencial computing approach. This means that as AGI systems are designed to mimic the information-seeking behavior of medical professionals, the overall patient experience, from diagnosis to post-treatment care, should be immersive and engaging, leveraging the full potential of experiencial computing.
>
> The process of becoming, as represented by the natural transformation $\beta: F_4 \to F_{4+}$, can be operationalized through phase transformations, representing the fluidity and transformation of identity. The field, undergoing transformations, metaphorically represents different states or aspects of becoming. As the phase varies, it signifies different configurations or states of identity, emphasizing the fluid and continuously transforming nature of identity.
>
> - In the context of breast cancer prevention, diagnosis, and treatment, this can be illustrated through the evolving identity of breast cancer care, as it

Lines of flight from structures and becoming of identities *(cont'd)*

transitions from traditional methods to technologically advanced, AI-driven approaches. Throughout these transformations, the magnitude of care's identity remains consistent, emphasizing the essence of providing the best possible care for patients. However, the phase changes, signifying the evolving nature of care practices and patient experiences. Just as the field undergoes phase transformations without changing its magnitude, breast cancer care evolves in its methods and approaches while maintaining its fundamental goal of patient well-being.

* **Traditional Breast Cancer Care (Initial Phase)**: At this phase, the identity of breast cancer care is rooted in traditional methods. Diagnostic procedures rely heavily on manual examinations and mammograms interpreted by radiologists. Treatments are based on generalized protocols, and patient experiences are largely determined by human interactions and standard procedures. When traditional methods start integrating digital tools, this marks the beginning of the "becoming" in breast cancer care.
* **Introduction of Digital Tools (Intermediate Phase)**: As digital tools are introduced, the identity of breast cancer care begins its transformation. AI-enhanced mammograms start providing more accurate detections, reducing false positives and negatives. Electronic health records streamline patient data management, and digital communication tools improve patient-doctor interactions. Next, the identity of breast cancer care undergoes a more profound change, with AGI and experiencial computing taking center stage.
* **Advanced AI-Driven Breast Cancer Care (Advanced Phase)**: At this phase, the identity of breast cancer care has evolved significantly. AGI-driven personalized treatment plans become the norm, ensuring that each patient receives care tailored to their unique genetic makeup and medical history. Virtual reality tools offer pain management solutions, and AI-driven chatbots address patient queries round the clock. The patient experience is no longer just about medical treatment but encompasses a holistic, technologically enhanced journey.[38]

Dynamics of lines of flight and becoming

The energy dynamics of transformation, change, and breaking free from established structures provides a mathematical framework to understand the tensions and interactions between identity and structure. This can be related to the dynamics of the functor transformations, where entities and systems undergo transformations, leading to the creation of new assemblages and the disruption of old ones. The kinetic energy term of the energy dynamics represents the active processes of becoming and lines of flight. This aligns with the dynamic and evolving nature of the ecosystem, where new cognitive processes and experiences are constantly being created and old ones are being disrupted. The potential energy term of the energy dynamics represents the tensions and interactions between identity and

Continued

Lines of flight from structures and becoming of identities *(cont'd)*

structure. This can be seen as the underlying forces or tensions that drive the process of deterritorialization and the creation of new structures.

A direct challenge and inversion signifies a radical shift or inversion in the established structures, emphasizing the Deleuzian concept of "lines of flight." The potential energy term captures the tensions between the current identity and the new structural configuration. As breast cancer care undergoes significant changes due to technological advancements, this term represents the tensions between traditional methods and innovative approaches.

- The introduction of groundbreaking AI tools or techniques can directly challenge established breast cancer care protocols. For instance, the widespread adoption of AI-driven diagnostic tools might challenge traditional diagnostic methods. The transformation $\beta: F_4 \rightarrow F_{4+}$ captures this direct challenge, representing the process of becoming as care practices undergo significant inversions.

Subtle, latent shifts in structure capture the more nuanced changes in established structures without a complete inversion. The kinetic term may involve intertwined dynamics of identity and structure. As breast cancer care evolves, the subtle shifts in care practices are represented by these cross terms.

- As care practices gradually incorporate AI tools and techniques, they undergo subtle shifts. For example, the gradual integration of AI-driven patient portals or chatbots represents a subtle enhancement in patient care without a complete overhaul. The transformation $\beta: F_4 \rightarrow F_{4+}$ embodies these subtle shifts, representing the continuous evolution of care practices.

A bifurcation or dominance in structure signifies a major shift in the established structures, leading to either a dominant new structure or a bifurcation into multiple paths. There will be a significant change in both the kinetic term and the potential term. As breast cancer care undergoes major shifts, energy dynamics capture the deep reconfiguration of the potential landscape and the dynamics of identity.

- Major advancements or breakthroughs in breast cancer care, such as the development of a revolutionary treatment method or the discovery of a new diagnostic technique, can lead to dominance or bifurcation in care practices. The transformation $\beta: F_4 \rightarrow F_{4+}$ captures this major shift, representing the process of becoming as care practices undergo significant reconfigurations.

Continuous transformation of identity entails the fluidity of identity, emphasizing the Deleuzian concept of "becoming." The kinetic term captures the dynamics of this transformation. As breast cancer care evolves, the active processes of becoming (like the integration of AI tools) are represented by this term.

- As AI tools become more integrated into breast cancer care (F_4's transformation), the continuous transformation of identity represents the evolving nature of care practices. The journey from traditional methods (Category C_5) to AI-driven

Lines of flight from structures and becoming of identities *(cont'd)*

approaches (Category C_{5+}) embodies this transformation.

Interaction between identity and established structures represents the more complex transformations associated with challenging or breaking free from established structures. The potential energy term captures the tensions and interactions between identity and structure. As AI tools challenge traditional breast cancer care methods, this term represents the tensions between old and new practices.

- The introduction of AI tools (like AGI-driven personalized treatment plans) can challenge established care protocols. The transformation β: $F_4 \rightarrow F_{4+}$ captures this shift, representing the process of becoming as care practices evolve.

Induced "lines of flight" in structures represent the breaking free from structures. The kinetic term captures the intertwined dynamics of identity and structure. As AI tools induce changes in breast cancer care, this term represents the evolving dynamics between traditional and new methods.

- As care practices transition from traditional methods to technologically advanced practices (like experiencial computing), they induce "lines of flight" in established structures. The transformation β: $F_4 \rightarrow F_{4+}$ embodies this shift.

Stabilization or destabilization of identity represents the intricate interplay between identity and structure. Both kinetic and potential terms capture the dynamics of stabilization or destabilization. As AI tools reshape breast cancer care, energy dynamics represents the potential for change and transformation in the system.

- The integration of AI tools can either stabilize (by improving diagnostic accuracy) or destabilize (by challenging established protocols) the identity of breast cancer care. The transformation β: $F_4 \rightarrow F_{4+}$ captures this dynamic interplay.[39]

[37] Examining the intra-actions between caregivers, practitioners, and digital technologies, Maslen and Harris (2021) illustrate how caregivers evolve, or engage in lines of flight, from traditional passive roles toward becoming active diagnostic agents, thereby embodying the concept of becoming. This transformation entails a shift from macroscopic, structured care methods based on established structures of medical knowledge to microscopic, transformative practices of digital media use and sensory work by caregivers to navigate and challenge the conventional boundaries of clinical diagnosis. Caregivers utilizing digital technologies for diagnostics exemplify the inversion from traditional passive caregiving roles to active diagnostic roles. The subtle yet profound effects of digital technology adoption by caregivers on the diagnostic process align with the latent operator. The emergence of caregivers as diagnostic agents signifies moments where local transformative processes become dominant.

[38] Maslen and Harris' (2021) work on "becoming a diagnostic agent" shows how caregivers break free from traditional, structured care methods to engage in transformative practices. Caregivers, through their interaction with digital media and sensory work, navigate and challenge conventional clinical diagnosis boundaries, embodying the concept of lines of flight. The evolution of caregivers into diagnostic agents exemplifies the continuous transformation of identity, or "becoming," as caregivers move toward digitally enhanced diagnostic processes. The transition of caregivers into active diagnostic roles can be seen as a natural transformation within the healthcare ecosystem, where the identity of caregivers evolves to incorporate technological capabilities.

Continued

Lines of flight from structures and becoming of identities *(cont'd)*

[39] In Maslen and Harris' (2021) exploration of the transformation of caregivers into active diagnostic agents through their interaction with digital technologies, the active engagement of caregivers with digital technologies and their evolution into diagnostic agents highlight the kinetic energy dynamics whereas the tension and interactions between the evolving identity of caregivers and the established structures of medical practice reflect the potential energy dynamics. In this study, the following transformational dynamics are particularly prevalent:

- Direct Challenge and Inversion: In a line of flight, indicative of a shift towards empowered, patient-centric care approaches, caregivers' use of digital technologies to perform diagnostic roles represents a form of inversion from the established medical authority to a more distributed, participatory form of healthcare.
- Continuous Transformation of Identity: The evolving role of caregivers as they integrate digital technologies into their diagnostic processes reflects the concept of "becoming" as caregivers adapt and redefine their roles through technology.
- Interaction Between Identity and Established Structures: Caregivers navigating the digital landscape to enhance their diagnostic capabilities represent a significant interaction between evolving caregiving identities and the traditional structures of healthcare.

CHAPTER 6

Dynamic relationalities of relational dynamics

As we saw in the previous chapter, central to Deleuze and Guattari's philosophy and DRT is the concept of "becoming"—a continuous transformation that challenges and reshapes established identities and structures. In category theory, this metamorphosis is represented as a natural transformation between functors, symbolizing the evolution of identity. This "becoming" is not a mere state change but a profound metamorphosis, indicating shifts in methodologies and paradigms. Visualizing this process can be likened to phase transformations, representing the fluidity of identity. Homotopy theory further enriches this perspective, introducing dynamics and providing a continuous path between functions, capturing the essence of the "becoming" process in dynamic systems.

We also saw how as assemblages evolve, the foundational principles governing their relational dynamics also undergo profound shifts, as we are witnessing (as of 2024) with generative AI-induced creative transformation. This is not a mere adaptation but a deep metamorphosis, a reterritorialization, that redefines the very essence of relational structure of assemblages, with implications for co-evolution of AI and Human Intelligence (HI) toward co-creative Machinic Generalized Intelligence (MGI).

As categories representing assemblages evolve, the functorial relationships that map and connect these categories must also transform. Guiding this intricate transformation is the concept of a natural transformation. It acts as a meta-layer of transformation, ensuring that as the categories (assemblages) evolve and their internal and external (functorial) relationships shift, there remains a coherent, overarching structure that preserves the integrity of the entire system. This natural transformation ensures that the evolution of relational dynamics is not chaotic but follows a structured path, even as it allows for profound changes.

While category theory provides a way to study the structure and relationships of mathematical objects and morphisms, differential topology focuses on the geometry and topology of smooth manifolds. Merging the principles of differential topology with category theory paves the way for a more holistic approach to understanding relational dynamics. In healthcare, differential topology can be applied to study continuous dynamics like decision-making processes, treatment implementations, and patient care routines. For instance, it can analyze how doctors adapt their decision-making based on new information or feedback. Similarly, it can explore the continuous learning and adaptation processes of healthcare professionals. Further, sheaf theory, which studies functions on spaces with local and global properties, can be applied to vector fields on manifolds using differential topology (as we will elaborate in the next chapter). Meanwhile, category theory can explore the relationships between these vector fields.

In this integrated perspective, each category can be visualized as a smooth manifold, where vector fields indicate the direction of transformations. Lie derivatives play a pivotal role in measuring the changes in these fields, effectively capturing the nuances of symmetries and transformations. Differential forms offer insights into the intensity of these transformations, while de Rham cohomology provides a means to evaluate the evolution of categories. When it comes to dynamics between categories, functors act as the mapping bridges between manifolds, facilitating the modeling of continuous transformations across categories. Concepts such as transversality and the differential nature of functors shed light on the intersections and rates of transformations between categories. Monads, in this context, serve as bundles that encompass manifold, vector field, and differential form, providing a panoramic view of intracategory dynamics. The principles of transversality and homotopy theory, which focus on nontangential intersections and continuous deformation, respectively, further enhance our understanding of transformations between categories. This amalgamation of differential topology and category theory offers a profound, dynamic insight into the world of relational dynamics.

Building upon category theoretical analysis, application of differential topological concepts enhances our understanding of singularity in Artificial General Intelligence (AGI). Two key theoretical constructs guide this exploration: the monad concept and homotopy theory. The monad concept in category theory is represented as a triad (T, υ, μ), comprising an endofunctor (T), and two natural transformations, unit (υ), and multiplication (μ). Offering a comprehensive perspective on intracategory dynamics, it provides insights into the enrichment process of our category framework. Homotopy theory, on the other hand, offers a perspective on the transformation dynamics between categories. In category theory, a natural transformation transforms one functor into another while preserving categorical structure. However, this transformation is static. Homotopy introduces dynamics, providing a continuous path between functions. The evolution from Category C_5 to C_{5+} is influenced by the synergy of these constructs, with advancements in generative AI, mixed reality, quantum computing, and more playing pivotal roles.

Dynamic relationality #6: Differential transformation

Combining differential topology with category theory can provide a dynamic perspective on the transformations and flows that occur within and between our categories. Consider the following:

- Dynamics of Relationalities Within Categories
    - **Smooth Manifolds**: Each category can be seen as a smooth manifold, where each point represents a particular configuration of objects and morphisms in the category. This allows one to model continuous transformations within each category.

- **Vector Fields**: A vector field can be assigned to each manifold (category), where each vector represents the "flow" or "direction" of transformation at a particular point in the category. This can model the dynamics of transformations within each category.
  - **Lie Derivatives**: A Lie derivative is a way to measure how much a vector field changes along another vector field. We can use Lie derivatives to analyze the transformation and flow of objects and morphisms within each category. This can provide a measure of the change of a tensor field along the flow of a vector field, modeling the symmetries and transformations within each category.
- **Differential Forms**: Using differential forms, we can measure quantities that depend on the orientation in each category. This can capture the "intensity" or "rate" of transformations within each category.
  - **De Rham Cohomology**: De Rham cohomology is a tool for studying the topology of a space using differential forms. It measures the extent to which a space is not simply connected, and it can be used to classify different types of spaces based on their topology. We can use de Rham cohomology to study the differential forms on each manifold (category). This can help analyze the structure and properties of each category, and how they change over time.
- Dynamics of Relationalities Across Categories
  - **Functors as Maps Between Manifolds**: We can consider functors as maps between manifolds (categories). This allows one to model the continuous transformations between different categories.
  - **Transversality**: Using the concept of transversality, we can study the intersections between different categories (manifolds). This can model the points where lines of flight intersect with the strata.
  - **Differential of Functors**: We can consider the differential of functorial maps to capture the rate of change of transformations between categories. This can provide a dynamic perspective on the functors linking our categories. Homotopy theory is a branch of algebraic topology that studies the properties of spaces that can be continuously deformed into each other.

Hence, by integrating these elements of differential topology with category theory, we capture the dynamic and continuous nature of transformations and flows within and between our categories, providing a richer and more nuanced perspective on our theoretical framework.

Within categories, the category theoretical construct of monad, consisting of the endofunctor, unit transformation, and multiplication transformation, serves as a powerful tool that encapsulates the intricate dynamics. Drawing parallels from differential topology: the endofunctor mirrors the transformative flow of vector fields on smooth manifolds and captures global structures akin to differential forms and de Rham cohomology. The unit transformation provides foundational structure, reminiscent of the local Euclidean descriptions on manifolds. Meanwhile, the multiplication transformation ensures the coherence of transformations, analogous to how Lie derivatives measure changes along flows. Together, these monadic components can bundle the rich dynamics and structures of differential topology, offering a comprehensive perspective on category dynamics.

Across categories, transversality is the way two submanifolds intersect in a differentiable manifold, such that they meet in a nontangential manner, in order to study the intersections between different categories (manifolds), such as modeling the points where lines of flight intersect with the strata. Homotopy theory provides a way to study the dynamics behind natural transformations as a continuous path in the space of functors from one functor to another, not just the initial and final states of the transformation, but also the entire "process" of the transformation.

These are the key ways in which the category theory of transformations becomes truly differential and dynamic—i.e., "dynamic relationalities of relational dynamics." So, next we discuss these two directions separately.

Differential forms to measure the flow of natural transformations

In the context of category theory, differential forms can be thought of as a way to measure the "rate of change" or "flow" of transformations within and between categories. This can be particularly useful in understanding the dynamics of complex systems, such as the healthcare ecosystem in the context of breast cancer prevention, diagnosis, and treatment.

- Let us consider the category C_4. In C_4, we have objects like "Adaptive Structuration," "Digital Niche Construction," and "Co-Shaping Technology and Organizational Practices." A differential form over C_4 could be defined as a measure of the rate of change of these objects with respect to certain variables. For instance, we could define a differential form that measures the rate of change of "Adaptive Structuration" with respect to the variable "time." This could represent the rate at which the adaptive structuration process is evolving over time. Mathematically, this could be represented as a derivative, $d(Adaptive\ Structuration)/dt$, where t represents time.
- This can be particularly useful in understanding the dynamics of complex systems, such as the healthcare ecosystem in the context of breast cancer prevention, diagnosis, and treatment. Differential forms over the category C_4 could be defined to measure the rate of change of the objects "Adaptive Structuration," "Digital Niche Construction," and "Co-Shaping Technology and Organizational Practices" with respect to certain variables. These differential forms could provide a dynamic perspective on the transformations and flows within the category C_4. They could help us understand how these transformations are evolving over time and how they are being influenced by various factors, such as time, patient population size, and resource investment.[1]

[1] Onno et al. (2023) provides a comprehensive analysis of the implementation and integration of Artificial Intelligence-based Computer Aided Detection (AI-CAD) in the global health context. The evolution of health care practices in response to the introduction of AI-CAD, such as changes in screening guidelines or treatment protocols, can be quantified using differential forms. For instance, the rate of adoption of AI-CAD technologies ($d(Adaptive\ Structuration)/dt$) could provide insights into how health practices adapt over time to technological advancements. Rapid deployment of AI-CAD solutions and their scalability directly impacts the number of patients served by these technologies, which can be measured by $d(Digital\ Niche\ Construction)/dp$. The investment in AI research and its integration into clinical practice represent a dynamic relationship where resources allocated to AI-CAD ($d(Co\text{-}Shaping\ Technology\ and\ Organizational\ Practices)/dr$) can indicate the extent to which technological investment influences organizational changes.

- **Adaptive Structuration**: This object could be seen as representing the evolving structure of healthcare practices and policies. A differential form over this object could measure the rate of change of these structures with respect to time. For example, we could measure how the adoption of new screening guidelines or treatment protocols changes over time. Mathematically, this could be represented as a derivative, $d(Adaptive\ Structuration)/dt$, where t represents time. This could provide insights into how quickly healthcare practices and policies are adapting to new research findings or technological advancements.
- **Digital Niche Construction**: This object could be seen as representing the development and implementation of digital technologies. A differential form over this object could measure the rate of change of these technologies with respect to the number of patients served. For example, we could measure how the use of digital mammography or telemedicine services changes as the number of patients increases. Mathematically, this could be represented as a derivative, $d(Digital\ Niche\ Construction)/dp$, where p represents the number of patients. This could provide insights into how the healthcare system is scaling up digital technologies to meet the needs of a growing patient population.
- **Co-Shaping Technology and Organizational Practices**: This object could be seen as representing the interplay between technological advancements and changes in organizational practices. A differential form over this object could measure the rate of change of this interplay with respect to the amount of resources invested. For example, we could measure how the integration of AI tools into clinical practice changes as more resources are invested in AI research and development. Mathematically, this could be represented as a derivative, $d(Co\text{-}Shaping\ Technology\ and\ Organizational\ Practices)/dr$, where r represents the amount of resources. This could provide insights into how investment in technology is driving changes in organizational practices.

- In the context of differential geometry and topology, **de Rham cohomology** is a powerful tool that uses differential forms to study the topology of a space. It provides a way to measure the extent to which a space is not simply connected, and it can be used to classify different types of spaces based on their topology. In the context of category theory, we can apply the concept of de Rham cohomology to study the differential forms on each manifold, which in our case represents a category. This can help us analyze the structure and properties of each category, and how they change over time.
  - In the context of the healthcare ecosystem, we can think of the manifold representing the category C_4, which includes states like "Adaptive Structuration" and "Digital Niche Construction," as a topological space. The differential forms on this manifold, as we discussed earlier, could represent the rate of change of these states with respect to certain variables, such as time, patient population size, or resource investment. These differential forms provide a dynamic perspective on the transformations and flows within the category C_4. de Rham cohomology of this manifold would provide a way to study the topology of the category C_4, or in other words, the global structure and properties of the healthcare ecosystem as it undergoes, say, an Electronic Health Record (EHR) implementation.

* For instance, the zeroth de Rham cohomology group, $H^0(C_4)$, would measure the extent to which the healthcare ecosystem *is* connected. If the ecosystem is fully connected, meaning that information and practices related to the EHR implementation can flow freely from one state to another, then $H^0(C_4)$ would be trivial. If the ecosystem is not fully connected, meaning that there are barriers to the flow of information or practices, then $H^0(C_4)$ would be nontrivial.[2]
  * In practical terms, this could mean looking at how different parts of the ecosystem, such as different healthcare practices or technologies, are interconnected. For example, a differential form that measures the rate of change of "Adaptive Structuration" with respect to time could be integrated over the entire healthcare ecosystem to give a measure of the overall rate of change. If this integral is zero, it would suggest that the healthcare ecosystem is disconnected in some way, with different parts evolving independently of each other.
* The first de Rham cohomology group, $H^1(C_4)$, would measure the extent to which the healthcare ecosystem is *not simply* connected. If the ecosystem has "holes" or "gaps," such as disparities in the adoption of the EHR system between different departments or clinics, then $H^1(C_4)$ would be nontrivial. If the ecosystem is simply connected, meaning that there are no such disparities, then $H^1(C_4)$ would be trivial.[3]
  * This could mean looking at how different parts of the ecosystem are interconnected in more complex ways, such as through feedback loops or network effects. For example, a differential form that measures the rate of change of "Digital Niche Construction" with respect to the number of patients could be integrated over a loop in the healthcare ecosystem, such as a cycle of patient care that includes diagnosis, treatment, and follow-up. If this integral is nonzero, it would suggest that the healthcare ecosystem has a nontrivial loop, indicating a more complex level of interconnection.
* H^0 focuses on the basic connectedness of the space, whereas H^1 provides insights into the higher-order topological features such as "holes" that could impede connectivity on a more complex level. Higher order de Rham cohomology groups, such as $H^2(C_4)$, $H^3(C_4)$, and so on, would measure more complex topological features of the healthcare ecosystem. However, these would likely be less relevant in a practical healthcare context, as they would involve integrating differential forms over higher-dimensional cycles, which may not have a clear interpretation in terms of healthcare practices or technologies.

---

[2] A trivial cohomology group typically means that there are no "holes" or "obstructions" in the space at the dimension being considered. For the zeroth de Rham cohomology group, being trivial suggests that there is exactly one connected component in the space. In practical terms, a trivial $H^0(C_4)$ indicates that the ecosystem is cohesive and interconnected, allowing for uniform evolution and implementation processes across different states or parts of the ecosystem. A nontrivial $H^0(C_4)$ suggests that the space has multiple connected components, with a dimension greater than one, reflecting the number of disconnected or independently evolving segments within the ecosystem.

[3] In topological terms, a simply connected space is one without "holes" that would prevent the space from being contractible to a point within that dimension. A nontrivial $H^1(C_4)$ indicates that the space is not simply connected, meaning there are loops or cycles that cannot be continuously shrunk to a point without leaving the space.

* In this way, de Rham cohomology provides a way to study the global structure and properties of the healthcare ecosystem as it undergoes the EHR implementation. This can provide valuable insights for healthcare administrators, helping them understand the potential challenges and opportunities associated with the EHR implementation and plan accordingly.[4]

Toward a new monadology of dynamic relationalities

In category theory, a monad is a structure that encodes a certain kind of self-contained behavior. We could use monads to model the addition of structure or effects to our categories as a result of transformations. A monad in category theory is a type of endofunctor (functors from a category to itself) that comes with two natural transformations. Technically, a monad is a triple (T, υ, μ) where T is an endofunctor, υ is a natural transformation called unit, and μ is a natural transformation called multiplication, subject to certain laws that mirror the laws of a monoid. The formal definitions are as follows:

- An **endofunctor** is a functor that maps a category to itself. We could consider each transformation within a category as being represented by an endofunctor. This would allow us to model the dynamics of transformations within each category in a way that preserves the category's structure. As the underlying functor of the monad, the functor T: C → C for a category C is an endofunctor from a category C to itself. This functor is often referred to as the "context" or "enrichment" functor, as it adds additional structure to the objects and morphisms in C. It describes how to transform objects and morphisms within the category. An endofunctor acts on an entire category, mapping all objects and morphisms within that category to other objects and morphisms within the same category. It doesn't act on individual objects separately but rather on the entire structure of the category.
- The **unit transformation** is a natural transformation from the identity functor to the endofunctor. The natural transformation $\upsilon: \text{Id} \to T$ (where Id is the identity functor on C) is the unit of the monad. It encodes the idea of embedding an object in its context. It associates to each object X of the category a mapping $\upsilon\_X: X \to T(X)$.
- The **multiplication transformation** is a natural transformation from the composition of the endofunctor with itself to the endofunctor. The natural transformation $\mu: T^2 \to T$ (where T^2 is the functor T applied twice) is the multiplication of the monad. It describes

---

[4] Abramowitz et al. (2023) illustrate the importance of efficient data flows and the challenges posed by complex healthcare ecosystems during public health emergencies. The practical challenges identified in the study, such as the lack of interoperability between different data collection and reporting systems, and the impact of vertical programs on routine data flows, can be understood as manifestations of the complex topological features of the healthcare ecosystem. When analyzed through the theoretical lens of de Rham cohomology, this practical scenario reveals deeper insights into the structure and dynamics of healthcare systems. For example, $H^0(\mathbf{C}_4)$ could help measure the connectedness within the healthcare system's response to the epidemic. The challenges in data interoperability and the creation of parallel data systems can be seen as "holes" or "gaps" in the system, which would be analyzed through $H^1(\mathbf{C}_4)$.

how to combine or simplify two contexts into a larger one. It associates to each object X of the category a mapping μ_X: T(T(X)) → T(X).

These components must satisfy two laws (associativity and unit), which ensure the coherence of the structure. The unit law ensures that if you embed an object in its context and then immediately apply the functor, that's the same as just applying the functor to the object. The associativity law ensures that if you apply the functor twice and then combine the contexts, that's the same as applying the functor once to the combined context.

In the context of a hospital implementing a new electronic health record (EHR) system, we can illustrate the monad in category theory as follows:

- Endofunctor T (EHR Implementation): The endofunctor T in our category $\mathbf{C}_4$ could represent the process of implementing the EHR system. When applied to an object like "Adaptive Structuration," it transforms it into a new state, which we could call "Adaptive Structuration with EHR." Similarly, when applied to "Digital Niche Construction," it transforms it into "Digital Niche Construction with EHR." The morphism f_{41}: "Adaptive Structuration" → "Digital Niche Construction" would then be transformed into a new morphism: "Adaptive Structuration with EHR" → "Digital Niche Construction with EHR."

- Unit Transformation υ (Planning): The unit transformation υ could represent the process of planning the EHR system implementation. For each object in $\mathbf{C}_4$, it provides a way to get from just "being in the state" to "being in the state with a plan for EHR system implementation." For instance, for the object "Adaptive Structuration," υ would provide a way to get from "Adaptive Structuration" to "Adaptive Structuration with Plan for EHR." Similarly, for "Digital Niche Construction," υ would provide a way to get from "Digital Niche Construction" to "Digital Niche Construction with Plan for EHR."

- Multiplication Transformation μ (Consolidation): The multiplication transformation μ could represent the process of consolidating multiple plans into a single, integrated plan for EHR system implementation.
  * Suppose that the hospital initially develops separate plans for different aspects of the EHR system implementation, such as:
    * A plan for training healthcare providers to use the new EHR system (Plan A)
    * A plan for integrating the new EHR system with existing IT infrastructure (Plan B)
    * Each of these plans could be seen as adding a layer of structure or "planning" to the state of "Adaptive Structuration." So, we might say that the hospital is in a state of "Adaptive Structuration with Plan A and Plan B."
  * Now, let's consider the "Digital Niche Construction" object. This could represent the process of integrating the new EHR system into the hospital's existing digital environment. The hospital might develop separate plans for different aspects of this process, such as:
    * A plan for customizing the EHR system to fit the hospital's specific needs (Plan C)
    * A plan for training IT staff to maintain and troubleshoot the EHR system (Plan D)

- These plans could be seen as adding a layer of structure or "planning" to the state of "Digital Niche Construction." So, we might say that the hospital is in a state of "Digital Niche Construction with Plan C and Plan D."
- The multiplication transformation μ would then simplify "Adaptive Structuration with Plan A and Plan B" to just "Adaptive Structuration with a consolidated Plan for EHR", and "Digital Niche Construction with Plan C and Plan D" to just "Digital Niche Construction with a consolidated Plan for EHR." The morphism f_{41}: "Adaptive Structuration" → "Digital Niche Construction" would then be transformed into a new morphism: "Adaptive Structuration with a consolidated Plan for EHR" → "Digital Niche Construction with a consolidated Plan for EHR."[5]

In the context of differential topology, we could think of a monad as a way of "bundling" together a manifold (corresponding to the category), a vector field (corresponding to the endofunctor), and a differential form (corresponding to the natural transformations), providing a unified way of studying the dynamics of transformations within each category.[6]

---

[5] Applying the monad framework to Cruz (2022) study of the intricate social life of biomedical data within digital health systems can help us understand how data representations in healthcare both reflect and shape the reality of care in nuanced ways. Similar to how an endofunctor maps objects and morphisms within a category to new ones within the same category, the process of implementing EHRs fundamentally changes how healthcare workers interact with patient data, transforming traditional healthcare practices into data-driven ones, which can be seen as adding a new layer of structure (the "EHR layer") to the existing healthcare practice. The unit transformation as a critical step that embeds the potential for transformation within the organization, akin to embedding an object within its context in the monad framework, could be seen in the planning phase of EHR system implementation, where the healthcare organization prepares to integrate digital technologies into its operations by mapping out how the EHR system will be implemented, how it will change existing processes, and how it will impact patient care. The monad's multiplication transformation, where multiple layers of structure or "contexts" are simplified into a single, integrated framework, making the overall system more efficient and effective, is mirrored by the way healthcare workers reconcile multiple data representations with the realities of patient care, as they consolidate data from various sources to get a holistic view of patient health or streamline multiple digital processes into a unified workflow.

[6] This association is a way of bridging the worlds of category theory and differential topology. By drawing these parallels, we can leverage the geometric and topological intuition provided by manifolds, vector fields, and differential forms to gain deeper insights into the abstract structures and transformations of categories, functors, and natural transformations:

- Associating a category with a manifold suggests that we're considering the category's objects as points in a space, and the morphisms as paths or trajectories between these points.
- Associating an endofunctor with a vector field suggests that we're viewing the functor's action as inducing a kind of "flow" or "dynamics" on the category, similar to how a vector field induces dynamics on a manifold.
- Associating natural transformations with differential forms suggests that we're viewing these transformations as providing a measure or "rate" of change within the category. Just as differential forms measure change on a manifold, the natural transformations measure the "coherence" or "consistency" of change induced by the monad's endofunctor.

- **Smooth Manifolds**: As we discussed earlier, a manifold is a space where each point can be locally described using Euclidean coordinates. It allows for continuous transformations. Similarly, a category consists of objects and morphisms (arrows) between them. Each object can be seen as a unique "state" or configuration, and morphisms represent transformations between these states. Unit transformation, as a natural transformation, maps each object of the category to its image under the monad. It serves as an "inclusion" or "embedding" of the object into the monadic context. The unit transformation's role of embedding an object into the monadic context can be likened to the local Euclidean descriptions of points on a manifold. Both provide a foundational structure upon which transformations are built.[7]
- **Vector Fields**: In the context of differential topology, we could think of each endofunctor as generating a vector field on the manifold representing the category, with each vector representing the "flow" or "direction" of the endofunctor at a particular point in the category. In any context, a manifold represents a complex system with each point signifying a specific state of the system. An endofunctor symbolizes a transformation within the system, altering its state and consequently its behavior and outputs. The vector field, generated by the endofunctor, indicates the direction and magnitude of this transformation at each point in the system. Each vector points toward the direction of change and its length signifies the extent of this change. Therefore, the vector field visually and mathematically depicts the transformation process within the system, showing how each state would change under the influence of the endofunctor. This understanding can provide valuable insights for system management, aiding in planning and implementing transformations.
    - In the context of the hospital implementing a new electronic health record (EHR) system, the vector field generated by the endofunctor T could represent the direction and magnitude of the changes brought about by the EHR implementation at each point in the healthcare ecosystem. Let's consider the manifold representing the category C_4, which includes states like "Adaptive Structuration" and "Digital Niche Construction." Each point on this manifold represents a specific state of the healthcare ecosystem. The endofunctor T, representing the EHR implementation, generates a vector field on this manifold. Each vector in this field corresponds to a point on the manifold and represents the "flow" or "direction" of the EHR implementation at that point.
        - For example, at a point representing a state of "Adaptive Structuration", the vector generated by T might point toward a new state, "Adaptive Structuration with EHR." The length of this vector could represent the magnitude of the

---

[7] Cruz (2022) exploration of how healthcare workers recognize, interpret, and act upon biomedical data within digital safety-nets can be likened to navigating the complex manifold of the digital health system. The local "Euclidean" understanding provided by healthcare workers' interpretations of data allows for the application of calculus—or, in this case, decision-making processes and care strategies—in navigating this complex space. The manifold analogy underscores the complexity and multidimensionality of the digital health system, highlighting the need for local understanding and interpretation (similar to local Euclidean descriptions) to navigate this space effectively. The application of category theory, particularly the concept of unit transformations, further emphasizes the importance of embedding individual data points into broader contexts to facilitate meaningful transformations in patient care and healthcare delivery.

change required to implement the EHR system, such as the extent of training needed for healthcare providers or the degree of change in administrative processes.
  - Similarly, at a point representing a state of "Digital Niche Construction," the vector generated by T might point toward a new state, "Digital Niche Construction with EHR." The length of this vector could represent the magnitude of the change required to integrate the EHR system into the hospital's digital environment, such as the extent of IT infrastructure modifications or the degree of change in data management practices.
  - In this way, the vector field provides a visual and mathematical representation of the EHR implementation process across the entire healthcare ecosystem. It shows how each state of the ecosystem would change under the influence of the EHR implementation and the extent of these changes. This can provide valuable insights for healthcare administrators, helping them understand the potential impacts of the EHR implementation and plan accordingly.[8]
- A **Lie derivative** is a mathematical operation that measures the change of a tensor field along the flow of a vector field. In other words, it describes how a given quantity changes as we move along the flow lines of a vector field. In the context of differential geometry and topology, Lie derivatives are used to study the symmetries and transformations of differentiable manifolds. They provide a way to compare the values of a tensor field at different points on a manifold, while taking into account the structure of the manifold and the flow of the vector field. This makes them a powerful tool for analyzing the dynamics of complex systems.
  - Now, let us consider the healthcare ecosystem represented by the category C_4, which includes states like "Adaptive Structuration" and "Digital Niche Construction." In this context, we can think of each state as a point on a manifold, and each process (like the EHR implementation or provider training) as a vector field on this manifold. For example, let us consider two vector fields on this manifold:
    - The vector field associated with the endofunctor T, which represents the process of EHR implementation. Each vector in this field indicates the direction and rate of change in the healthcare ecosystem due to the EHR implementation at a particular state.
    - The vector field associated with another endofunctor T′, which represents the process of training healthcare providers to use the new EHR system. Each vector in this field indicates the direction and rate of change in the healthcare ecosystem due to the provider training at a particular state.
  - The Lie derivative of the vector field associated with T with respect to the vector field associated with T′ would then measure how much the EHR implementation

---

[8] In Cruz (2022) study, when the EHR system is implemented, it initiates a transformation across the healthcare ecosystem, altering practices, workflows, and interactions, which can be envisioned as a vector field. Each endofunctor generating a vector field could symbolize a specific aspect of EHR implementation, such as data capture, data interpretation, or operational changes. The vectors point toward the direction of change—improved data accessibility, enhanced patient care, or possibly increased administrative burden—and their length signifies the extent of this change, whether it's a minor adjustment or a system-wide transformation.

process (as represented by T) changes as we move along the flow of the provider training process (as represented by T′).
- For instance, at a point representing a state of "Adaptive Structuration," the Lie derivative could indicate how much the direction and rate of change in the EHR implementation process are affected by the progression of the provider training process. If the Lie derivative is positive, this could indicate that the EHR implementation process is accelerating as provider training progresses. Conversely, if the Lie derivative is negative, this could indicate that the EHR implementation process is slowing down as provider training progresses.
- In this way, the Lie derivative provides a way to analyze the interplay between different processes within the healthcare ecosystem, helping administrators understand how these processes influence each other and plan accordingly.[9]

- **Differential Forms**: As we saw above, differential forms are tools to measure quantities on manifolds, especially those that depend on orientation. They can capture rates of change and provide a way to integrate over the manifold. On the other hand, natural transformations in a monad, specifically the unit and multiplication transformations, are fundamental structures in category theory that describe how one functor transforms into another in a way that is compatible with the morphisms in the category. Differential forms can be thought of as a way to measure the "rate of change" or "flow" of these natural transformations. These differential forms can provide a dynamic perspective on the transformations and flows within each category. They can help us understand how these transformations are evolving over time and how they are being influenced by various factors.
  - The unit transformation, as a specific natural transformation, is the transformation that maps each object to itself. A differential form associated with the unit transformation would measure the rate of change of the identity transformation (i.e., the transformation that leaves each object unchanged) as it evolves into the endofunctor transformation.
    - In the context of the hospital implementing a new electronic health record (EHR) system, the unit transformation represents the initial planning phase of adapting healthcare practices and policies. In the context of "Adaptive Structuration," this could involve the planning for the adoption of new screening guidelines or treatment protocols. A differential form associated with this transformation could measure the rate at which these new guidelines or protocols are being adopted over time. For instance, if we consider time as a variable, the differential form could represent how the adoption status of the new guidelines or protocols changes with respect to time.

---

[9] In the context of Cruz (2022) study, we can think of the "tensor field" as the structured practices and processes of healthcare delivery (e.g., patient care protocols, decision-making processes, administrative tasks), and the "vector field" as the influence of digital technologies and data analytics (e.g., the implementation of EHRs, the use of performance metrics). Just as Lie derivatives measure changes in a tensor field along the flow lines of a vector field, Cruz' exploration can be seen as measuring how the practices and processes of healthcare change as they are influenced by the flow of digital technologies and data analytics, i.e., the direction and manner, in which digital technologies are integrated into and transform healthcare practices, influencing decisions, patient care strategies, and administrative processes.

- In a more concrete sense, suppose we have a timeline that outlines when each new guideline or protocol should be adopted. We could then define a differential form that measures the rate of adoption of these guidelines or protocols per unit of time. This could provide insights into whether the adoption process is on track or if there are delays that need to be addressed.
- The multiplication transformation, as another specific natural transformation, is the transformation that maps the composition of the endofunctor with itself to the endofunctor. A differential form associated with the multiplication transformation would measure the rate of change of the composition of the endofunctor with itself as it evolves into the endofunctor transformation.
  - In the context of "Digital Niche Construction," the consolidation process could involve integrating multiple digital technologies into a unified system for breast cancer care. This might include merging separate systems for digital mammography and telemedicine services into a single, streamlined platform. A differential form associated with this transformation could measure the rate at which this consolidation process is occurring. For example, if we consider the number of separate systems as a variable, the differential form could represent how the number of separate systems changes as the consolidation process progresses.
  - In a more concrete sense, suppose we initially have several separate digital technologies being used in breast cancer care. As the consolidation process progresses, we could define a differential form that measures the rate of change in the number of separate systems per unit of time. This could provide insights into how quickly the healthcare system is able to integrate these separate technologies into a unified platform.
  - The multiplication transformation's role in ensuring the coherence of successive functorial applications can also be likened to how Lie derivatives measure the change of tensor fields along flows. Both concepts deal with the consistency and structure of transformations.[10]

---

[10] Referring back to Cruz (2022) study, differential forms could represent tools to measure the orientation-dependent changes within healthcare practices (e.g., clinical decision-making, patient care strategies, and administrative workflows) as they are influenced by digital technologies. A differential form associated with the unit transformation could measure how the introduction of EHRs changes healthcare practices (workflow, communication, and care delivery) from their original state, even when the intention is to retain the core identity of those practices. A differential form associated with the multiplication transformation would measure how the compounded use of EHRs and data analytics alters healthcare practices, reflecting the deeper integration and systemic changes that occur beyond initial implementation. As healthcare systems adapt to digital technologies, natural transformations in a monad, through unit and multiplication transformations, ensure that these adaptations are coherent with the system's overarching goals, such as improving patient care, enhancing efficiency, and maintaining ethical standards. The differential forms then measure how well these transformations maintain their coherence and alignment with healthcare objectives over time.

Homotopies

Homotopy theory can be used to study the "differential" of functors in a categorical sense. In the context of category theory, a homotopy is a continuous transformation from one morphism to another in the category of topological spaces (or other relevant categories where the notion of homotopy applies). In simpler terms, if you have two continuous functions $f, g: X \to Y$ between topological spaces, a homotopy is a family of continuous functions $\mathbf{H}: X \times [0,1] \to Y$ such that $\mathbf{H}(x, 0) = f(x)$ and $\mathbf{H}(x, 1) = g(x)$ for all x in X. This can be interpreted as a continuous deformation of f into g, with the second argument of $\mathbf{H}$ representing 'time.'

- Now, let's try to illustrate this with a conceptual example from a healthcare ecosystem. Consider the category whose objects are different patient care strategies for a particular medical condition (say, diabetes) and whose morphisms are transformation procedures from one care strategy to another. Each care strategy can be thought of as a complex function that takes in various inputs (patient health data, healthcare resources, etc.) and outputs a health outcome. Now, suppose we have two care strategies, f and g. Perhaps f represents the current standard of care, and g represents a new experimental strategy. A "homotopy" in this context would be a continuous transition plan from strategy f to strategy g. It would represent, for each "time" t between 0 (now) and 1 (the future), a care strategy $\mathbf{H}(t)$ that smoothly interpolates between f and g. This homotopy could reflect a gradual introduction of new treatments, a phased rollout of a new care protocol across different regions or patient populations, or any other changes that occur over time. The requirement of continuity would mean that at any given time, the care strategy is 'close' to the strategies at earlier and later times, i.e., changes are introduced in a gradual, controlled manner rather than abruptly.

In particular, the concept of a "natural transformation" in category theory is analogous to the concept of a homotopy in topology. A natural transformation provides a way to transform one functor into another while preserving the structure of the categories involved. This is similar to how a homotopy provides a way to deform one function into another in a continuous way. In the context of our model, you could consider a natural transformation between two functors as a kind of "differential" that captures how one functor changes into another. This could be used to model the dynamics of transformations between categories.

Homotopy theory can help make the dynamics behind natural transformations more explicit. In category theory, a natural transformation provides a way to transform one functor into another while preserving the categorical structure. However, this transformation is typically static and doesn't capture the "motion" or "path" of the transformation. Homotopy theory, on the other hand, is all about continuous transformations and deformations. A homotopy between two functions (or more generally, morphisms in a category) provides a continuous path from one to the other.

By considering natural transformations in the context of homotopy theory, one can introduce a notion of "dynamics" or "motion" into a model. For example, one might consider a "homotopy natural transformation," which provides a continuous path in the space of functors from one functor to another. This would capture not just the initial and final states of the transformation, but also the entire "process" of the transformation. This approach can make the dynamics behind natural transformations more explicit and can provide a richer and more nuanced understanding of the transformations in our model.

Let's consider the context of breast cancer prevention, diagnosis, and treatment and see how we can model the unit and multiplication transformations, both of which are natural transformations, and apply homotopy theory to understand their dynamics further:

Unit Transformation υ and Homotopy Theory: The unit transformation υ could represent the process of embedding a new practice or protocol within the existing healthcare system. For instance, let's say the hospital is considering implementing a new protocol for early detection of breast cancer that involves a novel genetic testing technique. The unit transformation υ would then represent the initial plan for implementing this new protocol within the existing healthcare practices.

- Now, let us consider a scenario where the hospital wants to transition from the current standard of care to this new protocol. We could define a homotopy that provides a continuous path from the current standard of care to the new protocol. This homotopy would represent a gradual transition plan that smoothly interpolates between the current standard of care and the new protocol over time. For instance, the hospital might gradually phase in the use of the new genetic testing technique while maintaining the existing standard of care throughout the transition. The continuity requirement of the homotopy ensures that the transition from the current standard of care to the new protocol is gradual and controlled, rather than abrupt. This can help in understanding how the planning phase of the new protocol evolves over time and how different plans can be smoothly transitioned.

Multiplication Transformation μ and Homotopy Theory: The multiplication transformation μ could represent the process of integrating multiple aspects of a single treatment plan. For instance, let us say a patient's treatment plan involves surgery, radiation therapy, and chemotherapy. The multiplication transformation μ would then represent the process of integrating these multiple aspects into a single, cohesive treatment plan.

- Suppose the hospital wants to transition from a state where these aspects are managed separately to a state where they are integrated into a single plan. We could define a homotopy that provides a continuous path from the separate-aspects state to the integrated-plan state. This homotopy would represent a gradual integration process that smoothly interpolates between the separate-aspects state and the integrated-plan state over time. For instance, the hospital might gradually improve coordination between the surgery, radiation therapy, and chemotherapy teams, leading to a more integrated approach to treatment. The continuity requirement of the homotopy ensures that the integration process is gradual and controlled, rather than abrupt. This can help in understanding how the integration process evolves over time and how different states of the process can be smoothly transitioned.

In both cases, homotopy theory provides a dynamic perspective on the transformations in the monad, capturing not just the initial and final states of the transformations, but also the entire process of the transformations. This can provide a richer and more nuanced understanding of the transformations in the monad, and can help in dynamically analyzing these transformations.[11]

Becoming of singularity in AGI as differential transformation

Let us apply differential topological concepts we have introduced in this chapter to expand our category theoretical analysis started earlier in the chapter to deepen our intuition and understanding of singularity concerning AGI (see Chapter 5).

To deepen our comprehension of the metamorphosis from Category $\mathbf{C}_5$ to Category $\mathbf{C}_{5+}$, where foundational elements like the FreeEnergyPrinciple, ActiveInference, and Phenomenotechnics aren't merely supplanted but rather expanded to encompass PerceptionCognition, AGI, and ExperiencialComputing, we turn to the two pivotal theoretical constructs of the monad concept and homotopy theory. While the former is encapsulated as a triad (T, υ, μ), which offers insights into the enrichment process of our category framework, the latter provides a lens to discern the intricate dynamics of the transformation β: $F_4 \rightarrow F_{4+}$ (see Fig. 5.1). In essence, the evolution from Category $\mathbf{C}_5$ to $\mathbf{C}_{5+}$ is potentially orchestrated by the interplay of a monadic structure and a homotopic natural transformation. Building on this foundation, we'll delve into the implications of contemporary advancements in areas like generative AI, mixed reality, Quantum computing, and more, emphasizing their role in shaping this categorical transformation and the broader implications for immersive, AI-driven human-computer interactions. Our exploration will be anchored in understanding the nuanced expansion of the core objects and morphisms from Category $\mathbf{C}_5$ to $\mathbf{C}_{5+}$.

Remember that the manifold $\mathbf{C}_5$, representing the category Creative-Experiencer, is akin to the assemblage space. It's a topological space that captures the entire landscape of possibilities for the assemblage. In the context of our categories, this manifold represents the entire space of configurations of the original objects and morphisms, such as FreeEnergyPrinciple,

---

[11] In the context of Hampel et al.'s (2022) study on the next-generation clinical care pathway for Alzheimer's disease (AD), homotopy theory allows us to see the interconnectedness of various aspects of AD care, from early detection and diagnosis to comprehensive treatment planning and management. The unit transformation represents the initial integration and embedding of biomarker-guided approaches within the existing clinical framework, where the homotopy would embody a series of intermediate steps or "paths" that healthcare systems might undertake over time, progressively incorporating biomarker-based early detection strategies into routine symptom-based practice. The multiplication transformation describes the integration of various therapeutic interventions, including emerging pharmacological treatments and nonpharmacological interventions (e.g., lifestyle adjustments, cognitive training), into a comprehensive care plan tailored to the individual's disease stage and biological markers, where the homotopy represents the process of evolving care strategies from a disjointed set of separate treatments to a fully integrated, personalized care plan.

ActiveInference, and Phenomenotechnics. The points on this manifold represent specific states or configurations of the assemblage. In the context of Category C_5 becoming Category C_{5+}, these points can represent various states of transformation, from the original objects and morphisms to the enriched ones like PerceptionCognition, AGI, and ExperiencialComputing. Consider how some of the currently emerging technologies might be instrumental in various transformations: [12]

- **Generative AI** can be seen as a mechanism that enriches the FreeEnergyPrinciple. By generating new data and scenarios, generative AI can enhance our understanding of how systems minimize discrepancies between their internal models and sensory data.
- **Neuromorphic Computing** directly relates to the ActiveInference object. By mimicking the human brain's architecture, neuromorphic chips can offer insights into how living systems actively seek out information.
- **Quantum Computing** can influence both the FreeEnergyPrinciple and ActiveInference. Quantum systems process information in fundamentally different ways, potentially offering insights into human cognition and AGI development.
- **Mixed Reality (AR/VR)** can be linked to Phenomenotechnics. By altering human perception and creating immersive experiences, AR/VR enriches the way technology shapes human experience.

Monad Concept for Category Extension/Enrichment: As we saw earlier, a monad is a mechanism to structure programmatic computations. It is a way to build computational steps and chain them together. In the context of Category C_5 becoming Category C_{5+}, the monad can be seen as a computational structure that captures the enrichment process.

- **Endofunctor T:** This represents the computational process that transforms Category C_5 into Category C_{5+}. It's the mechanism by which the original objects and morphisms of C_5 are expanded and transformed. In our context, T could be seen as the technological advancements (like generative AI, AR/VR, Quantum computing, etc.) that are driving the transformation.
    * The endofunctor can be visualized as generating a vector field on the manifold C_5. Each vector in this field indicates the direction and magnitude of transformation at a

---

[12] To give one example: in elucidating the convergence of neuromorphic and quantum computing with the Free Energy Principle and Active Inference, Fields et al. (2022) underscore a seminal shift toward realizing systems that embody biological cognitive processes. Neuromorphic computing, by mimicking brain-like architectures, serves as a critical technological underpinning for Active Inference, enabling artificial systems to minimize free energy through adaptive interactions with their environments. Meanwhile, quantum computing enables the simulation of complex probabilistic models inherent in the Free Energy Principle, thus enhancing the accuracy and speed of predictions related to active inference processes. This capability directly supports the development of AGI by providing a computational framework that can emulate the intricate and efficient decision-making processes observed in human cognition.

specific point (or state) in the category. Generative AI & Neuromorphic Computing can be seen as vectors pushing the FreeEnergyPrinciple toward a more comprehensive PerceptionCognition. The vector's direction indicates the transformation, while its length signifies the extent influenced by these technologies.[13]

- **Unit Transformation υ:** This captures the process of embedding an object from C_5 into the enriched Category C_{5+}. For instance, the FreeEnergyPrinciple, when enriched with insights from generative AI or neuromorphic computing, becomes a more comprehensive PerceptionCognition object in C_{5+}.
  * The differential form associated with the unit transformation would capture how each point (or state) in the category C_5 is evolving toward its transformed state under the influence of the endofunctor. For example, it would measure how quickly FreeEnergyPrinciple is transforming into PerceptionCognition.[14]
- **Multiplication Transformation μ:** This represents the process of flattening or combining multiple enriched structures. For example, combining insights from Quantum computing and generative AI might lead to a new form of ActiveInference in C_{5+} that's more powerful and encompassing than its C_5 counterpart.
  * The differential form for the multiplication transformation would capture the rate of change when the endofunctor is applied successively. This can be visualized as measuring the acceleration of the transformation process.[15]

Homotopy Theory for Natural Transformation Dynamics: A natural transformation provides a mapping between two functors, in this case, F_4 and F_{4+}. A homotopy between these functors would provide a continuous family of functors that transition from F_4 to F_{4+}. This family of functors captures the evolving dynamics of the transformation, showing how

---

[13] The attention mechanism in deep learning, as reviewed by Niu et al. (2021), exemplifies a concrete instantiation of the endofunctor T. By incorporating attention mechanisms, foundational elements like the Free Energy Principle and Active Inference undergo a conceptual upgrade, enhancing their interpretative and decision-making capabilities. Morphisms, representing the dynamic processes between these elements, are similarly transformed to reflect more complex, adaptive interactions. This categorical transformation encapsulates a shift toward more advanced cognitive architectures, facilitating nuanced perception, cognition, and AGI within an enriched category, C_{5+}.

[14] The integration of machine learning and neuromorphic computing into human–machine interfaces (HMIs), as explored by Zhu et al. (2020), exemplifies a practical embodiment of the unit transformation υ in the context of Category C_5 transitioning to C_{5+}. This transformation highlights the evolution of HMIs from basic tactile interfaces to advanced systems capable of sophisticated perception and cognition-like processes. By applying the FreeEnergyPrinciple—originally focused on reducing surprise or uncertainty in systems—to the enhancement of HMIs, this transition minimizes the entropy in human-computer interactions. The augmented HMIs become a more comprehensive PerceptionCognition object within Category C_{5+}, marking a pivotal shift in the capabilities of HMIs to understand, predict, and adapt to user intentions, thereby mirroring human cognitive processes more closely.

[15] The exploration by Dunjko and Briegel (2018) demonstrates how the integration of quantum computing's processing capabilities with machine learning's (ML) adaptive learning can significantly advance the computational models and cognitive architectures within an enriched Category C_{5+}, where the multiplication transformation μ symbolizes the unification and enrichment of systems or ideas to forge a more comprehensive structure. Combining quantum computing's ability to quickly process complex data sets with ML's iterative learning from data can lead to the development of new forms of Active Inference that are more efficient, adaptive, and capable of addressing real-world complexities.

each object and morphism in $\mathbf{C}_5$ is transformed under the lens of F_4 to its counterpart in $\mathbf{C}_{5+}$ under the lens of F_{4+}.

- **Homotopy of** *β_AdaptiveStructuration*: This homotopy would represent the continuous transformation of the FreeEnergyPrinciple into PerceptionCognition. It would capture the evolving understanding of adaptive structuration processes in organizations or systems and how they relate to perception and cognition processes. The homotopy would provide a path showing how minimizing uncertainty and adapting to environmental changes influence the way agents perceive and process information, evolving over time.[16]
- **Homotopy of** *β_DigitalNicheConstruction*: This homotopy would capture the continuous transformation of ActiveInference into AGI. It would show the evolving dynamics of how agents actively seek information and adapt to their niches and how this understanding informs the development of AGI systems. The path would highlight the transition from understanding adaptive behavior in biological systems to designing AGI algorithms that mimic this behavior.[17]
- **Homotopy of** *β_CoShapingTechnologyOrganizationalPractices*: This homotopy would represent the continuous transformation of Phenomenotechnics into ExperiencialComputing. It would capture the evolving interplay between human perception, technology, and organizational practices, showing how these dynamics inform the design of immersive computing experiences.[18]

---

[16] Keding and Meissner (2021) show how the introduction and integration of AI into decision-making processes in strategic R&D investment effectuate a transformation from reliance on the FreeEnergyPrinciple, where decisions are made to minimize uncertainty through human cognition, toward an enhanced PerceptionCognition framework, where AI systems offer structured, trustworthy advice, thereby reducing uncertainty in decision-making processes.

[17] The development of AGI systems is shaped by the underlying adaptive behaviors observed in biological systems, with an additional layer of complexity introduced by the cultural and symbolic habitus that frames these technologies. Romele (2024) argues that digital technologies, particularly those driven by AI algorithms, function as powerful habitus machines, embedding and embodying social classifications within individuals through algorithmic curation. This process not only tailors personalized services but also shapes individuals' perceptions, desires, and interactions with the world, effectively creating information bubbles that limit exposure to a homogenized set of stimuli. The transition from Active Inference to AGI thus encompasses both the algorithmic and the cultural–symbolic dimensions, where the evolving dynamics of agents adapting to their niches are intertwined with the societal perceptions, expectations, and imaginaries surrounding AI technologies.

[18] Baabdullah et al. (2022) demonstrates the transition from Phenomenotechnics to Experiencial Computing, underlining the vital roles of personalization, responsiveness, and ubiquitous connectivity in enhancing AI-powered chatbot interactions. Their findings on readability and transparency as pivotal elements for immersive computing experiences underscore the gradual evolution of technology and organizational practices toward more intuitive and user-centered computational experiences.

PART III

Organizations and creative transformation

CHAPTER 7

Creative transformation along a molar/molecular stack

In **Part 2**, we discussed in detail the creative transformations of dynamic relationalities—within and across assemblages, on and across (stacked) strata, lines of flight, becoming, and reterritorialization in co-evolution of Machinic Life-Experience Ecosystems (MLXEs). We saw how by considering the two pivotal theoretical constructs of the monad concept and homotopy theory, one can encapsulate and make the dynamics of the entire continuous process of transformation more explicit, thereby providing a much richer and nuanced understanding of the underlying dynamic relationalities of creative transformations.

In this chapter, we delve into how Dynamic Relationality Theory (DRT) underscores the intricate interplay between local and global dynamics, represented by molecular and molar lines of segmentation. Molar lines, shaped by overarching socio-political forces, influence the broader structures of systems. In contrast, molecular lines delve into microtransformations, capturing the nuances of cognitive landscapes and sense-making processes. The fluidity of an MLXE is an expression of the reciprocal influence between these lines: while molar lines mold local dynamics, molecular lines subtly shape global systems. We also discuss how through sheaf analysis, we can contextualize this interplay, understanding how local cognitive processes intertwine with global dynamics, offering a comprehensive view of the dynamic relationalities of MLXE relational dynamics.

Dynamic relationality #7: Molar/molecular lines of segmentation between strata

Returning to Fig. 3.1, the distinction between molar and molecular lines of segmentation in the categories C_3, C_4, C_5, and C_6 presents a complex problematic of understanding the relationship and influence between local and global systems. Molar lines of segmentation, represented by C_3 and C_4, are characterized by large-scale, macroscopic structures and processes. They represent the global systems, where the collective identity of the ecosystem is shaped by larger socio-political forces and the strata signify varying degrees of codification or rigidification. On the other hand, molecular lines of segmentation, represented by C_5 and C_6, are characterized by microtransformations that disrupt and recreate the cognitive landscape of the creative-experiencer and the sense-making processes in the Experience-verse. They represent the local systems, where the fluid and dynamic nature of cognitive processes and experiences are driven by the constant state of flux within the assemblage itself.

The local and global systems are not separate, but rather intertwined in a complex web of interactions and influences. The local systems are embedded within the global systems, and their dynamics are shaped by the larger socio-political forces, and structures represented by the molar lines of segmentation. At the same time, the global systems are influenced by the micro-transformations occurring within the local systems, as represented by the molecular lines of segmentation. Understanding this complex interplay between local and global systems, and the influence they exert on each other, requires a nuanced approach that takes into account both the molar and molecular lines of segmentation, and the dynamic and fluid nature of the MLXEs.

A sheaf is a mathematical concept that provides a way to organize and relate local information within a global context. It offers a framework for understanding how local systems (say, the objects of C_5) relate to and are influenced by global systems (the objects of C_3). The sheaf approach provides a framework for understanding how the local cognitive processes of the Free Energy Principle, Active Inference, and Phenomenotechnics in C_5 interact with and are influenced by the more global processes of Social Actors and Networks, Individual-Technology-Environment Relations, and Black Boxes in C_3.[1] By considering the objects of C_5 as sheaves over C_3, we can study the complex interdependencies and interactions between the elements of these categories, shedding light on the intricate dynamics of an MLXE. This approach can help us better understand how the cognitive processes of the creative-experiencer in C_5 are influenced by the broader stakeholder practices in C_3. Additionally, it may provide insights into how local cognitive processes can contribute to the larger dynamics of social actors and networks, individual-technology-environment relations, and the exploration of black boxes, thereby enriching our understanding of the complex relationships within an MLXE.

---

[1] In Finkler's (2004) study of the Mexican healthcare delivery system, with its blend of biomedicine and traditional healing practice, the global (molar) structures of biomedicine, characterized by standardized practices, technologies, and terminologies, are confronted with local (molecular) realities, including traditional beliefs, economic constraints, and social norms, which leads to a hybrid form of medical practice that integrates global biomedical approaches with local cultural understandings and practices. The Free Energy Principle and Active Inference are manifested as local healthcare systems seek to minimize the discrepancy between global medical standards and local practices by updating their beliefs and practices based on local needs and cultural contexts. Phenomenotechnics further elucidates the transformation of global medical technology and knowledge through its integration with local socio-cultural phenomena, leading to a unique, hybrid form of healthcare delivery. This dynamic exchange between global influences and local adaptations underscores the interconnectedness of social actor-networks and individual-technology-environment relations, revealing the "black boxes"—complex, hidden factors that shape healthcare practices at the local level while being influenced by global trends.

Relating global transformations to local transformations

To develop a sheaf approach for understanding the interaction between, say, local processes in C_5 and global processes in C_3:

1. Define a topological space (plane of immanence) by identifying all possible configurations (assemblages) in C_3 as the underlying structure on which sheaves are built.
2. Define a base for the topology as the collection of open sets (regions of space where certain interactions are dominant) in the topological space that generates the topology.
3. Define a presheaf, which is a functor from the category of open sets of the topological space in C_3 to the category of objects in C_5, with inclusions as morphisms, and specify how these objects "restrict" or "localize" as we move from one open set to another.
4. Define a sheaf by ensuring that the presheaf satisfies certain "locality" and "gluing" conditions, that is, global data can be restricted to get local data and local data can be consistently pieced together to form global data, respectively.
5. Examine the stalks defined as the direct limit of the sections of the sheaf over all open sets containing a point X representing all the local information, that is, experiences and understandings of different communities of practice come together to form a coherent picture of the ecosystem as a whole, and this global picture is informed by and reflects the local realities at each point in stakeholder practices.[2]

The topological space and the base referred to in Step 1 and 2 are the same as discussed earlier in Chapter 3. In defining a presheaf in Step 3, we assign to each open set in the topological space an object from C_5, which represents a particular way of interpreting the interactions within that set. For instance, consider the open set U, which represents the Collaborative SocialActorNetworks & Technology-Dependent IndividualTechnologyEnvironmentRelations in the context of breast cancer prevention, diagnosis, and treatment (please see "Healthcare Case: Details of Category Theory Analysis" in the Appendix and Fig. 1 therein). To this open set, we might assign the object Free Energy Principle from C_5 (see Fig. 7.1). This assignment reflects the idea that the interactions within U, such as the collaboration between healthcare professionals, patients, and caregivers facilitated by technology, can be understood in terms of minimizing discrepancies between internal world models (such as predictive models of disease progression) and actual sensory data (such as patient symptoms and diagnostic test results).

Now, consider a subset V of U, representing a more specific set of interactions within U, such as the use of mammography technology for early detection of breast cancer. The inclusion V ⊆

---

[2] Finkler's (2004) study highlights how global biomedical practices, as part of a broader "plane of immanence," become locally reinterpreted within the specific cultural and institutional context of Mexico. Here, the "open sets" could be seen as the various domains within the Mexican healthcare system, where the interactions between global biomedical knowledge and local cultural practices become most apparent, generating a topology that supports the emergence of localized medical practices. Global biomedical knowledge is adapted and applied within local contexts, i.e., a presheaf, becoming a sheaf if and when global information restricts to local data (locality) and local understandings can be pieced together to inform a coherent global strategy (gluing). The "stalks" represent the localized interpretations and applications of global biomedical practices at specific points within the Mexican healthcare ecosystem.

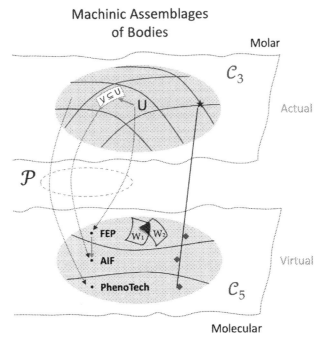

FIGURE 7.1   Relating global transformations to local transformations.

U is a morphism in the category of open sets, reflecting the fact that V is part of U in the topological structure (shown by the red arrow in the Molar stratum). However, as we move from U to V, the interpretation of interactions might change, reflecting the more specific context of V. This is where restriction maps come into play. We might assign to V the object Active Inference from $\mathbf{C}_5$, reflecting a shift in focus toward actively seeking out information (such as mammography results) to improve internal models (such as predictive models of breast cancer). The restriction map $\mathscr{P}(U) \to \mathscr{P}(V)$ is a morphism in $\mathbf{C}_5$ that captures this shift in interpretation as we move from U to V (shown by the other red arrow in the Molecular stratum). Thus, the presheaf, as a functor, not only assigns objects to open sets but also provides a way to track how these objects "restrict" or "localize" as we move from one open set to another.

The presheaf $\mathscr{P}$ can be seen as a mapping of the rhizomatic connections between different territories in the plane of immanence, assigning to each territory a particular way of making sense of the interactions (see Fig. 7.1). The morphisms in the presheaf, which include both the morphisms assigned to inclusions and the restriction maps, capture the transformations of these perspectives as we move from one territory to another. This reflects the dynamic and fluid nature of the plane of immanence, and the diverse perspectives and understandings that emerge from the deterritorialization and reterritorialization processes.[3]

In step 4, we check whether the presheaf behaves well with respect to the topology of the base category $\mathbf{C}_3$, that is, presheaf satisfies the sheaf condition, which consists of two parts: the locality condition and the gluing condition. In the context of a digitalized health ecosystem, the sheaf condition would ensure that our understanding of the interactions and processes at the global level (across the entire ecosystem) is consistent with our understanding at the local level (within individual communities of practice). It would ensure that we can "piece together" the experiences and understandings from different communities of practice to form a coherent picture of the entire ecosystem (locality condition), and that we can "break down" our understanding of the entire ecosystem to understand what is happening within individual communities of practice, with these local understandings agreeing on overlaps (gluing condition).

- The locality condition ensures that if local interpretations of interactions agree on overlaps, there is a unique global interpretation that restricts to this local data. For U = Open Set 3 (Collaborative SocialActorNetworks & Technology-Dependent IndividualTechnologyEnvironmentRelations), consider two local settings: W_1 (hospital setting) and W_2 (telemedicine setting). In W_1, healthcare professionals might use the Free Energy Principle to guide decision-making in the diagnosis and treatment of breast

---

[3] In Finkler's (2004) study, presheaf can be used to understand how different configurations of social actors' networks, technology-environment relations, and the transparency or opacity of medical technologies (black boxes) influence the localization of global health practices. Through inclusion morphisms and restriction maps, the presheaf captures the dynamic translation of global medical knowledge into local, culturally resonant practices, illustrating the fluid interplay between global influences and local realities in healthcare delivery. We can consider three plausible presheaves:

- When hierarchical social structures dominate and there's a heavy reliance on technology for diagnosis and treatment (Open Set 1 in $\mathbf{C}_3$), global biomedical practices are localized through blending technology-dependent relations with traditional hierarchical social structures to adapt treatments to individual patient contexts, as healthcare professionals actively infer patient needs and treatment outcomes based on the available technological inputs and established hierarchical protocols (Active Inference object in $\mathbf{C}_5$).
- When collaborative networks among social actors are emphasized and there's still a dependency on technology for managing healthcare (Open Set 3 in $\mathbf{C}_3$), adaptation of global biomedicine to local contexts by blending different knowledge bases and practices minimizes uncertainty between expected healthcare outcomes and actual patient experiences (Free Energy Principle object in $\mathbf{C}_5$), ensuring culturally sensitive healthcare delivery.
- When collaborative social actors' networks interact with transparent black boxes, signifying technologies or methodologies whose workings are open and accessible to all healthcare stakeholders (Open Set 7 in $\mathbf{C}_3$), the blend of global medical technologies and local traditional healing practices is openly discussed, understood, and integrated into the healthcare delivery process, as both healthcare professionals and patients engage with and understand the technologies at play in a creative experiential process (Phenomenotechnics object in $\mathbf{C}_5$), fostering a shared understanding and collaborative innovation.

cancer, constantly adapting and learning in response to new patient data, research findings, and technological advancements. In W_2, the same principle guides decision-making in remote patient monitoring and virtual consultations, adapting strategies based on patient feedback and remote monitoring data. If these local interpretations align with the Free Energy Principle, the locality condition ensures a unique global interpretation for U that restricts to the Free Energy Principle in both W_1 and W_2. This could represent a global strategy for breast cancer prevention, diagnosis, and treatment applicable across different settings and technologies.

- The gluing condition ensures that a global interpretation can be restricted to give local interpretations on any open cover, and these local interpretations will agree on overlaps. Suppose we have a global interpretation for U consistent with the Free Energy Principle, representing a global strategy for breast cancer prevention, diagnosis, and treatment that might involve a combination of regular mammograms, genetic testing for high-risk individuals, personalized treatment plans based on genetic and lifestyle factors, and remote patient monitoring and follow-up care for patients undergoing treatment. The gluing condition ensures this global interpretation can be restricted to give local interpretations in W_1 and W_2, both consistent with the Free Energy Principle and agreeing on overlaps. This represents how the global strategy adapts to specific needs and constraints of different settings, maintaining consistency with the Free Energy Principle and ensuring coherence across different settings and technologies.

Two points are in order. First, however counterintuitive it might sound, both conditions involve restriction (from global to local) and gluing (from local to global), but they emphasize different aspects of this process: the locality condition focuses on the gluing aspect (forming a global section from local data), while the gluing condition focuses on the restriction aspect (breaking down a global section into local data). Second, to verify that a presheaf satisfies the locality and gluing conditions, one needs to check the conditions for every open cover of every open set in the topological space, especially in a more formal mathematical context, or if the behavior of the presheaf is complicated or unpredictable. However, in practice, especially in applied contexts, it is often sufficient to check the condition for a representative selection of open sets and their covers.[4]

---

[4] In the context of Finkler's (2004) study, the presheaf constructed from Category C_3 to C_5 transitions into a sheaf if global biomedical practices, when adapted to the local Mexican context, remain coherent and can be pieced together from localized cultural interpretations, by satisfying:

- Locality Condition: If local interpretations of biomedicine that incorporate traditional healing practices in overlapping communities within Mexico agree, say, on the importance of understanding patient's cultural background in treatment, there's a coherent, global approach to integrating biomedicine with traditional practices.
- Gluing Condition: We can take localized practices of biomedicine integration from various communities and piece them together to form a coherent, global strategy for the healthcare system, reflecting both the universal principles of biomedicine and the localized, culturally-specific interpretations and practices of traditional healing.

In Step 5, examining the stalks over the points in C_3 provides insights into how the local processes in C_5 (represented by the Free Energy Principle, Active Inference, and Phenomenotechnics) interact with the global processes in C_3 (represented by the different types of interactions in the healthcare system). A stalk over a point in the topological space, i.e., a specific scenario or situation within the ecosystem, represents all the information from the sheaf that is "visible" or understood by the different communities of practice in C_5. It is formed by taking the direct limit (also known as the colimit) of the sections over all open sets containing the point, which essentially "glues" together the sections over different open sets to form a coherent whole. For example, examining the stalk over a point representing a patient's consultation could reveal how the Free Energy Principle (representing the patient's adaptation and learning), Active Inference (representing the healthcare professional's decision-making), and Phenomenotechnics (representing the co-creation of the healthcare experience) all come together to shape the outcome of the consultation.[5]

The preceding sheaf approach also serves as a framework for understanding how the local processes of sense-making, series and structure, and poiesis and enchantment in C_6 interact with and are influenced by the more global processes of adaptive structuration, digital niche construction, and the co-evolution of technology and organizational practices in C_4. This approach can help us better understand how the generative and transformative processes of experience creation and consumption in C_6 are influenced by the broader co-evolutionary processes in C_4.

For instance, for the open set representing interactions associated with the AI and Data Analytics Field, where AI is being welcomed and incorporated into healthcare practices, driving

---

[5] Using Finkler's (2004) study, let us illustrate the concept of forming a stalk at a point X: the treatment of a common ailment like diabetes, which can be approached both through biomedicine (medication, technology) and traditional practices (dietary advice, herbal remedies). We consider open sets in C_3 that reflect different aspects of healthcare delivery and interaction:

- Open Set 1, representing the structured, biomedical approach to treating diabetes, including the use of medications and monitoring technologies within a hierarchically organized healthcare system, is assigned to Active Inference, reflecting the active engagement of healthcare professionals and patients in managing diabetes through technology and medication.

- Open Set 7, representing a more collaborative healthcare approach, integrating traditional healing practices with biomedicine, making the process transparent to patients, is assigned to object Phenomenotechnics, symbolizing the integration of traditional knowledge and practices with biomedicine, enhancing patient experience and outcomes.

To form the stalk at point X, we take the direct limit of the sections over all open sets containing the point, where sections represent specific healthcare practices or interpretations within each open set, e.g., for Open Set 1, its section could detail the regimen of insulin therapy combined with lifestyle advice based on biomedical guidelines, and for Open Set 7, its section might describe an integrated treatment plan that includes both the biomedical approach and traditional practices like herbal remedies known to benefit diabetes management. The direct limit of these sections, forming the stalk at point X, represents a unified, comprehensive approach to diabetes treatment that is both globally informed by biomedical practices and locally adapted to include traditional healing.

co-evolution of technology and organizational practices, the presheaf might assign an object that represents a particular way of understanding this change at the molecular (local) level, as a process of organization and structuring of experiences, that is, "SeriesStructure." This means that within this set of interactions, the prevalent way of making sense of the change is through identifying a series of steps or stages. The presheaf also includes morphisms that capture how the interpretation changes when we move from one open set to another. For instance, if U represents all interactions where receptiveness of AI and Data Analytics drives changes in organizational practices, and V represents a subset of interactions where AI specifically drives changes in healthcare practice, $\mathscr{P}(V \subseteq U)$ could be the morphism from SeriesStructure to PoiesisEnchantment in $\mathbf{C}_6$, indicating that the structuring of experiences represented by V contributes to the creation and transformation of meaningful experiences represented by U.

Examining the stalks of $\mathbf{C}_4$ allows us to study all the different ways in which a specific aspect of the healthcare ecosystem (such as the use of AI in healthcare or the focus on patient-centered care) could be understood and experienced across the healthcare ecosystem, taking into account the perspectives of different communities of practice and the influence of different technologies and organizational practices. For example, for open set U from above, we can compare and contrast how the point (assemblage) of "AI being incorporated into healthcare practices" (such as the use of AI for personalized medicine) vis-a-vis the point (assemblage) "Analyzing population-based data using AI" (such as use of AI to analyze large datasets for insights into population health) manifest processes of sense-making, structuring of experiences, and creation of enchanting experiences in specific ways, and yet how the experiences and understandings of different communities of practice come together to form a coherent picture of the healthcare ecosystem as a whole.

Transformations along stacked strata: Sheaf morphisms

Although the sheaf approach by itself provides a rigorous mathematical framework to study the relationships between local (molecular) and global (molar) processes at a fairly general level, we can further elaborate this approach to conceptualize Deleuze and Guattari's concepts of induction, transduction, and translation, through which these molar and molecular lines interact and influence each other. Induction is a dynamic process, where a preindividual field of potentials is actualized into a structured individual entity, for example, global potentials in the healthcare ecosystem are developed into specific local processes in different communities of practice. Translation is a process characterized by a shift in scale or context, where a local entity is transformed as it is transferred or reinterpreted in a broader, global context, taking on new meanings or functions, for example, the adoption of a specific clinical practice by other communities of practice within the ecosystem, or the scaling up of a local technological innovation to be used across the ecosystem. Transduction is a dynamic process of structuration where a change in one part of a system leads to changes in other parts, creating a new, structured whole, for example, transformation and propagation of an activity across a domain, where the activity at one location occasion the activity at the next, from within the system, rather than being imposed from outside.[6]

The processes of induction, translation, and transduction can be modeled by "sheaf morphisms," which are mappings between two sheaves that respect the sheaf structure (see Fig. 7.2). In other words, a sheaf morphism transforms one sheaf into another in a way that preserves the relationships between the objects and morphisms in each sheaf. Induction can be modeled by sheaf morphisms that map global processes in the topological space (C_3) to local processes in the category of objects (C_5). This embodies the concept of induction as a process where global, overarching patterns or principles are inferred from specific, local observations or data.[7] Translation, on the other hand, can be modeled by sheaf morphisms that map local processes in C_5 to global processes in C_3. For translation, this represents the process by which local insights or findings are expanded or applied to the global context.[8] Transduction can be modeled by sheaf morphisms that map between different global processes in C_3 or different local processes in C_5. For transduction, this illustrates the process where information or signals undergo conversion from one form to another, occurring either within the global context (between different global processes) or within the local context (between different local processes).[9]

We propose the following as a framework for understanding the interaction between local processes in C_5 and global processes in C_3, effectively modeling the induction, transduction, and translation processes in ecosystems:

1. Define the topological spaces for both the molar and molecular levels, capturing the essential features and relationships of each level.

---

[6] Studies from the sociology of health/medicine literature can be used to illustrate how induction, translation, and transduction operate within the complex assemblages of healthcare innovation, policy-making, and cultural adaptation through the nuanced interplay between local and global processes in the medical/health geography domain. As an empirical illustration of induction, Wu's (2012) examination of multiple-embryo transfer regulation in Taiwan's IVF policy showcases how global standards and practices in assisted reproductive technology (ART) are adapted to fit local regulatory frameworks and societal norms. Exemplifying processes of translation, the work of Spicer et al. (2014) on scaling up health innovations in Ethiopia, India, and Nigeria discussed how local health innovations, effective in improving maternal and neonatal health, are positioned for broader adoption and integration into national policies and practices. As an empirical depiction of transduction, Foley's (2014) study on the global diffusion of the Roman-Irish bath from its origins in Ireland to a worldwide phenomenon highlights how a local hydrotherapeutic innovation influenced and transformed health and wellness practices globally, facilitating new forms of therapeutic assemblages.

[7] Wu (2012) depicts induction through the selection and adaptation of global forms (e.g., American guidelines, European trends) to create a specific regulatory approach for ART in Taiwan. This reflects the sheaf morphism mapping global processes (the various international ART guidelines) to local processes (Taiwan's unique adaptation and implementation), signifying the actualization of global potentials into structured local entities.

[8] Scaling-up process of health innovations discussed by Spicer et al. (2014) requires navigating between local realities (e.g., community uptake, government adoption) and global ambitions (e.g., donor support, international recognition), mirroring the sheaf morphism that transforms local processes into global phenomena. The translation is vividly demonstrated as local health interventions are reinterpreted and applied across wider geographic and cultural contexts, thereby acquiring new meanings and functions.

[9] The movement and adaptation of the Roman-Irish bath concept through various socio-material networks, as documented in Foley (2014), demonstrate the dynamic structuration inherent in transduction. Foley's narrative reveals how changes in one locale (the invention and cultural embedding of the bath in Ireland) lead to modifications and new structured wholes globally, illustrating sheaf morphisms that map transformations within the system rather than imposing changes from outside.

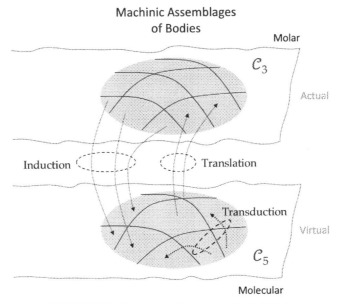

FIGURE 7.2 Transformations along stacked strata.

2. Specify the local systems (molecular lines) in one topological space (e.g., the space associated with C_5) and the global systems (molar lines) in another topological space (e.g., the space associated with C_3).
3. Define the sheaves over each topological space, representing the data or properties associated with the local and global systems.
4. Use the concept of sheaf morphisms to model the process of induction, transduction, and translation between the local and global systems. Sheaf morphisms represent how local data or properties are affected by or influence the global structures, allowing you to capture the interplay between molar and molecular lines.
5. Investigate the properties of these sheaf morphisms to gain insights into the dynamics and interactions between the local (molecular) and global (molar) processes.

For Steps 1–3, we have already seen how a topological space, its associated open sets and base sets, and a sheaf can be specified in C_3 (and they can be defined in an analogous fashion for C_5). The process of defining the sheaves involves specifying the open sets in each topological space, defining a base for these open sets, and making plausible assignments of the sheaf to each open set. This process allows us to track the interactions between the local and global processes in the healthcare ecosystem, and to understand how these interactions influence each other. In the global (molar) level, the sheaf $\mathscr{P}$ is defined over the topological space C_3, with open sets representing different scenarios or states in the healthcare ecosystem. Each open set is associated with a particular field in the healthcare ecosystem, and the sheaf maps each open set to an object in C_5 that represents a local process in that field. In the local (molecular) level, the sheaf $\mathscr{Q}$ is defined over the topological space C_5, with open sets representing different processes in different communities of practice. Each open set is associated with a particular community, and the sheaf maps each open set to an object in C_3 that represents a global process in the healthcare ecosystem.

For Steps 4 and 5, we can define the sheaf morphisms implicated in processes of induction, translation, and transduction. In Deleuze and Guattari's philosophy, **induction** is a process where a global potential is actualized into a local process. In our context, this can be modeled as a sheaf morphism from $\mathscr{P}$ to $\mathscr{Q}$. This sheaf morphism consists of a collection of morphisms $\mathscr{P}(U) \to \mathscr{Q}(V)$ for each open set U in $\mathbf{C}_3$ and V in $\mathbf{C}_5$, such that for every inclusion of open sets $U' \subseteq U$ and $V' \subseteq V$, the diagram commutes. This ensures that the sheaf morphism respects the topological structure of the ecosystem and represents a meaningful process of induction.[10] Going in the opposite direction, **translation** is a process where a local entity is transformed as it is transferred to a global context. In our context, this can be modeled as a sheaf morphism from $\mathscr{Q}$ to $\mathscr{P}$. This sheaf morphism consists of a collection of morphisms $\mathscr{Q}(V) \to \mathscr{P}(U)$ for each open set V in $\mathbf{C}_5$ and U in $\mathbf{C}_3$, such that for every inclusion of open sets $U' \subseteq U$ and $V' \subseteq V$, the diagram commutes.[11] In the lateral direction, **transduction** is a process where a

---

[10] In Wu's (2012) analysis of the regulatory trajectory of multiple-embryo transfer, the global space encompasses various international regulatory models and ethical guidelines (such as those from the UK, the US, and European countries), while the local space is defined by Taiwan's unique socio-political, medical, and cultural context. The sheaves over these spaces can be seen as the specific regulatory practices and ethical considerations pertinent to IVF technologies within each jurisdiction. The transformation of global regulatory models into locally adapted regulations in Taiwan exemplifies the concept of sheaf morphisms, representing the induction process between the local and global systems. The selection of global standards (e.g., the UK's ethical guidelines, the American guideline, and the European trend) and their adaptation or resistance within the local Taiwanese context (e.g., adding one more embryo to the American guideline or setting a specific limit in the Human Reproduction Law) can be seen as instances of induction where global potentials are actualized into local processes. Wu highlights how global standards do not simply impose themselves on local contexts; rather, they undergo a complex process of negotiation, adaptation, and sometimes transformation based on local needs, stakeholder power dynamics, and specific socio-cultural contexts.

[11] In the context of Spicer et al. (2014), the local systems at the molecular level ($\mathbf{C}_5$) are the innovative maternal and newborn health (MNH) interventions being piloted in Ethiopia, India, and Nigeria, whereas the global systems at the molar level ($\mathbf{C}_3$) refer to the policies, strategies, and scaling-up frameworks developed by governments and international partners to enhance maternal and neonatal health outcomes on a larger scale. For $\mathbf{C}_3$, a sheaf could represent the collection of all potential policy and scaling-up strategies applicable to MNH interventions globally. For $\mathbf{C}_5$, a sheaf could encapsulate the diverse local practices, innovations, and their adaptations within each targeted community. The translation process is embodied by sheaf morphisms that map the local adaptations and successes of MNH innovations to inform and shape global health policies and scaling-up strategies. Examining the properties of these sheaf morphisms could reveal how local community engagement, the involvement of government in program design, and the use of evidence to advocate for policy change are critical for successful translation from local successes to global policies. Each step taken by implementers—from designing scalable innovations, planning for scale-up, advocating for government adoption, to promoting community acceptance—serves as a microcosm of the translation process.

local (global) process influences another local (global) process. In our context, this can be modeled as a sheaf morphism from $\mathscr{C}$ to $\mathscr{C}$. This sheaf morphism consists of a collection of morphisms $\mathscr{C}(U) \rightarrow \mathscr{C}(V)$ for each pair of open sets U and V in $\mathbf{C}_5$, such that for every inclusion of open sets $U' \subseteq U$ and $V' \subseteq V$, the diagram commutes.[12]

Recall that the Open Set Collaborative SocialActorNetworks & Technology-Dependent IndividualTechnologyEnvironmentRelations in $\mathbf{C}_3$, which can be associated with the Healthcare Technology Innovators Field, is mapped by presheaf $\mathscr{P}$ to the object Free Energy Principle in $\mathbf{C}_5$, reflecting the constant adaptation and learning as mechanisms for reducing discrepancies and improving the accuracy of internal models, in response to new technologies and collaborations. Turning to $\mathbf{C}_5$, we have Open Set Entropy-Reducing FreeEnergyPrinciple & Data-Driven ActiveInference, which can be associated with the Predictive Analytics Community, where the entropy-reducing Free Energy Principle is a key aspect of predictive analytics, as healthcare professionals interpret and make sense of complex medical data to make clinical decisions, and the data-driven Active Inference represents the standardized analytical methods that guide these decisions. The object in $\mathbf{C}_3$ that this open set is mapped by the presheaf $\mathscr{C}$ can be IndividualTechnologyEnvironmentRelations, as the process of predictive analytics involves individuals interacting with technology and their environment to reduce uncertainty and make sense of complex medical data.

A sheaf morphism from $\mathscr{P}$ to $\mathscr{C}$ representing the process of **induction** can be interpreted as follows: The global potential of FreeEnergyPrinciple in the Healthcare Technology Innovators Field, represented by $\mathscr{P}$(Collaborative SocialActorNetworks & Technology-Dependent IndividualTechnologyEnvironmentRelations), is actualized into the local process of IndividualTechnologyEnvironmentRelations in the Predictive Analytics Community, represented by $\mathscr{C}$(Entropy-Reducing FreeEnergyPrinciple & Data-Driven ActiveInference). Here, FreeEnergyPrinciple represents a global potential in the healthcare ecosystem. It is a potential for healthcare professionals and patients to minimize discrepancies between their internal world models and actual sensory data, facilitated by advanced technologies. This potential exists at the level of the entire ecosystem, but it can be actualized in different ways in different communities of practice. On the other hand, IndividualTechnologyEnvironmentRelations represents a specific process within the Predictive Analytics Community. It is a process where individuals interact with technology and their environment to interpret and make sense of complex medical data. Thus, the process of induction, in this context, could be understood as the actualization of the global potential of FreeEnergyPrinciple within the Healthcare

---

[12] Foley's (2014) study presents the Roman-Irish Bath as a therapeutic innovation that initially emerged from a specific locale but eventually diffused globally. This diffusion mirrors the establishment of topological spaces at both molar and molecular levels, where the molar level represents the global spread and acceptance of hydrotherapy practices, and the molecular level reflects the localized innovations and adaptations that made the Roman-Irish Bath relevant and effective across diverse contexts. In Foley's narrative, the sheaves could be interpreted as the various attributes and properties of the Roman-Irish Bath that were relevant at different points in its diffusion—ranging from its medical efficacy and cultural significance to its architectural design, as multiple layers of meaning and utility that are context-dependent. Roman-Irish Bath's evolution and adaptation over time and space illustrate sheaf morphisms, which model how local adaptations of hydrotherapy (e.g., modifications to the bath's design to include steam or adapt to different cultural contexts) influenced and were integrated into the broader global practice of hydrotherapy. Examining these historical pathways can reveal the mechanisms through which healthcare practices evolve, adapt, and integrate into different cultural and medical contexts, driven by the continuous interplay between innovation, adaptation, and adoption.

Technology Innovators Field into the local process of IndividualTechnologyEnvironmentRelations within the Predictive Analytics Community. This could happen, for example, when a new predictive analytics technology is introduced into the Predictive Analytics Community that enables healthcare professionals to minimize discrepancies between their internal world models and actual sensory data, leading to changes in their data interpretation practices.

The commutative diagram ensures that the induction process is consistent across different levels of abstraction, from global to local and vice versa. In the context of the healthcare ecosystem, this means that the global potential of the Free Energy Principle within the Healthcare Technology Innovators Field can be consistently actualized into the local process of IndividualTechnologyEnvironmentRelations within the Predictive Analytics Community, regardless of the specific path taken: (1) Direct Induction: This is where the global potential is directly actualized into a local process. For example, a new predictive analytics technology could be introduced into the Predictive Analytics Community. This technology, based on the Free Energy Principle, could enable healthcare professionals to minimize discrepancies between their internal world models and actual sensory data. This could lead to changes in their data interpretation practices, enhancing the accuracy and efficiency of their predictions. (2) Indirect Induction: In this scenario, the global potential is first actualized in a different local context within the Healthcare Technology Innovators Field, specifically the Telemedicine Implementation Community. Here, a new approach to delivering healthcare services remotely might be developed that reduces discrepancies between patients' internal world models and actual sensory data, aligning with the Free Energy Principle. This approach could involve new ways of communicating with patients, understanding their experiences, and integrating their feedback into care plans remotely. Once this approach has been developed and refined within the Telemedicine Implementation Community, the insights and methodologies could then be transferred to the Predictive Analytics Community, influencing how predictive analytics are used in this community.[13]

A sheaf morphism from $\mathscr{E}$ to $\mathscr{P}$ represents the process of **translation** where the local process of IndividualTechnologyEnvironmentRelations is transformed as it is transferred to the global context of FreeEnergyPrinciple. Consider a local entity, which is a specific clinical practice using

---

[13] Returning to Wu's (2012) analysis of IVF policy-making in Taiwan, in the global context, the sheaf $\mathscr{P}$ over the topological space C_3 symbolizes the overarching policy frameworks and societal norms governing IVF practices, where each open set represents different scenarios or states in the healthcare ecosystem related to IVF. This global perspective encompasses the entire healthcare ecosystem's potential responses to IVF, offering insight into how these global potentials are actualized into local processes within the field of reproductive healthcare. On the local level, the sheaf $\mathscr{E}$ over C_5 represents the specific IVF practices, protocols, and community responses within Taiwan. Each open set corresponds to different local processes in various communities of practice, such as medical professionals, patient advocacy groups, and individual patients navigating the IVF landscape, illustrating how the global policies are interpreted, adapted, and enacted at the community level. The sheaf morphism from $\mathscr{P}$ to $\mathscr{E}$ highlights how, for example, the global potential of improving access to IVF services, as part of a broader reproductive health policy framework represented by $\mathscr{P}$, is actualized into specific local processes such as the development of patient-centric IVF protocols, public health campaigns, and insurance coverage adjustments within the Taiwanese healthcare system, represented by $\mathscr{E}$. Regarding the application of commutative diagram to identify direct versus indirect paths of induction, the consideration of the British model in the 1990s and the American guidelines in the early 2000s reflects an attempt to directly actualize global standards within the local Taiwanese IVF regulatory framework. However, the unique local context—such as the absence of pressure from religious or bioethical groups that was present in Britain or the affinity toward the American model due to technological admiration and training lineage—resulted in the adaptation rather than direct adoption of these models.

a data-driven active inference method for predicting patient outcomes within a specific hospital (Base Set Predictive Analytics Community). This practice has been developed and refined within this local context, and it involves specific routines and procedures (Open Set Entropy-Reducing FreeEnergyPrinciple & Data-Driven ActiveInference). Now, let us consider the process of translation, where this local entity is transferred to a global context. This could involve integrating the insights gained from this predictive analytics practice into a broader healthcare technology innovation field (Base Set Healthcare Technology Innovators Field). As this practice is transferred to this broader context, it needs to be adapted to fit different settings and circumstances. This could involve sharing the insights and methodologies with other healthcare professionals in the field, adapting the method to work with different patient populations and healthcare systems and developing new protocols for integrating the method into existing workflows. This adaptation and learning process is not just about understanding the technical aspects of the method but also about interpreting and making sense of the new types of data that the method provides and understanding how this data can inform clinical decisions. This could involve interpreting complex patterns in the data, understanding the implications of different types of results, and making decisions about how to act on this information. This aligns with the Free Energy Principle's focus on minimizing discrepancies between internal world models and actual sensory data.

The commutative diagram ensures that the translation process is consistent across different levels of abstraction, from local to global and vice versa. In the context of the healthcare ecosystem, this means that the insights and methodologies derived from the local practice of using a data-driven active inference method for predicting patient outcomes can be consistently integrated into the broader healthcare technology innovation field, regardless of the specific path taken: (1) direct translation, whereby the local practice is directly integrated into the broader healthcare technology innovation field, for example, presenting the insights and methodologies at a conference or publishing them in a peer-reviewed journal, where they can be accessed by healthcare professionals from different settings; (2) indirect translation, in which the local practice is first adapted for use in a different local context, such as another hospital or a different patient population, and then, the adapted practice is integrated into the broader healthcare technology innovation field.[14]

To motivate **transduction**, in addition to the previously discussed Open Set in C_5, Entropy-Reducing FreeEnergyPrinciple & Data-Driven ActiveInference (EFDA), representing the local process of IndividualTechnologyEnvironmentRelations in the Predictive Analytics Community, by which healthcare professionals interpret and make sense of complex medical data to make clinical decisions to recommend a specific treatment plan, such as surgery,

---

[14] In Spicer et al.'s (2014) study of the scale-up of health innovations in Ethiopia, India, and Nigeria, the translation process can be seen in how a local entity such as a new approach to community health worker training in Ethiopia (representing an open set in C_5) could be adapted and integrated into national health worker training programs (an open set in C_3). The sheaf morphism reflects this process, ensuring that the adaptation maintains the core insights and effectiveness of the original innovation while being applicable across a broader context. The adaptation and learning process Spicer et al. describe can be directly related to the sheaf morphism's function of ensuring consistency across levels of abstraction. The commutative diagram in the sheaf theory ensures that the translation process from local practices to global integration is coherent and maintains the integrity of the original innovation, whether through direct translation (e.g., specific innovations are directly advocated for and adopted by policy-makers) or indirect translation (e.g., innovations influence broader health system strengthening efforts and policy formulations indirectly through the demonstration of success, evidence synthesis, and stakeholder engagement).

chemotherapy, or radiation therapy, let us bring in another Open Set Entropy-Reducing Free-EnergyPrinciple & Direct-Experience Phenomenotechnics (EFDEP), which represents the local process of SocialActorNetworks in the Patient-Centered Care Community, for example, healthcare professionals using direct-experience phenomenotechnics and the entropy-reducing Free Energy Principle to interpret and understand patient experiences to deliver personalized care based on a patient's unique experiences and preferences as an instance of SocialActorNetworks. The sheaf morphism from $\mathscr{C}$(EFDA) to $\mathscr{C}$(EFDEP) represents the process of transduction where the local process of IndividualTechnologyEnvironmentRelations influences the local process of SocialActorNetworks. This could occur in several ways: (1) healthcare professionals increasingly recognize the importance of personalized medicine in treatment, this could drive the development of new social networks, for example, patient support groups or personalized care teams, for patient-centered care. (2) Healthcare professionals' feedback on the effectiveness and limitations of current predictive analytics tools could inspire the development of more personalized and patient-centered care practices. (3) The Patient-Centered Care Community could adapt its social networks to accommodate the Predictive Analytics Community's evolving insights from healthcare data.

The commutative diagram ensures that the transduction process is consistent across different levels of abstraction, from one local process to another. This means that the insights and methodologies developed within the Predictive Analytics Community can be consistently transferred to the Patient-Centered Care Community regardless of the specific path taken: (1) Direct Transduction: This is where the local process directly influences another. For example, healthcare professionals in the Predictive Analytics Community could directly influence the Patient-Centered Care Community by sharing their insights and methodologies. This could lead to changes in the way patient-centered care is delivered, such as the development of new patient support groups or personalized care teams. (2) Indirect Transduction: In this scenario, the local process first influences a different local context within C_5, and then these insights and methodologies are transferred to the Patient-Centered Care Community. For instance, feedback from healthcare professionals in the Predictive Analytics Community could inspire the development of more personalized and patient-centered care practices in another community of practice within C_5, such as the Telemedicine Implementation Community. These adapted practices could then be transferred to the Patient-Centered Care Community, influencing how patient-centered care is delivered in this community.[15]

As with the sheaf approach in general terms, we can use the methodology involving sheaf morphisms to model the process of induction, transduction, and translation so as to

---

[15] In Foley's (2014) exploration of the Roman-Irish Bath's historical diffusion, the transformation of the bath's concept from purely therapeutic to incorporating social and leisure aspects in some cultures demonstrates how local processes influence one another. Similarly, the feedback loop between the efficacy of hydrotherapy practices and their evolution over time exemplifies the sheaf morphism from local processes back to the global understanding of hydrotherapy, enriching it with new insights and adaptations. Direct Transduction in the context of the Roman-Irish Bath can be understood through the process where a local innovation—the development of a novel therapeutic practice—directly influenced broader medical and wellness communities across different geographies, aligning with broader health trends focused on natural and holistic treatments. Indirect Transduction can also be observed in how the Roman-Irish Bath through the feedback loop between the innovation's effectiveness and its societal acceptance encouraged the development of new social networks around health and wellness, such as spa and hydrotherapy resorts, which catered to a growing interest in personal well-being and natural treatments.

understand how changes in local processes in C_6 (like SenseMaking or SeriesStructure) influence global processes in C_4 (like AdaptiveStructuration or DigitalNicheConstruction), and vice versa.

Let the open set, or territory of assemblages, U (Decentralized AdaptiveStructuration & Technology-Receptive DigitalNicheConstruction) in C_4, represent the global potential of SenseMaking in the Remote Healthcare Delivery Field, where patients and healthcare professionals share in decision-making facilitated by technology, and bottom-up technology changes are accepted and integrated into a healthcare niche. Also let the open set V (Data-Driven SenseMaking & Routine-Based SeriesStructure) in C_6 represent the local process of AdaptiveStructuration in the Clinical Decision-Making Community, where healthcare professionals interpret and make sense of complex medical data to make clinical decisions with standardized care pathways and protocols guiding these decisions. Then, the sheaf morphism from $\mathscr{P}(U)$ to $\mathscr{Q}(V)$ represents the process of **induction** where the global potential of SenseMaking is actualized into the local process of AdaptiveStructuration. An example for this would be a new technology being introduced into the Clinical Decision-Making Community that enables healthcare professionals to make better sense of complex medical information, leading to changes in their decision-making practices.

Next, if we take open set V in C_6, which represents the local process of AdaptiveStructuration in the Clinical Decision-Making Community, and the open set U in C_4, which represents the global potential of SenseMaking in the Remote Healthcare Delivery Field, the sheaf morphism from $\mathscr{Q}(V)$ to $\mathscr{P}(U)$ represents the process of **translation** where the local process of AdaptiveStructuration is transformed as it is transferred to the global context of SenseMaking. For example, when the practice of using a new AI-driven diagnostic tool for early detection of breast cancer within a specific hospital is scaled up to be used across multiple hospitals or even nationwide, it needs to be adapted to fit different settings and circumstances which will involve training healthcare professionals in different locations to use the tool, adapting the tool to work with different IT systems, and developing new protocols for integrating the tool into existing workflows.

Now, in addition to the open set V in C_6, which represents the local process of AdaptiveStructuration in the Clinical Decision-Making Community, let the open set V' (Data-Driven SenseMaking & Adaptive SeriesStructure) in C_6, where interpreting and understanding data generated by advanced technologies goes hand in hand with practices and processes constantly being adapted in response to new technologies, represent the local process of DigitalNicheConstruction in the Technological Innovation Community. Then, the sheaf morphism from $\mathscr{Q}(V)$ to $\mathscr{Q}(V')$ represents the process of **transduction** where the local process of AdaptiveStructuration influences the local process of DigitalNicheConstruction. This can be seen in healthcare ecosystems, as clinicians increasingly recognize the importance of personalized medicine in breast cancer treatment, this drives the development of new technologies for genetic testing and targeted therapies.

Transversality as the pathway between global and local transformations

In the previous section, we introduced transduction as a process where local practices in one community influence another, ensuring consistency across different levels of abstraction. In the healthcare context, we saw how the Predictive Analytics Community's practices, which focus on interpreting complex medical data, can influence the Patient-Centered Care Community, which emphasizes personalized care based on patient experiences, potentially leading to the development of new patient support groups or personalized care teams, either directly, with insights and methodologies being shared, or indirectly, where practices are first adapted in another community before being transferred. Exactly how that transductive "influence" works was not fully explained, but now we will show that we can view the functors F_3 (Flows) and F_4 (Event) as a way to bridge the sheaf-based representation of the relationship between C_3 and C_5, with the additional insights provided by C_4 (collective assemblages of enunciation) as the "circuitry" that drives the process of transduction.

Following Fig. 7.3, since F_3 maps the processes in C_3 (the stakeholder) to the processes in C_4 (the ecosystem) and F_4 maps, these processes in C_4 to processes in C_5 (the creative-experiencer), by analyzing the composition of functors F_3 and F_4 ($F_4 \circ F_3$), we can establish an indirect link between C_3 and C_5, which incorporates the additional insights provided by C_4. It is crucial to recognize; however, that the composition of functors $F_4 \circ F_3$ and the presheaf $\mathcal{Q}$ operate at different levels of abstraction: $F_4 \circ F_3$ operates at the global level (objects and morphisms in categories), while $\mathcal{Q}$ operates at the local level (open sets in a topological space). As we saw earlier, a sheaf morphism from $\mathcal{Q}$ to $\mathcal{Q}$, where $\mathcal{Q}$ is a presheaf from the open sets of C_5 to C_3, consists of a collection of morphisms $\mathcal{Q}(U) \to \mathcal{Q}(V)$ for each pair of open sets U and V in C_5. For each open set U in C_5, we can consider the object $\mathcal{Q}(U)$ in C_3

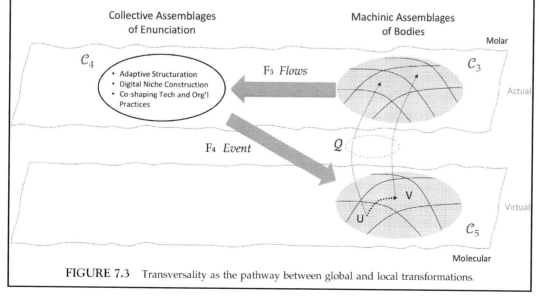

FIGURE 7.3 Transversality as the pathway between global and local transformations.

Continued

Transversality as the pathway between global and local transformations *(cont'd)*

and its image under $F_4 \circ F_3$ in $\mathbf{C}_5$. This gives us a way to connect the local processes represented by the open set U with the global processes represented by the objects and morphisms in $\mathbf{C}_4$.

In the context of Deleuze and Guattari's philosophy, the topological space of the healthcare ecosystem can be understood as a plane of immanence, a space of potentialities where each point represents a particular assemblage of technology, healthcare, and organizational change. This plane is not static but is constantly evolving, with new points (assemblages) emerging and existing ones transforming or dissolving. Let U, the Predictive Analytics Community, represent the assemblage of radiology department, predictive analytics technology, and the associated organizational changes. It is a territory within the plane of immanence where the entropy-reducing Free Energy Principle and data-driven Active Inference are prevalent. Let V, the Patient-Centered Care Community, represent the assemblage of oncology department, patient-centered care practices, and the associated organizational changes. It is another territory within the plane of immanence where the entropy-reducing Free Energy Principle and direct-experience Phenomenotechnics are prevalent. The morphism $\mathscr{O}(U) \to \mathscr{O}(V)$ in $\mathbf{C}_5$ represents the ideal patient journey from diagnosis in radiology (U) to treatment in oncology (V). This journey is not a simple linear path but a complex trajectory that traverses the plane of immanence, connecting different assemblages and communities of practice.

Transduction can be seen as a process that involves the transformation and propagation of an activity across a domain, as a process of structuring and organization that occurs from within the healthcare ecosystem. The morphisms in $\mathbf{C}_3$ and $\mathbf{C}_5$ provide pathways for these local processes to influence each other, and the mappings of the composition of functors F_3 and F_4 provide a way to integrate insights from the global level ($\mathbf{C}_4$) into these local processes. The composition of functors F_3 and F_4 ($F_4 \circ F_3$) maps the stakeholders and their interactions in $\mathbf{C}_3$ through the healthcare ecosystem in $\mathbf{C}_4$ to the patient's experiences and actions in $\mathbf{C}_5$. Through this process of transduction, insights from $\mathbf{C}_4$ can be integrated into the local processes represented by $\mathscr{O}$. This allows for the structuring and organization of the healthcare stakeholder-ecosystem from within, with morphisms in $\mathbf{C}_3$, $\mathbf{C}_4$, and $\mathbf{C}_5$ providing pathways for local processes to influence each other.

The morphism $\mathscr{O}(U) \to \mathscr{O}(V)$ is a path in the healthcare ecosystem that connects those two territories of U and V. It represents the transduction of technological acceptance and organizational change from the realm of predictive analytics to patient-centered care. This transduction is not a simple transfer but a complex process of deterritorialization and reterritorialization, where old practices are dismantled and new practices are formed. In this process, the Predictive Analytics Community (U) undergoes deterritorialization as the insights gained from predictive analytics disrupt the traditional ways of diagnosis. These insights then reterritorialize in the Patient-Centered Care Community (V), leading to the formation of new patient-centered care practices. The validation of the model involves comparing the actual patient journey with this ideal trajectory. If the actual journey matches the ideal trajectory, then the diagram commutes, and the model is validated. This would demonstrate that the composition of functors F_3 and F_4 effectively channels the process of transduction from the global level ($\mathbf{C}_3$) to the local level ($\mathbf{C}_5$), as intended. This validation process itself can be

Transversality as the pathway between global and local transformations *(cont'd)*

seen as a form of rhizomatic mapping, a concept from Deleuze and Guattari's philosophy, where the map is open and connectable in all of its dimensions and capable of being dismantled and reversed.

In the preceding subsection, we also explored the concept of **induction** within an ecosystem, using the framework of sheaf theory. We considered how global potentials, such as the Free Energy Principle within the Healthcare Technology Innovators Field from C_3 or SenseMaking in the Remote Healthcare Delivery Field from C_4, can be actualized into local processes, such as IndividualTechnologyEnvironmentRelations within the Predictive Analytics Community from C_5 or AdaptiveStructuration in the Clinical Decision-Making Community from C_6, respectively. Although, this can occur directly, with a new technology or approach being introduced into a community, or indirectly, with insights and methodologies first being developed and refined in another field, we have not elaborated on how this inductive "actualization" comes about. Now, we will briefly explain how to view the functors F_4 (Event) and F_5 (Sense) as a way to bridge the sheaf-based representation of the relationship between C_4 and C_6, with the additional insights provided by C_5 (machinic assemblages of bodies) as the "machinery" that drives the process of induction.

Since F_4 maps the processes in C_4 (the ecosystem) to the processes in C_5 (the creative-experiencer) and F_5 maps these processes in C_5 to the processes in C_6 (the experience-verse), by analyzing the composition of functors F_4 and F_5 (i.e., $F_5 \circ F_4$), we can establish an indirect link between C_4 and C_6, which incorporates the additional insights provided by C_5. Let's elaborate on the sheaf-based representation of the relationship between C_4 and C_6, in conjunction with Deleuze and Guattari's discussion of molar and molecular lines. In particular, C_4 represents the collective assemblage of enunciation, where different strata signify varying degrees of codification or rigidification. F_4, representing the Event, signifies lines of flight or processes of escape from existing structures. In C_5, the concepts of the free energy principle, active inference, and phenomenotechnics represent the cognitive processes. These cognitive processes, or "creative events," are territorialized as Sense in their respective Experience-verses. In C_6, the Experience-verse is perceived as a collective assemblage of enunciation. In other words, the sheaf morphism from $\mathscr{P}$ to $\mathscr{C}$, through the composition of functors F_4 and F_5, drives the process of induction, enabling the transformation of global potentials into local entities.

Consider the open set U (Decentralized AdaptiveStructuration & Technology-Receptive DigitalNicheConstruction) in Remote Healthcare Delivery Field from C_4, e.g., a woman actively uses a digital health platform to monitor her breast health and report symptoms, and the platform, in response, facilitates remote consultation and diagnostic tests; and the open set V (Data-Driven SenseMaking & Routine-Based SeriesStructure) in Clinical Decision-Making Community from C_6, e.g., healthcare professionals interpret biopsy results to diagnose a woman with breast cancer and devise a personalized treatment plan following standardized care pathways. The process of induction ($\mathscr{P}(U) \to \mathscr{C}(V)$) represents how the potential for SenseMaking is actualized into the process of AdaptiveStructuration as we move from the global context of the Remote Healthcare Delivery Field to the local context of the Clinical Decision-Making Community, e.g., healthcare professionals adapt their decision-making practices in response to new information from AI-based tools and changing circumstances, such as patient's health status and preferences. The processes in

Continued

Transversality as the pathway between global and local transformations *(cont'd)*

C_5, through the composition of functors F_4 and F_5, serve as the "machinery" that drives the process of induction, enabling the transformation of global potentials into local entities. The integration of digital mammography into routine screening practices, a form of AdaptiveStructuration, aligns with the FreeEnergyPrinciple in C_5, where healthcare professionals minimize discrepancies between their understanding of a patient's health and actual mammography data, which then maps to SenseMaking in C_6, illustrating how professionals interpret mammography images to diagnose breast cancer.[16]

---

[16] In Foley's (2014) study of the diffusion of the Roman-Irish Bath, we see a clear example of transversality—the process by which local innovations (e.g., the specific architectural and therapeutic innovations introduced by Richard Barter) influence broader practices within the wellness and healthcare communities. The circuitry, represented by the interactions and discourses that facilitated the bath's diffusion, and the machinery, seen in the cultural and societal adoption and adaptation of the baths, together illustrate the complex processes of transduction and induction that drive changes in healthcare practices and societal health trends.

- The concept of "circuitry," involving the functors F_3 (Flows) and F_4 (Event) as bridges in the sheaf-based representation, can be likened to the network of influences and interactions that enabled the Roman-Irish Bath's diffusion. The initial proponents of the Roman-Irish Bath, such as Dr. Richard Barter, interacted with broader medical and wellness communities, sharing insights and promoting the health benefits of these baths. This interaction represents the mapping of processes from C_3 (stakeholders) to C_4 (the ecosystem), where the idea of the bath moved from individual proponents to a collective discourse. As the concept gained traction, it influenced the practices of healthcare professionals and institutions (C_5, the creative-experiencer), who integrated the Roman-Irish Bath into their offerings. The societal discourse around health and wellness provided the circuitry for this transduction, transforming individual innovations into widely adopted practices.

- The "machinery" driving the process of induction, represented by the composition of functors F_4 (Event) and F_5 (Sense), can be understood through how the Roman-Irish Bath influenced not just the medical community but also the broader cultural and societal perceptions of health and wellness. The transformation from a novel therapeutic practice to a staple of wellness culture shows how global potentials (the concept of hydrotherapy) were actualized into local entities (specific implementations of Roman-Irish Baths in various locales). The adaptations of the Roman-Irish Bath to suit local preferences and technologies represent the machinery of induction in action. Feedback from users and observations of the bath's benefits in different contexts fueled further innovation and spread, showcasing the adaptive structuration of this health practice across the experience-verse.

Diagrammatic logic of organizations in ecosystems

The previous chapter emphasized the interplay between local (molecular) and global (molar) dynamics, with sheaf theory contextualizing how local cognitive processes intertwine with global dynamics. In this chapter, we discuss how DRT offers a diagrammatic perspective to decipher the complex architecture of Machinic Life-Experience Ecosystems (MLXEs). Assemblages within categories resemble smooth manifolds, where manifold points depict unique configurations of entities and their interrelations. Functors interlink these categories, illustrating entity-assemblage dynamics. An organization's evolutionary path in the ecosystem is charted by a sequence of functors. Diagrams, acting as abstract maps, spotlight relationships within assemblages and extend to larger entities of organization, which as assemblage configurations navigate between their current and aspirational states. Grasping these transformations empowers enterprises to steer the MLXEs in which they are implicated organizationally toward envisioned goals.

Category Theoretical interpretation of organizations and ecosystems

Creating separate categories for entities and assemblages only to map them to each other might artificially compartmentalize the system, potentially obscuring the dynamic interrelationships and transformations that occur across different levels of the system. Instead, we focus on how the functors in our framework capture the processes by which entities come together to form assemblages—the essence of "*agencements*" or *agencial* assemblages, and how these assemblages interact and transform each other. An enterprise or organization is constituted by a "path" or "patches" in the ecosystem, represented by a sequence or array of mappings.[1] In turn, if an assemblage in a category is consistently mapped to another

---

[1] A sequence is an ordered list of functors that represent a process or transformation pathway, where the order of functors matters because each functor's output becomes the input for the next. An array implies a systematic arrangement of processes, which could be ordered but not necessarily sequentially dependent.

assemblage in another category under a given functor, this indicates that both assemblages are part of the enterprise or organization (*and/or vice versa*).

Each functor represents a transformation or process that occurs within the healthcare ecosystem. Different sets of functors could represent different paths or patches in this ecosystem, corresponding to the operations and interactions of different enterprises, organizations, or communities. In the simplest scenario, one enterprise might be *primarily* involved in the transformation from individual interactions (C_1) to networked individuation (C_2), represented by the functor F_1.[2] Another enterprise might be primarily focused on the transformation from ecosystem co-evolution (C_4) to perception and cognition (C_5), represented by the functor F_4. These different sets of functors (implicated in a path or patches) would represent the different roles and activities of the enterprises within the healthcare ecosystem. This interpretation allows for a flexible and dynamic understanding of the healthcare ecosystem, where enterprises, organizations, or communities are not fixed entities but rather are defined by their interactivity, as represented by the functors. It also allows for the possibility of change and evolution, as enterprises can shift their primary focus and activities over time, corresponding to changes in their associated functor sets.

For example (see Fig. 8.1), the transformational focus of many Preventive Healthcare Organizations (PHO) is from individual behavior insights to networked healthcare services and ecosystem-wide health strategies, i.e., F_1 (Lived Journey) → F_2 (Engagement) → F_3 (Flow). In general, a PHO may employ F_1 to track individual lifestyle data related to breast cancer risk, use F_2 to connect individuals with a network of healthcare providers for community outreach, and leverage F_3 to influence broader healthcare strategies and policies. Meanwhile, Health Policy Advocacy Groups (HPAGs) dedicate a transformational focus from understanding patient and provider experiences to influencing co-innovation practices and platform policies, that is, F_4 (Event) → F_5 (Sense) → F_6 (Enactment). An HPAG may work with F_4 to analyze the impact of technological ecosystems on patient care, harness F_5 to shape the discourse around patient-centered care, and employ F_6 to promote enactment of patient-centric policies at co-innovation levels. Digital Health Startups (DHS), on the other hand, have a transformation focus from innovation and design to platform development and service implementation, that is, F_6 (Enactment) → F_7 (Resourced Capabilities) → F_8 (Interactional Creation). Typically, a DHS might use F_6 to co-create innovative apps for breast cancer self-management, apply F_7 to leverage their capabilities in creating a digital platform for patient engagement, and utilize F_8 to cycle back and refine individual user interactions with their app.[3] Last but not least, Integrated Care Networks (ICN) have a transformational focus from enhancing patient–caregiver engagement to evolving service platforms and reinforcing individual care pathways, for example, F_2 (Engagement) & F_7 (Resourced Capabilities) → F_8 (Interactional Creation).[4] An ICN might leverage F_2 to deepen

---

[2] Note that we are *not* saying an enterprise is only involved in one set of transformations, rather we want to draw attention to the empirical fact that many organizations have a *primary* focus on a particular set of transformations by virtue of strategic choice and historical legacy.

[3] Note that both HPAG and DHS have F_6 (Enactment) as part of their functor sequences, but it has totally different transformative effects since it maps different assemblages in each organization.

[4] ICN is an example where an *array* of functors can be ordered, but they are *not sequential*, thus we talk about a *patch*, instead of a *path*, of functors defining the focus of the organization.

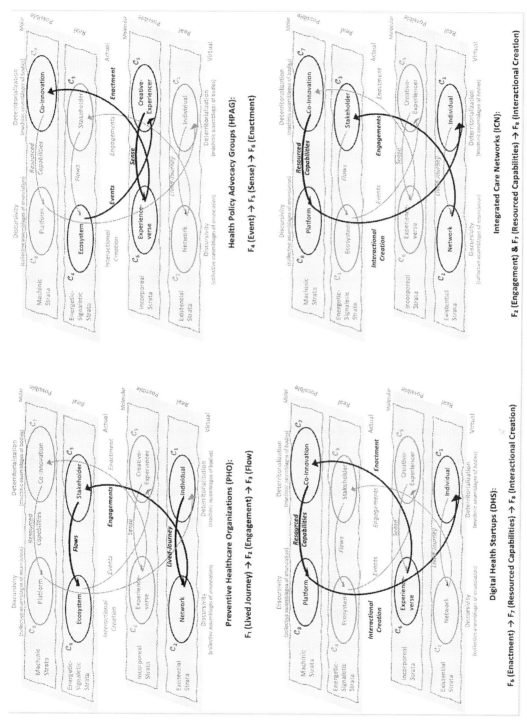

FIGURE 8.1 Category theoretical interpretation of organizations.

patient–caregiver interactions within the network, use F_7 to develop resources that support integrated care platforms, and employ F_8 to tailor individual patient experiences through enhanced interactional creation.

In each scenario, organizations use a sequence or array of functors to navigate the complex landscape of breast cancer care. The paths or patches they take through the category theoretical framework represent strategic decisions to focus on specific aspects of patient care, policy, innovation, or integrated service delivery. Each sequence or array of functors defines the roles, contributions, and evolutionary path of the organization within the MLXE.[5]

Different assemblages mapped by different instantiations of the same functor sequence could represent different enterprises, organizations, or communities. In this context, an "instantiation" of a functor sequence could be thought of as a specific way in which the functors are applied to the objects in the categories, resulting in a specific path or trajectory through the category theoretical framework. Each enterprise, organization, or community could then be seen as a unique trajectory through the framework, represented by a unique sequence of functor applications. This sequence would capture the specific ways in which the entities within the enterprise or organization interact and transform each other, as well as their relationships with the larger ecosystem. This approach allows for a high degree of flexibility and adaptability, as the functor sequences can be adjusted and reconfigured to reflect changes in the structure and dynamics of the enterprises or organizations. It also provides a powerful tool for analyzing and comparing different enterprises or organizations, by examining the similarities and differences in their functor sequences.

When different organizations focus on the same transformations but "instantiate" functor sequences in unique ways, it reflects their distinctive approaches, methodologies, and strategic priorities. Whereas PHO_1 might focus on digital platforms to engage individuals with gamified health tracking to promote early detection and preventive measures for breast cancer, PHO_2 might emphasize community health programs that use data analytics to tailor preventive care and wellness workshops to high-risk populations. Both organizations use the sequence $F_1 \to F_2 \to F_3$, but PHO_1 "instantiates" this path by creating interactive platforms for user engagement, while PHO_2 instantiates it through local community interventions that are informed by data analytics. Likewise, while DHS_1 might develop mobile applications that allow for personalized health monitoring and direct communication with healthcare providers, DHS_2 might create a virtual health assistant platform that leverages AI to guide

---

[5] In Polk et al.'s (2023) case study on Memorial Sloan Kettering Cancer Center (MSK), paths and patches illustrate the strategic and operational efforts by MSKCC to integrate precision oncology into their practices, particularly through the establishment and evolution of the Early Drug Development (EDD) service, alongside other initiatives. For example, its *path* from $F_1 \to F_3$ reflects MSK's journey from its initial engagement with genomic testing technologies, where individual patients' genomic profiles begin to impact the broader understanding of cancer treatments (F_1), through transitioning from utilizing genomic data to enhancing stakeholder relationships, including the integration of genomic data into personalized care plans (F_2), to the adaptive structuring of cancer treatment protocols based on genomic insights, influencing broader healthcare practices and policy (F_3). Similarly, the *patch* for organizational and clinical trial innovation reflects the strategic and operational reconfigurations within MSK to support precision oncology, for example, developing a tissue-agnostic approach to cancer treatment based on genomic understanding (F_5), implementation of genotype-matched clinical trials (F_6); and reconfiguration of MSK's clinical research enterprise to align with precision oncology goals (F_3).

patients through their healthcare journey. Both follow $F_6 \to F_7 \to F_8$, DHS_1 "instantiates" it by focusing on mobile solutions for individual patient use, whereas DHS_2 applies it to create a broader AI-powered platform for patient guidance. Similar considerations would apply to Health Policy Advocacy Groups and ICNs.

Enrollment and agencing in organizations and ecosystems

We can think of the "belonging" relation as being embedded in the structure of the categories and functors themselves. In this view, an entity "belongs" to an assemblage if it is part of the assemblage's structure as represented in the category. Similarly, an assemblage "belongs" to an enterprise, organization, or community if it is part of the structure of these larger entities as represented by the functors. In practical terms, this means that the "belonging" relation is not something that needs to be explicitly modeled with additional categories or functors but is something that can be inferred from the structure of the categories and functors in our framework.

A diagram in a category can be seen as a representation of an assemblage, capturing the structure of the assemblage in terms of the entities that participate in it and the relationships between these entities. The entities are represented by the objects in the diagram, and the relationships between entities are represented by the morphisms in the diagram. For example, an entity might choose to participate in an assemblage in a particular way, which could be represented by a specific morphism in the diagram.

Likewise, a diagram could represent a collection of assemblages and the functors connecting them, effectively capturing the structure of an enterprise, organization, or community within our ecosystem. The "gluing" of these diagrams together, or the way they fit into the larger structure of the ecosystem, could then be captured by the concept of a "limit" or "colimit," to represent how a structure is composed of and distributed across existing category theoretical notions. Limits and colimits are universal constructions that "summarize" or "combine" structures from multiple categories into a single object, capturing their common features and relationships. A limit is an object that represents the "best" way to combine multiple objects under a given condition, while a colimit represents the "best" way to separate or distribute multiple objects. Limits and colimits are ways of combining objects and morphisms in a category to form new objects and morphisms, and they can be used to capture the ways in which the assemblages and enterprises or organizations interact and overlap with each other.

Entities can be seen as actively participating in the diagrams and limits/colimits in the categories.

- An entity's role in an assemblage can be understood in terms of its position within the diagram and the morphisms that connect it to other entities and in terms of its contribution to global structures. For example, an entity might contribute to the limit of a diagram by participating in multiple assemblages that are part of the diagram. Similarly, an entity might contribute to the colimit of a diagram by bridging between different assemblages in the diagram.
- An entity's knowledge of the diagrams and limits/colimits in the categories can be modeled as a functor that maps from the category of entities to the category of

diagrams or limits/colimits. This functor would capture the relationship between each entity and the diagrams or limits/colimits it is aware of. However, as we saw earlier, creating separate categories for entities and assemblages might artificially compartmentalize the system. In reality, the entities and their knowledge of the diagrams and limits/colimits would likely be represented within the same categories (for example, the Active Inference object in Creative Experiencer category), with the relationships between them captured by the morphisms and functors in our system.

- In terms of agency and intentionality, entities can be seen as actively shaping the diagrams and limits/colimits in the categories. For example, an entity might choose to participate in certain assemblages and not others, thereby influencing the structure of the diagrams. Similarly, an entity's actions and decisions can influence the formation of the limits and colimits in the categories, thereby shaping the overall structure of organizational enterprises. An entity as active participant, that is, influencing the structure of the diagrams and formation of the limits and colimits in the categories, can be modeled by a natural transformation between functors, which would capture the changes in the diagrams or limits/colimits that result from the entity's actions.

Based on the foregoing considerations, the concept of **Complex Adaptive Relational Event-Sense (CARE) Architecture** presents a novel approach to understanding and structuring organizations. This framework, deeply rooted in category theory, transcends traditional models by viewing organizations not as static entities but as dynamic assemblages of various interconnected elements. CARE Architecture captures the essence of organizational life through a complex web of relationships and interactions, employing diagrams, functors, and the concepts of limits and colimits to map and understand these intricate connections. This approach allows for a comprehensive view of organizations as living, evolving systems, characterized by adaptability, responsiveness to change, and continuous evolution, positioning CARE Architecture as a pivotal tool for navigating the complexities of modern organizational dynamics.

CARE Architecture is grounded in the principles of complexity theory, relational thinking, and adaptive systems. The architecture is designed to capture the multifaceted interactions and evolutionary dynamics within organizations and between them and their broader environments. At its core, CARE Architecture views an organization not as a static entity but as a living, evolving assemblage. This assemblage is composed of various elements—individuals, networks, stakeholders, ecosystems, and more—each represented as distinct but interconnected categories within a larger categorical structure. These categories encapsulate different aspects of organizational life, ranging from individual experiences and network dynamics to broader ecosystem interactions and innovative platforms.

In CARE Architecture, relationships and interactions are key. The architecture uses category theory to map these elements, employing diagrams to represent the complex web of connections and interactions. Each element is seen both in its individual capacity and as part of the larger whole, allowing for a nuanced understanding of both micro-level and macro-level dynamics. Furthermore, CARE Architecture is adaptive and responsive. It recognizes the constant flux in organizational and environmental contexts, embracing change as a fundamental aspect of existence. The architecture is designed to accommodate and leverage this change, using it as a driver for innovation and growth.

In essence, CARE Architecture provides a sophisticated, holistic approach to understanding and designing organizations. It combines theoretical depth with practical applicability, offering a robust framework for navigating the complexities of modern organizational life and fostering environments conducive to co-innovation, resilience, and continuous evolution.[6]

Dynamic relationality #8: Diagrammatics

For Deleuze and Guattari, a diagram is an abstract, nonrepresentational "map" that captures the relationships, potentials, and forces at play within an assemblage. It is a way of thinking about and analyzing the complex network of connections and interactions between components that make up an assemblage. Diagrams can be used to trace the lines of flight or deterritorialization that allow for the emergence of new forms and possibilities within the assemblage.

Define a diagram $D: \mathcal{I} \to \mathbf{C}$.

Let $\mathcal{I}$ have eight objects ($C_1, C_2, C_3, C_4, C_5, C_6, C_7, C_8$) representing the constituents of the CARE architecture. Define the diagram D as follows:

$D(C_1) = \mathbf{C}_1$ (representing Individual)
$D(C_2) = \mathbf{C}_2$ (representing Network)
$D(C_3) = \mathbf{C}_3$ (representing Stakeholder)
$D(C_4) = \mathbf{C}_4$ (representing Ecosystem)
$D(C_5) = \mathbf{C}_5$ (representing Creative-Experiencer)
$D(C_6) = \mathbf{C}_6$ (representing Experience-verse)
$D(C_7) = \mathbf{C}_7$ (representing Co-Innovation)
$D(C_8) = \mathbf{C}_8$ (representing Platform)

The indexing category $\mathcal{I}$ provides the structure or shape of the diagram, while the target category C is the category in which the diagram resides. The objects of $\mathcal{I}$ are mapped to objects of **C** by the functor D, and the morphisms of $\mathcal{I}$ are mapped to morphisms of **C**, such that the functor preserves the composition and identity morphisms of the indexing category. A

---

[6] In Polk et al.'s (2023) study, the case of the Early Drug Development (EDD) unit at MSK can be used to illustrate the concepts of belonging, agencing, and complex adaptive relational enterprise (CARE) architecture. EDD, as an entity, being part of the Division of Solid Tumor Oncology and cutting across traditional oncological categories with a tissue-agnostic approach, belongs to MSK's larger organizational ecosystem, not merely as a matter of administrative alignment but in a deeply operational and epistemic sense. EDD oncologists, or "molecular champions," embody the agency of entities within assemblages by specializing in specific oncogenes or molecular pathways and acting as liaisons between the EDD and organ-based services, engaged not just in clinical decision-making but also in research and pharmaceutical pipelines. The operational and strategic activities of the EDD—such as crafting a tissue-agnostic service, running basket trials, and establishing working groups—are guided by a diagrammatic representation of cancer biology. The EDD's formation, the development of genomic profiling platforms, and the establishment of new centers can be understood in terms of limits and colimits, where the organization combines various elements (e.g., technological platforms, clinical expertise, research initiatives) to form a coherent response to the challenges and opportunities of precision oncology. The EDD's approach to trial prioritization, collaboration with other departments, and the design of innovative research strategies reflect the dynamics of a complex adaptive system, where the evolutionary path of the organization is charted through a sequence of transformations.

diagram can be thought of as a way to represent a pattern of relationships between objects and morphisms in a category. An indexing category serves as a template or blueprint for the diagram we want to create in the target category. The functor D maps the objects and morphisms of the indexing category to the objects and morphisms of the target category, thereby creating the desired diagram in **C** following the structure specified by $\mathcal{I}$.

The index category $\mathcal{I}$ is a separate category containing objects like C_1, C_2, C_3, etc., which represent different aspects of the CARE architecture. The diagram D: $\mathcal{I} \to$ **C** is a mapping between the index category $\mathcal{I}$ and the categories $\mathbf{C}_1$, $\mathbf{C}_2$, $\mathbf{C}_3$, etc. The diagram D maps the objects in $\mathcal{I}$, which represent the constituents of the CARE architecture, to the corresponding categories. In this mapping, the categories capture the structure and relationships between their objects and morphisms, which are related to different aspects of the MLXE (see Fig. 8.2).

C_1 is an object in the index category $\mathcal{I}$, while $\mathbf{C}_1$ is a category that captures the "Individual" aspect of the MLXE. In the context of constructing the limit object for the CARE architecture, $\mathcal{I}$ is an index category whose objects represent the constituents of the CARE architecture (and the expanded version of it). In this case, $\mathbf{C}_1$ is an object in $\mathcal{I}$ that represents the "Individual" aspect of the "expanded" $\mathbf{C}_9$—CARE architecture (**this is the target category C, we talked about earlier, where the diagram resides, here we have started numbering it**). C_1, as an object in $\mathcal{I}$, does not have its own objects like $\mathbf{C}_1$ does. Instead, C_1 is a way to index or refer to the corresponding category $\mathbf{C}_1$ in the diagram D: $\mathcal{I} \to$ **C**. In other words, the object C_1 in $\mathcal{I}$ rather serves as a way to connect or refer to $\mathbf{C}_1$ in the context of the diagram and the construction of the limit object for the CARE architecture.

On the other hand, $\mathbf{C}_1$ is a category that captures the structure and relationships between its objects and morphisms related to the individual aspect of the MLXE. The diagram D: $\mathcal{I} \to$

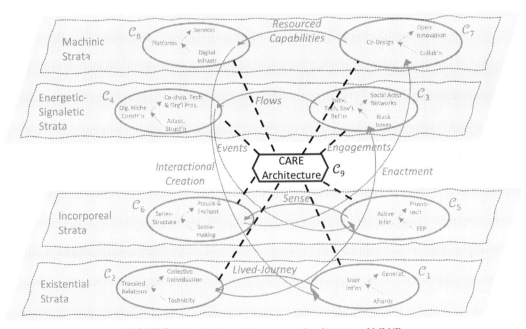

FIGURE 8.2 CARE architecture as the diagram of MLXE.

C maps the objects in $\mathcal{I}$ to the corresponding categories in **C**. In this mapping, $D(C_1)=C_1$, which means that the object C_1 in the index category $\mathcal{I}$ is associated with the category C_1 that represents the individual aspect of the MLXE.

The diagram is an abstract construction that involves mapping between the index category $\mathcal{I}$ and the categories like C_1, C_2, etc. This diagram is used to relate different categories and their structures within the context of constructing the limit object for the CARE architecture. It is not directly represented in the graphical texts and arrows. The graphical texts and arrows help us visualize the structure and relationships within individual categories, while the diagram D: $\mathcal{I} \rightarrow$ **C** is a more abstract construction that relates different categories and their structures within the context of constructing the limit object for the CARE architecture.

Category theory can be thought of as a diagrammatic practice because it often involves the use of diagrams to represent abstract mathematical concepts and relationships. Diagrams in category theory help to visualize the relationships between categories, objects, and morphisms, making it easier to grasp the underlying structure and interconnections. However, it is important to keep in mind that the diagrams in category theory are not the same as the actual mathematical constructs they represent. The diagrams are visual aids that help us to understand and reason about the abstract mathematical concepts. The actual constructs in category theory, such as categories, objects, and morphisms, are defined rigorously in terms of sets, functions, and other mathematical entities. So, while category theoretical analysis and encoding can involve diagrammatic practice as a way of visually representing and reasoning about the concepts, the diagrams themselves are not the same as the actual mathematical structures they represent.[7]

Limit/colimit object of a system of categories

A limit is a construction that, in a sense, encapsulates the "most general" object with specific properties or relationships, whereas a colimit represents the "most specific" or "least constrained" object that still maintains those properties or relationships. Both limits and colimits are defined by a universal property, which characterizes the unique relationship between the limit/colimit object and the objects in the categories involved. Suppose we have a group of people living in different cities and working for the same company. The company wants to hold a meeting with all of these employees. The limit represents the optimal meeting

---

[7] In Polk et al.'s (2023) study, the indexing category $\mathcal{I}$ organizes the conceptual architecture of the CARE model, delineating distinct aspects of healthcare innovation and practice as separate entities such as C_1 through C_8, each of which encapsulates a specific dimension of EDD's operational and strategic initiatives, from individual molecular champions (C_1) through the broader digital infrastructure platforms enabling precision oncology (C_8). The functor D: $\mathcal{I} \rightarrow$ **C** operates as the transformative mechanism that maps these conceptual entities from the index category $\mathcal{I}$ into the target category **C**, where they are instantiated in the context of MSK's EDD. This mapping elucidates how abstract theoretical constructs are practically applied to orchestrate a complex ecosystem of cancer research and treatment. Through this lens, the EDD's endeavors—from fostering interdisciplinary collaboration to implementing the MSK-IMPACT platform—are understood not merely as operational activities but as manifestations of a broader, diagrammatically structured endeavor to navigate and reshape the oncological landscape. The diagram D: $\mathcal{I} \rightarrow$ **C**, therefore, is more than a representation; it is a strategic tool that facilitates the identification of new pathways for innovation, collaboration, and patient care by highlighting the interdependencies and potential lines of flight within the ecosystem.

location to minimize travel distance, while the colimit represents a virtual meeting platform that enables communication between employees in different cities. These limits and colimits can be seen as capturing the essence of the CARE Architecture as a whole, combining the different aspects and their relationships across the existing category theoretical notions.

We can create a new category C_9 representing the CARE architecture, with each of its eight constituents as objects. To build a limit or colimit of the CARE architecture, we will consider a diagram D: $\mathcal{I} \to \mathbf{C}$, where $\mathcal{I}$ is an index category and $\mathbf{C}$ represents the categories. For each object in $\mathcal{I}$, we define a functor $L_n: \mathbf{C}_n \to \mathbf{C}_9$, with $n \in [1, 2, \ldots, 8]$. Since $\mathbf{C}_9$ represents the entire CARE architecture, each L_n functor would map the core aspects or features of its respective $\mathbf{C}_n$ to a corresponding feature or aggregation in $\mathbf{C}_9$ that embodies the properties and relationships of that aspect as it exists within the larger architecture. The functors are defined as follows:

$L_1(\mathbf{C}_1) =$ The aggregate object in $\mathbf{C}_9$ that represents the integration of individual contributions and experiences within the CARE architecture.

$L_2(\mathbf{C}_2) =$ The network object in $\mathbf{C}_9$ that encapsulates the networked relationships and interactions.

$L_3(\mathbf{C}_3) =$ The stakeholder object in $\mathbf{C}_9$ that represents the stakeholders' collective roles and influences.

$L_4(\mathbf{C}_4) =$ The ecosystem object in $\mathbf{C}_9$ that embodies the overall ecosystem structure and dynamics.

$L_5(\mathbf{C}_5) =$ The creative-experiencer object in $\mathbf{C}_9$ that encompasses the creative and experiential aspects as they manifest in the architecture.

$L_6(\mathbf{C}_6) =$ The experience-verse object in $\mathbf{C}_9$ that captures the realm of experiences and interpretations.

$L_7(\mathbf{C}_7) =$ The co-innovation object in $\mathbf{C}_9$ that synthesizes the collaborative and innovative efforts.

$L_8(\mathbf{C}_8) =$ The platform object in $\mathbf{C}_9$ that reflects the foundational platforms and infrastructures supporting the architecture.

These functors are conceptual tools that map the individual aspects of the CARE architecture as represented in categories C_1 through C_8 to their integrated form in the overarching category C_9. They are essentially the "glue" that connects the detailed structures of each aspect to the holistic view of the CARE architecture, ensuring that the contributions of each part are recognized and synthesized into the overall structure.

Limit construction (most general object)

The limit object (CARE_limit) is the most general object with morphisms from each object in C_9 to the corresponding object in the diagram D. These morphisms satisfy the universal property, which states that for any other object Z with morphisms to the objects in the diagram, there exists a unique morphism from Z to CARE_limit, making the triangles commute.

Limit objects offer a way to explore the emergent properties and behaviors of assemblages. As the universal object that best approximates the diagram's elements, a limit object captures the essence of an assemblage's dynamics, revealing how its components interact and evolve over time. In this context, limit objects can be seen as points of convergence, where the behavior and properties of an assemblage stabilize, despite the inherent complexity of its constituent elements.

Colimit construction (least constrained object)

The colimit object (CARE_colimit) is the least constrained object with morphisms from the objects in the diagram D to the corresponding objects in C_9. These morphisms satisfy the universal property, which states that for any other object Z with morphisms from the objects in the diagram, there exists a unique morphism from CARE_colimit to Z, making the triangles commute.

Colimit objects provide a way to understand the potential expansion and diversification of assemblages. As the least constrained object that "extends" or "pushes out" the diagram's elements, a colimit object captures the potential growth or evolution of an assemblage, revealing how its components could interact and evolve in new ways. In this context, colimit objects can be seen as points of divergence, where the behavior and properties of an assemblage could expand or diversify, despite the inherent complexity of its constituent elements. This makes colimit objects particularly useful for exploring the potential for innovation, adaptation, and growth within an assemblage.

Limits and colimits as universal properties

A universal mapping property is a characterization of a particular object or morphism in a category that captures its unique and "best-fitting" behavior with respect to other objects and morphisms. An object or morphism satisfying a universal mapping property is often the solution to an optimization or minimization problem in the category and is unique up to unique isomorphism. Limits and colimits are examples of universal constructions in category theory, and they can be defined using universal mapping properties. Limits and colimits are examples of universal constructions that can be defined using universal mapping properties, where the limit is the "best-fitting" object for a given diagram with respect to all other objects in a category, and the colimit is the "least-fitting" object for the same diagram.[8]

The universal property of the limit (CARE_limit) states that for any object Z in C_9 with morphisms from Z to the objects in the diagram D, there exists a unique morphism from Z to CARE_limit, making the triangles commute. The universal property of the colimit (CARE_colimit) states

---

[8] Let us return to Polk et al.'s (2023) study on EDD's operational model at MSK. The limit object in this scenario, represented by the EDD's integrative efforts, could be seen as the "most general" object that synthesizes the specific properties or relationships across the CARE architecture. This encapsulation reflects the universal property of limits, characterizing the unique and optimized relationship between the limit object (EDD's operational model) and the objects (various components of precision oncology) within the involved categories. For example, the molecular champions' role, the MSK-IMPACT platform, and the Ecosystems Project collectively form a cohesive structure that mirrors the limit object's function—harmonizing diverse inputs into a unified, functional entity that advances precision oncology. Conversely, the colimit object could be represented by the open, expansive nature of EDD's initiatives, such as its outreach to pediatric and young adult oncology, and its collaboration with preclinical or basic science units. This "least constrained" object still maintains the core properties of the CARE architecture but allows for the exploration of new directions and possibilities in cancer treatment, embodying the colimit's universal property. This showcases the EDD's ability to extend the application of precision oncology beyond traditional boundaries, reflecting a dynamic, evolving landscape that remains grounded in the foundational principles of the CARE architecture.

that for any object Z in $\mathbf{C}_9$ with morphisms from the objects in the diagram D to Z, there exists a unique morphism from CARE_colimit to Z, making the triangles commute.

Higher-order categories to represent categories and functors

CARE architecture category, as constructed through the diagram object, floats above or supervenes on $\mathbf{C}_1$ to $\mathbf{C}_8$. The diagram object D maps the objects in $\mathcal{I}$ to their respective constituents in the CARE architecture. By defining functors and natural transformations that interrelate these constituents, we effectively create a higher-level category that supervenes on the lower-level categories $\mathbf{C}_1$ to $\mathbf{C}_8$. In this context, supervenience describes the relationship between the CARE architecture and its underlying constituents ($\mathbf{C}_1$ to $\mathbf{C}_8$). The CARE architecture emerges from the complex interactions between these constituents, and any change in the underlying constituents would have an effect on the CARE architecture itself. This higher-level category helps us understand and analyze the MLXE by focusing on the interactions between its various aspects and their contributions to the overall structure and function of the system.

A higher category could be used to capture the relationships between the categories and the functors that establish a trajectory between them. The objects in this higher category would be the categories in our framework, and the morphisms would be the functors. For example, consider a 2-category. The objects in this 2-category would be the categories in our framework. The 1-morphisms would be the functors between these categories. The 2-morphisms would then be natural transformations between these functors. This structure would capture the relationships between the categories and the ways in which they are connected by the functors. In this higher category, a diagram could be defined that represents the particular configuration of functors that establish a trajectory between the categories. The limit or colimit of this diagram would then represent a particular state or configuration of the entire system. This approach would allow us to capture the complex, multilayered relationships between the entities, assemblages, and organizations in our system, as well as the dynamic processes of change and transformation that occur within and between these entities.

The higher category that considers the categories as objects and functors as morphisms is a way of capturing the relationships between the categories and the ways in which they are connected by the functors. This is a separate concept from the diagrams and limits/colimits within the individual categories. However, the diagrams and limits/colimits within the individual categories could also be considered as part of the structure of the higher category. In this sense, they could be seen as being "embedded" in the higher category. The higher category provides a way of looking at the entire system as a whole, including the relationships between the categories and the structures within the categories.

The original diagram and limit/colimit that we defined for the categories in our framework can still be useful. They provide a way to capture the structure of each category and the relationships between the objects within each category. The higher category and the new diagram that we define within it are a way to incorporate the functors into the framework and to capture the relationships between the categories. In the original diagram, the focus is on the structure of each category and the relationships between the objects within each category. In the new diagram, the focus is on the relationships between the categories and the ways in which they are connected by the functors.

The original diagram and its associated limit/colimit provide a perspective on the structure within each category and the relationships between the objects in each category. This is a view of the system at the level of individual categories. The second diagram in the higher category, along with its associated limit/colimit, provides a perspective on the relationships between the categories themselves and how they are connected by the functors. This is a view of the system at a higher level, where the focus is on the interconnections between categories. These two perspectives are complementary and both are important for understanding the dynamics of our system. The first diagram and limit/colimit capture the "micro" level dynamics within each category, while the second diagram and limit/colimit capture the "macro" level dynamics between categories.

The diagrams and limits/colimits within the individual categories could also be considered as part of the structure of the higher category. In this sense, they could be seen as being "embedded" in the higher category. The higher category provides a way of looking at the entire system as a whole, including the relationships between the categories and the structures within the categories. Where does this higher order category lie, that is, where does the CARE architecture diagram itself as a limit or colimit object category for the rest of the ecosystem reside? **It is the platform category**, and as a collective assemblage of enunciation. So, while the true CARE architecture is just inherent to the entire framework and embedded into all the objects and morphisms in all the categories, our Co-Innovation Platform strata (Abstract Machinic Phyla) is where the CARE Architecture is manifested, articulated, for everyone to see, and even if they do not see it, they are steered in their existential territories by these co-innovation platforms. The problem of strategic misalignment of limit/colimit objects (i.e., perceived CARE architecture) can be solved by a conscious design of the machinic phyla, or co-innovation platforms. The problem of strategic misalignment is raised in Chapter 10. The problem with limit/colimit object as a free-floating category is that it also does not capture the functors between the categories. So, we need a higher order category to capture both the diagrams of categories and the functors between them. And that is accomplished (successfully or not) in the strata of machinic phyla (co-innovation platforms).[9]

---

[9] In Polk et al. (2023), D: $\mathcal{I} \rightarrow \mathbf{C}_9$ maps categories that encompass various operational and strategic initiatives, from the role of molecular champions to the establishment of genomic platforms from an index category $\mathcal{I}$ into a target category $\mathbf{C}_9$, instantiating abstract theoretical constructs into the practical orchestration of MSK's EDD ecosystem. In a 2-category framework, the objects are the categories ($\mathbf{C}_1$ to $\mathbf{C}_8$), 1-morphisms are the functors (such as $F_1, F_2, ..., F_8$) that map between these categories, and 2-morphisms are natural transformations between these functors, establishing a higher-order view of the relationships and transformations within the EDD's operational model. A diagram in this 2-category context can be understood as a specific configuration of categories and functors that outline the trajectory of operations and strategic initiatives within the EDD. The limit of such a diagram would represent the cohesive structure harmonizing diverse inputs (such as molecular champions' roles, the MSK-IMPACT platform, and the Ecosystems Project) into a unified entity advancing precision oncology. Conversely, a colimit object might represent the open, expansive nature of EDD's initiatives, embodying the dynamic, evolving landscape of cancer treatment that remains grounded in foundational principles yet open to exploration and innovation. Natural transformations in this framework might include the translation of genomic insights into clinical trial designs (molecular champions acting as liaisons), the integration of diverse treatment modalities (connecting pediatric and adult oncology), and the collaborative efforts across MSK (establishing working groups for interdisciplinary research).

Organizations as convergence of diagrams and limit/colimit objects

We define an enterprise, organization, or community as a configuration of assemblages whose components converge and act on a current state and future state of a diagram of which they are part. This definition emphasizes the shared understanding and collective vision of the members of the enterprise, which aligns well with many theories of organizational behavior and strategy. In the context of our category theoretical framework, the "current state" could be represented by a particular configuration of objects and morphisms in the categories, and the "future state" could be represented by a different configuration that the enterprise is aiming to achieve. This definition also implies a process of change or transformation, as the enterprise moves from its current state to its future state. This aligns with our description of enterprises as involving "complex adaptive relational event-sense architectures." In terms of stack (sheaf) analysis, this definition could inform the definition of the gluing conditions.[10] The gluing conditions could be defined in such a way that they reflect the shared understanding and collective vision of the members of the enterprise, as well as the transformations that occur as the enterprise moves from its current state to its future state.

In the context of an enterprise, a diagram in the category theoretical model can serve as a representation of the current state of the enterprise. The diagram thus provides a snapshot of the enterprise's current configuration and the relationships and interactions that exist within it.

The limit object of the diagram can then serve as a representation of the future state of the enterprise. In category theory, a limit object is an object that best approximates or 'summarizes' the objects and morphisms in a diagram. In the context of an enterprise, the limit object can be interpreted as the "ideal" or "goal" state of the enterprise that the members are collectively working toward. It represents the state of the enterprise where all the relationships and interactions between the aspects of the enterprise are optimally aligned and coordinated (see Fig. 8.3).[11]

The operational role of the diagram and limit object is thus to provide a framework for understanding and guiding the dynamic evolution of the enterprise. The diagram provides a map of the current state of the enterprise, highlighting the areas that need to be addressed or improved. The limit object provides a vision of the future state of the enterprise, guiding the actions and decisions of the members. The process of moving from the current state to the future state involves transforming the diagram into the limit object, which requires adjusting the relationships and interactions within the enterprise to better align with the "ideal" configuration represented by the limit object.

---

[10] As we will explain further in Chapter 10, *stack* is a generalization of a sheaf that allows for "twisted" or "nontrivial" gluing conditions. Stacks can be used to model situations where the local-to-global principle of sheaves is too restrictive, such as when the local data cannot be simply "glued together" to form the global data.

[11] The arrows pointing "out" symbolize the process of converging diverse elements from multiple categories into a unified vision, aligning with the enterprise's future state. This directionality underscores the limit object's role in synthesizing inputs to guide strategic alignment and decision-making, not mere combination.

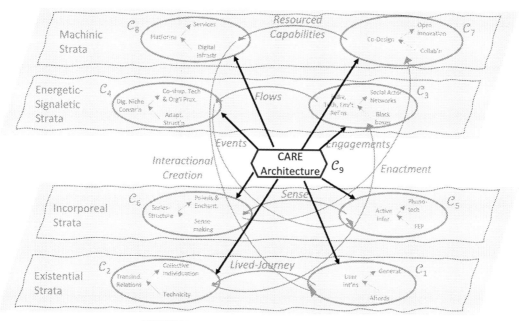

FIGURE 8.3 CARE architecture as limit object.

The colimit object of the diagram, on the other hand, can serve as a representation of the potential expansion or diversification of the enterprise. In category theory, a colimit object is an object that "extends" or "pushes out" the objects and morphisms in a diagram. In the context of an enterprise, the colimit object can be interpreted as the "growth" or "expansion" state of the enterprise that the members are collectively working toward. It represents the state of the enterprise where new relationships and interactions are being created that extend beyond the current configuration of the enterprise (see Fig. 8.4).[12]

The operational role of the diagram and colimit object is thus to provide a framework for understanding and guiding the dynamic growth of the enterprise. The diagram provides a map of the current state of the enterprise, highlighting the areas that can be expanded or diversified. The colimit object provides a vision of the expansion state of the enterprise, guiding the actions and decisions of the members. The process of moving from the current state to the expansion state involves transforming the diagram into the colimit object, which requires creating new relationships and interactions that extend beyond the current configuration of the enterprise, rather than adjusting the existing ones to better align with an "ideal" configuration. This process reflects the enterprise's orientation toward growth

---

[12] The arrows pointing "in" represent the colimit's role in facilitating the enterprise's expansion by distributing or separating elements into broader contexts. This direction supports the enterprise's growth by guiding the integration of new interactions, contrary to the intuition of separation.

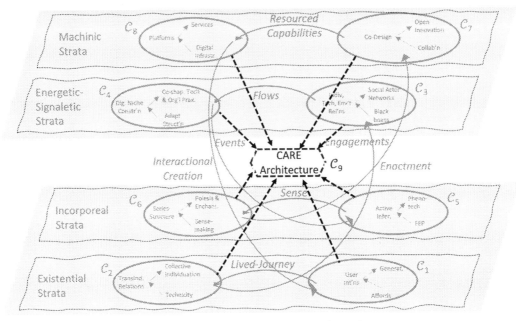

FIGURE 8.4 CARE architecture as colimit object.

and diversification and its ability to adapt and respond to new opportunities and challenges in its ecosystem.[13]

Case: Diagrammatic logic of organizations in healthcare ecosystems

The digital health ecosystem is a rich amalgam of interactions, decisions, and outcomes. By employing category theory and its diagrammatic practices, we can gain deeper insights into the structure and dynamics of this ecosystem, paving the way for more personalized and

---

[13] In the context of Polk et al.'s (2023) study, EDD is a complex assemblage, whose components converge and act on a current and future state of a diagram, as underscored by its strategic role in crafting MSK's precision oncology profile, its tissue-agnostic service, and the crafting of a genomic and bioinformatic ecosystem. The operational and strategic initiatives of EDD, such as the roles of molecular champions and the establishment of the MSK-IMPACT platform, can be seen as part of a diagram, representing the current state of the enterprise. The process of moving from the current state to the future state involves transforming the diagram toward the limit object, evident in the organization's efforts to align its operations with the goal of advancing precision oncology. Conversely, the colimit object, represented by EDD's outreach and collaborative initiatives, showcases the potential for expansion and diversification. The EDD's activities, including the formation of working groups and the management of its clinical trial portfolio, demonstrate the application of category theory's concepts of limits and colimits to understand and manage the dynamic interactions and transformations within the organization. By structuring its operations around the principles of tissue agnosticism and leveraging the capabilities of precision oncology platforms, EDD embodies the process of organizational morphogenesis, where the enterprise actively navigates and reshapes its trajectory toward envisioned goals.

effective care solutions. Let us illustrate the concepts using the context of breast cancer prevention, diagnosis, and treatment.

- **Assemblages and Functors:** In the digital health ecosystem for breast cancer care, various entities come together to form assemblages. These could be patients, healthcare providers, wearable devices, telemedicine platforms, and AI tools. The functors, like F_1 ("Lived Journey") and F_2 ("Engagements"), capture the transformation processes within this ecosystem. For instance, F_1 might represent the journey of a patient from diagnosis to treatment, facilitated by AI tools and telemedicine platforms.
- **Entities and Their Distributed Roles in the Ecosystem**: A wearable device, which is used for health monitoring, does not belong to a singular assemblage dedicated to "health monitoring tools." Instead, its function and interactions are distributed across the categories and functors. For example, the wearable device's basic functionalities, such as heart rate monitoring or step counting, are captured within C_1 (representing Complex); its ability to interact with other devices or platforms, and its role in a network of health monitoring tools is represented in C_2 (representing Relational); how data from the wearable device contributes to a patient's health journey, influencing decisions and interactions with healthcare providers is entailed by functor F_1, "Lived Journey." Likewise, a radiologist, as an entity within our category-theoretical framework, does not reside in a singular category dedicated to "breast cancer diagnosis." Instead, their expertise and interactions are distributed across multiple categories and functors. Similarly, a telemedicine platform's role and interactions are distributed across the categories and functors, rather than belonging to a specific assemblage of "remote consultation tools."
- **Intentionality and Active Participation:** A patient's decision to undergo a specific treatment, influenced by their research and consultations, can shape the trajectory of their care pathway. This decision-making process can be captured by a natural transformation between functors, representing the dynamic interplay of personal choices and medical recommendations. An oncologist, as an entity, might choose to collaborate with a nutritionist to provide holistic care to a breast cancer patient. This active decision influences the structure of the care team diagram.
- **Diagrams:** A diagram can represent the assemblage of a patient's care team, including doctors, nurses, and family members. The morphisms in this diagram capture the relationships and interactions among these entities. The diagram captures the dynamic interplay of forces within the breast cancer care ecosystem. It maps out the potentials and relationships between patients, healthcare providers, and technologies. This diagram captures the structure of the CARE architecture in the context of breast cancer care. The entire process of breast cancer prevention, diagnosis, and treatment can be visualized using category theoretical diagrams. These diagrams provide a holistic view of the patient's journey, capturing the intricate web of interactions, decisions, and outcomes.
- **CARE Architecture and Higher-Order Categories**: The CARE architecture, while inherent to the entire framework, is manifested in the Co-Innovation Platform strata. This platform serves as a collective assemblage of enunciation, guiding and steering the entities within the breast cancer care ecosystem. The Co-Innovation Platform strata is

where the CARE architecture's strategic alignment is articulated. By consciously designing this platform, we can ensure that all entities—be it healthcare professionals, patients, or diagnostic tools—are aligned in their goals and approaches to breast cancer care.

- **Limits and Colimits**: The "gluing" of diagrams, representing the integration of different care teams for comprehensive patient care, can be captured by the concept of a "limit" in the category. Limits and colimits in category theory offer a way to "summarize" or "combine" structures from multiple categories. In the context of breast cancer prevention, diagnosis, and treatment, they can represent the convergence of various diagnostic tools, treatment methods, and preventive measures. Consider a patient's journey through breast cancer care. The limit might represent the optimal combination of treatments and interventions tailored to the patient's unique needs, while the colimit could represent all potential treatment pathways available.
- **Enterprise as a Configuration of Assemblages**: A hospital specializing in breast cancer care could be seen as a trajectory through the ecosystem, represented by a sequence of functorial mappings. This sequence captures the hospital's unique approach to patient care, from prevention to post-treatment support. The limit object represents the convergence of the best practices and interventions for breast cancer care. In contrast, the colimit object showcases the potential diversification of treatments and approaches, allowing for innovation and adaptation.

CHAPTER 9

Organizational morphology and development

In the preceding chapter, we explored the diagrammatic logic of organizations within ecosystems, depicting organizations as assemblages interconnected by functors to illustrate evolutionary paths, and introduced CARE Architecture, emphasizing adaptability and evolution through diagrams showing intricate relationships and strategies toward goals, concluding with discussions on diagrams, limits, and colimits in representing organizational states. In this chapter, we present a category theory-based framework for analyzing organizational morphology, focusing on the stability of operational processes (stable vs. variable) and strategic orientation (goal-oriented vs. expansion-focused) while also exploring organizational morphogenesis through category theory concepts like diagrams, limits/colimits, and gauge transformations to shed light on organizational change dynamics.

A category theoretical typology of organizational morphology

In this section, we introduce a framework for understanding organizational morphology through the lens of category theory, particularly focusing on the stability of the functor set and the emphasis on limit or colimit objects. The former dimension refers to whether the set of functors (processes or transformations) that an organization uses to navigate its ecosystem is stable or variable. The latter dimension refers to whether an organization is oriented toward achieving a future state that includes new relationships and interactions (limit object focus) or toward expansion and diversification (colimit object focus). This approach transcends traditional organizational analysis paradigms by embedding the fluid, interconnected nature of modern enterprises within a rigorous, mathematical structure that offers both clarity and depth in analyzing organizational dynamics.

Stability of the Functor Set: The stability of the functor set dimension is paramount as it captures the essence of an organization's operational consistency and adaptability. Stable functor sets signify organizations that rely on established, consistent processes, reflecting a traditional, hierarchical structure aimed at maintaining control and predictability. This stability is not merely a reflection of rigidity but signifies an organization's commitment to proven strategies that ensure reliability and efficiency. On the other hand, variable functor sets characterize organizations that exhibit flexibility and innovation, adapting their operational processes in response to the ever-changing business environment. This adaptability is crucial in today's fast-paced world, where technological advancements and shifting market dynamics demand a dynamic approach to organizational strategy and operations. The ability to modify processes and strategies, encapsulated in the variable functor set, is indicative of an entrepreneurial spirit that drives growth and innovation.[1]

- Organizations with stable functors use established, consistent processes. They are often traditional and hierarchical, focusing on maintaining stability and control. These organizations have a fixed set of transformations that they apply consistently. The specific set of functors being used at any given moment, and the sequence or array in which they are applied, represent a specific "instantiation" of the functor sequence. This "instantiation" remains relatively constant over time, reflecting the organization's commitment to stability and control. For example, an enterprise primarily involved in the transformation from individual interactions to ecosystem co-evolution, represented by a set of specific functors, would consistently apply these transformations, maintaining a stable functor set. Examples might include traditional corporations with well-defined roles and structures. Since structure refers to the established, consistent patterns or frameworks that guide behavior and interactions within a system, predominance of stable functor sets in an organization implies privileging of "Structure."
- Organizations with variable functors are adaptable, changing their processes in response to environmental shifts. They are often innovative and entrepreneurial, focusing on learning and adaptation. These organizations have a flexible set of functors that they can draw from, and they can change the specific functors they use, or the sequence or array in which they are applied, as needed. Each time they make such a change, they create a new "instantiation" of the functor sequence. For instance, an enterprise might shift its focus *from* transforming individual interactions to ecosystem co-evolution *to* transforming ecosystem co-evolution to perception and cognition, represented by different functors. This shift would represent a new "instantiation" of the functor

---

[1] The stability of the functor set within organizations embodies a theoretical interplay between operational consistency and strategic adaptability, essential for navigating dynamic environments. Helfat and Winter (2011) delve into the distinction between operational and dynamic capabilities, suggesting that the stable versus variable nature of functor sets may reflect an organization's ability to manage and adapt to continuous change. Felin et al. (2012) underscore the importance of microfoundations, including individual actions and organizational structures, in shaping routines and capabilities, implying that stability and variability in functor sets are foundational to organizational agility and performance. Teece et al. (2016) further refine this discussion by linking dynamic capabilities to organizational agility, emphasizing the strategic necessity of balancing stability with the capacity to reconfigure operational processes in response to uncertainty and innovation demands. Therefore, stability in functor sets does not imply a resistance to change but rather a structured approach to leveraging proven processes, whereas variability signifies a strategic openness to reconfiguration and innovation. This balance is critical for organizations aiming to sustain competitiveness in rapidly evolving markets, where the ability to pivot operations and strategy becomes a pivotal factor in long-term success. Effective management of this balance—through leveraging both stable and variable functor sets—can serve as a core competency, enabling organizations to navigate the complexities of modern business landscapes with agility and foresight.

sequence or array, reflecting the organization's adaptability. Examples might include startups or innovative companies that are constantly experimenting with new business models and strategies. Since agency refers to the capacity of individuals or groups to act independently and make their own free choices, prevalence of variable functor sets in an organization implies privileging of "Agency."

Focus on Limit or Colimit Objects: Equally important is the focus on limit or colimit objects, which delves into the organization's developmental trajectory concerning its future state and expansion potential. The limit object focus directs attention toward organizations striving to achieve a specific "ideal" state through the integration and optimal alignment of relationships and interactions. This future-oriented perspective emphasizes goal-directed behavior, where every organizational activity is aligned toward achieving a coherent, predefined outcome. Conversely, the colimit object focus encapsulates organizations inclined toward exploration and diversification, aiming to extend their operational and strategic boundaries beyond the current state. This orientation is indicative of an organization's desire to explore new opportunities, foster innovation, and adapt to emergent challenges, thereby ensuring long-term sustainability and relevance in a constantly evolving ecosystem. While both limit object focus and colimit object focus involve creating new relationships and interactions, the key difference is whether these are oriented toward a specific "ideal" state (limit object focus) or toward open-ended expansion and diversification (colimit object focus). This distinction is independent of the stability of the functor set, which refers to whether the processes or transformations that the organization uses to navigate its ecosystem are stable or variable.[2]

- Organizations with a limit object focus are oriented toward achieving a future state that includes new relationships and interactions. This future state, represented by the limit object, is seen as an "ideal" configuration that the organization strives to achieve. This does not necessarily mean only optimizing the current state, but rather it involves a transformation toward a state where all the relationships and interactions between the aspects of the organization are optimally aligned and coordinated. This could involve creating new relationships and interactions, but the key point is that these are all oriented toward a specific, predefined "ideal" state. Since a corporation typically has a

---

[2] The focus on limit or colimit objects in organizational strategy underscores the nuanced balance between pursuing a well-defined, ideal future state and embracing open-ended exploration for growth and innovation. This duality mirrors the theoretical frameworks of exploration and exploitation, where organizations navigate the tension between leveraging existing competencies and venturing into new territories to foster long-term resilience and competitiveness (Lavie et al., 2010). The strategic orientation toward limit objects aligns with the pursuit of coherence and alignment within an organization's existing framework, aiming for an optimized configuration of relationships and interactions that embody an envisioned ideal state. This approach is reflective of a theory-based view of strategy, where firms are posited as economic actors formulating and testing hypotheses about value creation within a structured, goal-oriented framework (Felin and Zenger, 2017). Conversely, the emphasis on colimit objects represents a strategic inclination toward diversification and expansion, resonating with dynamic managerial capabilities that enable firms to modify their resource base and operational strategies in response to evolving environmental dynamics (Helfat and Martin, 2015). Together, these perspectives encapsulate a comprehensive approach to strategic management, where the deliberate balance between stability and adaptability, informed by a deep understanding of the organization's internal and external ecosystems, guides the pursuit of strategic objectives.

specific goal or "ideal" state that it is working toward, an organizational form where limit object focus dominates can be called a "Corporation."

- Organizations with a colimit object focus, on the other hand, are oriented toward expansion and diversification. They aim to create new relationships and interactions that extend beyond the current configuration, represented by the colimit object. Unlike the limit object focus, the colimit object focus does not have a specific "ideal" state that the organization is striving to achieve. Instead, the organization is open to exploring a variety of possible future states, and the goal is to expand and diversify the organization's configuration in ways that can adapt to changes and opportunities in the environment. Since a collective often focuses on expansion and diversification, creating new relationships and interactions that extend beyond the current configuration, an organizational form where colimit object focus is prevalent implies a "Collective."[3]

Overall, these dimensions allow for a nuanced categorization of organizations that goes beyond superficial traits, delving into the fundamental operational and strategic orientations that define an organization's essence. By employing these dimensions, DRT offers a comprehensive framework that captures the complex interplay between an organization's operational consistency, adaptability, goal orientation, and expansionary aspirations. Moreover, this category-theoretical approach to organizational morphology facilitates a deeper understanding of how organizations navigate their ecosystems, transform over time, and how they can strategically architect their evolution. It provides a robust analytical tool for dissecting the intricate mechanisms underlying organizational behavior and strategy, thus offering valuable insights for both scholars and practitioners aiming to decode the complexities of organizational dynamics in the contemporary landscape.

Thus, organizations can be located on a 2-dimensional space with Stable/Variable Functor Set and Limit/Colimit Object Focus as the two dimensions (See Table 9.1). The ideal types of

TABLE 9.1 A category theoretical typology of organizational morphology.

	Limit object focus	Colimit object focus
Stable functor set	Structur*ed* Corporation (e.g., traditional corporations)	Structur*ing* Collective (e.g., open-source communities)
Variable functor set	Agenc*ed* Corporation (e.g., startup companies)	Agenc*ing* Collective (e.g., entrepreneurial networks)

---

[3] Coleman's (1990) social theory, which offers insights into the social structures shaping behavior and societal outcomes, blending individual actions with collective dynamics, distinguishes corporate actors (corporations) from collectives (collectivity). Corporate actors are institutional entities like corporations and trade unions, capable of collective decision-making and societal influence. Collectives, in contrast, are decentralized, communal organizations with shared interests or values, operating through less formal structures. Our choice of corporation versus collective to refer to the ideal types of organizational forms with limit object focus versus colimit object focus is inspired, but *not* determined, by Coleman's terminology.

Structured Corporation and Agencing Collective can be situated in opposite corners of this two-dimensional space and associated with traditional corporations (with a strong focus on strategic planning and long-term goals) and entrepreneurial networks (or communities of practice), respectively. Based on these dimensions, we can identify four types of organizations, which we will discuss in the following sections.

Structured corporations as limit object focused organizations with stable functor sets

Structured Corporations with a Stable Functor Set and Limit Object Focus operate with a consistent set of processes and aim toward a specific "ideal" state.[4] This "ideal" state, represented by the limit object, is often a shared vision or goal, and the focus is on integration and coherence. These organizations are characterized by centralized control mechanisms and a focus on optimization and alignment of activities. Examples include traditional corporations with a strong focus on strategic planning and long-term goals.[5]

- A large pharmaceutical company like Pfizer or Roche consistently applies functors such as F_1 (Lived Journey), which involves the development of patient-focused treatment plans, and F_4 (Event), which could represent structured clinical trials leading to the development of new medications. They aim to achieve an "ideal" state, such as bringing a new breast cancer drug to market that optimally aligns with regulatory standards and maximizes patient outcomes (See Fig. 9.1).

Despite the stability in the processes used, these organizations are oriented toward achieving a future state that includes new relationships and interactions. This future state is not just an optimization of the current state but a state that includes new relationships and interactions that align and coordinate the activities to achieve an "ideal" configuration. This balance between stability and orientation toward a future state is a key characteristic of many successful organizations.

The "-ed" suffix in Structured Corporation communicates multiple potential implications: (1) presence of a stable functor set predetermines the organization's pursuit of a specific limit object or "ideal" state; (2) corporation's focus on achieving a particular limit object necessitates the establishment of a stable functor set; (3) stable processes provide a reliable

---

[4] Structured Corporations, with their centralized control and focus on achieving an ideal future state through stable processes, resonate with Deleuze and Guattari's (1977, 1987) concepts of the Despotic Machine and State Apparatus of Capture. This parallel is seen in the organization's emphasis on strategic planning and long-term goals, mirroring the State Apparatus's pursuit of an organized, hierarchical order that captures and directs social flows toward a centralized vision or "ideal" state. The balance between stability and pursuit of a future state reflects the dynamic between despotic control and the desire to shape society.

[5] Integrated care in the emergency department (ED), as discussed in Nugus et al.'s (2010) study (see Chapter 3), provides an illustration. The ED operates within a structured, yet highly adaptive environment, requiring centralized control mechanisms to manage the dynamic and complex interactions between various departments and external organizations. The focus on optimizing and aligning activities to achieve integrated care despite the inherent unpredictability mirrors the characteristics of Structured Corporations, where there is an emphasis on integration and coherence within a defined operational framework.

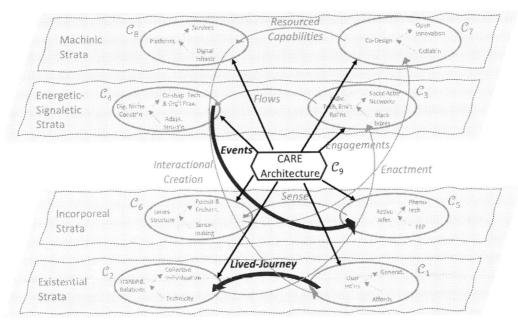

FIGURE 9.1 Structured corporations with a stable functor set and limit object focus.

foundation for pursuing the "ideal" state, while the pursuit of this state continuously refines and reinforces the stability of the operational processes; (4) an evolutionary strategy where the organization's structure and its objectives are adaptations to competitive pressures, market demands, and other environmental factors; (5) a deliberate design or architecture within the corporation that accommodates dynamic strategic planning and goal orientation within a stable operational framework.

Agenced corporations as limit object focused organizations with variable functor sets

Agenced Corporations with a Variable Functor Set and Limit Object Focus are adaptable, innovative, and entrepreneurial.[6] They constantly experiment with new strategies, models, and approaches to navigate their ecosystem, adjusting to external factors such as market trends, customer needs, and technological advancements. Despite the variability in their processes, these organizations aim toward a specific "ideal" future state,

---

[6] Agenced Corporations, characterized by their adaptability and entrepreneurial spirit, can be likened to Deleuze and Guattari's (1977) concept of Capitalist Machines. The constant experimentation with new strategies and adaptation to market trends and customer needs reflects the capitalist machine's focus on continuous innovation and economic expansion. Despite the variability in processes, the aim toward an "ideal" future state mirrors the capitalist machine's relentless pursuit of growth and transformation. This balance between flexibility and strategic direction exemplifies the capitalist machine's dynamic nature and its drive toward reshaping its environment.

represented by the limit object. This state includes new relationships and interactions that align and coordinate their activities to achieve an "ideal" configuration, guided by the organization's strategic goal or vision. In practice, these organizations might aim to enter new markets, develop new products, or establish new partnerships. They might also implement new business models or operational processes that align with their strategic objectives. Examples of such organizations include startups or innovative companies with a clear goal or "ideal" state.[7]

- A biotech startup such as Guardant Health utilizes F_3 (Flow) to adaptively respond to new research in diagnostic technologies and F_7 (Resourced Capabilities) to secure funding and partnerships for the development of innovative testing methods. They are driven to achieve a future state where their new diagnostic tool is integrated into standard care practices, optimizing early detection of breast cancer (See Fig. 9.2).

These organizations are dynamic and adaptive, effectively navigating their ecosystem while continuously evolving and innovating. This balance between flexibility and orientation toward a future state is a key characteristic of many successful organizations.

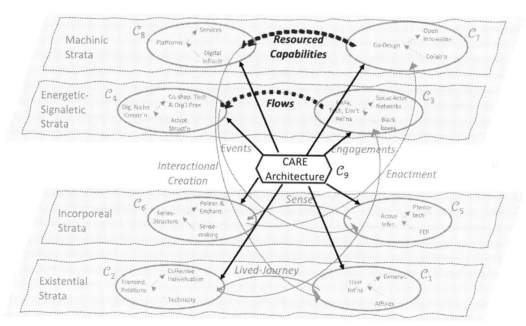

FIGURE 9.2 Agenc*ed* corporations with a variable functor set and limit object focus.

---

[7] Organizing precision medicine at MSK Cancer Center, as discussed in Polk et al.'s (2023) study (see previous section in this chapter), is a good example. MSK's engagement with genomics and precision oncology highlights an organization that is adaptable, innovative, and continuously evolving. Their effort to build an infrastructure for genomic practices and create conditions for actionable results aligns with the characteristics of Agenc*ed* Corporations, which prioritize adaptability and orientation toward a specific "ideal" future state through innovative strategies and models.

The "-ed" suffix in Agen*ced* Corporation signifies a number of possibilities: (1) variable nature of the functor set allows the corporation to explore diverse strategies, models, and approaches, thereby discovering and refining its "ideal" state through experimentation and adaptation; (2) corporation's strategic vision actively molds its agency, dictating the need for a variable set of processes to achieve its objectives; (3) an ongoing, iterative process of becoming, where agency and goal orientation are in constant dialog; and (4) successful application of agency within the bounds of strategic vision, where adaptability serves the pursuit of the "ideal" state without compromising the coherence of the organizational vision.

Structur*ing* collectives as colimit object focused organizations with stable functor sets

Structur*ing* Collectives with a Stable Functor Set and Colimit Object Focus are communal and decentralized.[8] They have established traditions and practices, represented by a stable set of processes or transformations. However, these organizations are oriented toward expansion and diversification, aiming to create new relationships and interactions that extend beyond their current configuration. This might manifest in various ways, such as forming new relationships and alliances, exploring new territories or resources, or incorporating new ideas or practices into their traditional ways of life. Examples of such organizations might include grassroots community organizations, cooperatives, or open-source software communities.[9]

- A patient advocacy group like the Breast Cancer Research Foundation regularly uses F_2 (Engagement) to involve patients and survivors in awareness campaigns and F_6 (Enactment) to implement community support programs. They foster new relationships and interactions, like partnering with diverse stakeholders to address the broader determinants of health and influence policy for better breast cancer care (See Fig. 9.3).

These organizations are dynamic and adaptive, able to navigate their ecosystem effectively while preserving their core identity and values. This balance between stability and openness to change is a key characteristic of many successful organizations.

The "-ing" suffix in Structur*ing* Collective suggests several potential interpretations: (1) inherent structure facilitates a controlled expansion, ensuring that while the collective seeks growth, it remains grounded in its established practices and traditions; (2) pursuit of new

---

[8] Structur*ing* Collectives, with their communal and decentralized nature, show similarities to Deleuze and Guattari's (1987) concept of Tribal Machines. Both are characterized by established traditions and practices, with a stable set of processes signifying a collective identity. The orientation toward expansion and diversification, as seen in forming new relationships and incorporating new practices, echoes the tribal machine's fluidity and openness to external influences. This balance between maintaining core traditions and openness to change reflects the tribal machine's adaptability while preserving a distinct communal identity.

[9] Multisited therapeutic assemblages for youth mental health support, as discussed in Trnka's (2021) study (see Chapter 3), can serve as an example. These collectives are described as communal, decentralized, and focused on expansion and diversification, aiming to create new relationships and interactions beyond their current configuration. The emphasis on forming new alliances and exploring new territories or resources mirrors the dynamic and adaptive nature of Structur*ing* Collectives that navigate their ecosystem effectively while preserving their core identity and values.

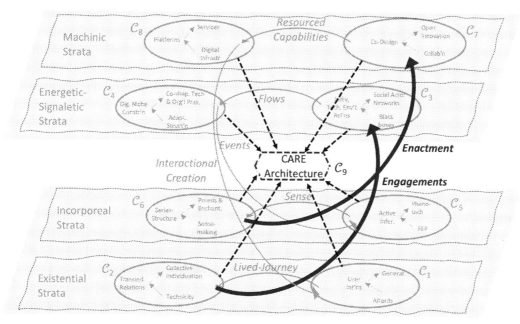

FIGURE 9.3 Structur*ing* collectives with a stable functor set and colimit object focus.

opportunities and relationships encourages the collective to maintain and even strengthen its core operational stability to support these ambitions effectively; and (3) collective's established practices provide a resilient framework that supports and is invigorated by its diversification and growth ambitions.

Agenc*ing* collectives as colimit object focused organizations with variable functor sets

Agenc*ing* Collectives with a Variable Functor Set and Colimit Object Focus are disruptive, transformative, and focused on expansion and diversification.[10] They are characterized by their adaptability and the constant evolution of their processes or transformations (functors) to navigate their ecosystem. This could include various strategies, tactics, and operational models that can change based on the evolving landscape, technological advancements, and other external factors. Despite this variability in the processes used, these organizations aim toward a future state that includes new relationships and interactions that extend beyond their current configuration, represented by the colimit object. This future state is not just an extension of the current state, but a state that includes new relationships and interactions that extend beyond the current

---

[10] Agenc*ing* Collectives, with their focus on disruption, transformation, and expansion, share characteristics with Deleuze and Guattari's (1987) concept of War Machines. The adaptability and constant evolution of processes in these organizations parallel the war machine's nature of challenging and destabilizing existing structures. The orientation toward a future state of expansion and diversification reflects the war machine's goal of extending influence and creating new power configurations. This balance between adaptability and a vision for transformative expansion captures the essence of the war machine's capacity for innovative and revolutionary change.

configuration. In practice, this might manifest in various ways. For example, the organization might aim to extend its influence or control over new territories, develop new alliances or partnerships, or acquire new capabilities or resources. It might also aim to disrupt or destabilize the existing structures or relationships, thereby creating new configurations of power or influence. Examples of such organizations might include entrepreneurial networks or communities of practice that are constantly evolving and innovating.[11]

- A collaborative research network like the Global Alliance for Genomics and Health dynamically utilizes F_5 (Sense) to incorporate emerging genomic data into personalized care plans, and F_8 (Interactional Creation) to innovate patient-engagement platforms based on real-world data. Their orientation is toward a future state that expands the boundaries of genomics in breast cancer, fostering growth in data-sharing networks that can lead to novel treatment pathways and more inclusive care models. They aim to diversify the field of genomics and create a more interconnected global research landscape (See Fig. 9.4).

These organizations are dynamic and expansive, able to navigate their ecosystem effectively while continuously evolving and innovating. This balance between adaptability and

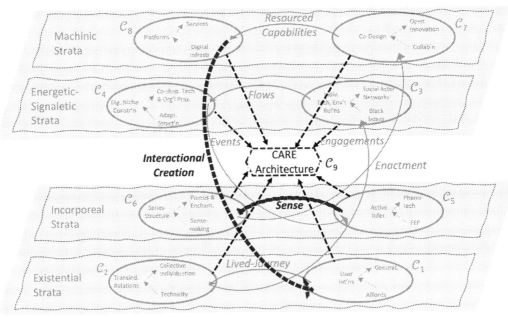

FIGURE 9.4 Agenc*ing* collectives with a variable functor set and colimit object focus.

---

[11] Scaling-up health innovations in Ethiopia, India, and Nigeria, as discussed in Spicer et al.'s (2014) study (see Chapter 7), provides a good context. The description of scaling-up health innovations involves disruptive, transformative efforts focused on expansion and diversification. The process of catalyzing scale-up, navigating complex ecosystems, and embracing a variable set of processes for innovation and expansion reflects the characteristics of Agenc*ing* Collectives. These collectives are adaptable and focused on creating new configurations of power or influence, aligning with the described efforts to scale up health innovations effectively.

orientation toward a future state of expansion and diversification is a key characteristic of many successful organizations.

The "-ing" suffix in Agenc*ing* Collective affords multiple possible implications: (1) ongoing, active process of leveraging adaptability to venture into new territories, form new alliances, and disrupt existing structures; (2) collective's aspiration toward expansion and diversification necessitates an adaptable, evolving array of strategies and operations; (3) perpetual state of becoming, where the organization's capacity for independent action and change is both the means and the outcome of its expansion and diversification efforts.

Toward a theory of organizational morphogenesis

As we have seen previously, in category theory, diagrams represent objects and morphisms (arrows) between them. Limit and colimit objects capture certain universal properties of these diagrams. The stability of the functor set and the focus on limit or colimit objects determine how these diagrams evolve over time, which can be thought of as "organizational morphogenesis." In the realm of organizational dynamics, the interplay between meaning-making (signification) and boundary crossing (transversality) offers insights into the structure and function of complex systems. By delving into the mathematical underpinnings of these concepts (see "The Unified Gauge Theory of Signification and Transversality" in the **Appendix**), we can illuminate the intricate web of relationships and interactions that define category theory, particularly as it pertains to diagrams, limits/colimits, and the roles of entities within these structures. This exploration seeks to apply insights from signification and transversality to the foundational elements of category theory, providing a comprehensive understanding of organizational morphogenesis.

Gauge-Invariant Lagrangian and Gauge Transformations: A gauge-invariant Lagrangian describes the dynamics of a system that remains unchanged under certain transformations, known as gauge transformations. These transformations can be thought of as "redundant" descriptions of the same physical state. The combined symmetry group for Signification and Transversality indicates that there are two sets of gauge transformations at play, each corresponding to a different aspect of the theory. The gauge transformations of the combined symmetry group for Signification and Transversality can be thought of as operations that act on the diagrams. Specifically,

- **Signification Gauge Transformations**: These transformations can be seen as operations that change the meaning or interpretation of the objects and morphisms in the diagram without altering their fundamental structure. In the context of an ecosystem, this could correspond to a shift in the roles or functions of certain entities without changing the overall network of interactions. When viewed through the lens of signification, diagrams can be seen as a manifestation of the coding, decoding, and overcoding processes. The entities within the diagram assign and interpret meaning, while the overarching structure of the diagram centralizes or dominates these meanings. The $GL(2,C)$ group, consisting of 2×2 invertible matrices with complex entries, represents a two-dimensional complex vector space. Each entity within the diagram occupies a position in this vector space, with its

Toward a theory of organizational morphogenesis *(cont'd)*

relationships to other entities defined by the dynamics of coding and decoding. This space can be thought of as a realm where different regimes of signs operate.

- **Coding State**: Represents the process of organizing signs into structured codes. In the context of an ecosystem, this could correspond to established norms or conventions.
- **Decoding State**: Represents the breakdown or deconstruction of these codes. This could correspond to challenging or critiquing established norms.
- **Overcoding State**: Represents the restructuring or domination of codes. This could correspond to a dominant narrative or ideology that overrides other codes.

- **Transversality Gauge Transformations**: These transformations can be seen as operations that introduce or remove connections (morphisms) between objects in the diagram. In an ecosystem, this could represent the emergence or dissolution of interactions between entities. The Braid Group, particularly B3, captures the essence of transversality, which is about crossing established boundaries and creating new connections. In the context of diagrams, this represents the dynamic interactions and boundary-crossing between entities. The strands of the braid, representing entities, continuously cross boundaries, challenging established relationships and forging new connections.

  - **Boundary State**: This state represents established boundaries or disciplines. In an organization, this could signify well-defined departments or teams with clear roles and responsibilities.
  - **Crossing State**: This state embodies the act of crossing or transgressing boundaries. In an organizational context, this could represent interdisciplinary teams or projects that bring together expertise from different domains.
  - **Interdisciplinary State**: Representing a dynamic interplay of disciplines, this state signifies organizations that thrive on collaboration and innovation, constantly seeking to break down silos and foster cross-functional teamwork.

- The gauge-invariant **Lagrangian** provides a principle that dictates how these transformations can be applied to the diagrams in a way that preserves certain properties. In other words, not all gauge transformations are allowed; only those that satisfy the conditions imposed by the Lagrangian can be applied. The Lagrangian captures the dynamics of signification, subjectification, and boundary transgression. It represents the tension between signification processes and transversal actions. The kinetic terms describe the "energy" associated with signification and boundary transgression, while the potential terms capture the tension inherent in interdisciplinary innovation.[12]

Operationalizing the Dynamics: The dynamic interplay of forces in an ecosystem can be captured by the way in which gauge transformations of signification and transversality act on the diagram. Specifically,

- The application of a gauge transformation can lead to the convergence or divergence of certain components of the ecosystem, represented by the objects in the

Toward a theory of organizational morphogenesis *(cont'd)*

diagram. For instance, a strong overcoding process might lead to a convergence around a dominant narrative, while a strong decoding process might lead to divergence and the emergence of alternative narratives.

- The focus on limit or colimit objects can determine the "equilibrium" states of the ecosystem. These objects represent stable configurations that the ecosystem tends to evolve toward. For example, in a system dominated by a strong overcoding process, the limit object might represent a homogenized, centralized structure. In contrast, in a system with active boundary crossings, the colimit object might represent a diverse, decentralized structure.
  * Limits and colimits in category theory allow for the combination of objects and morphisms, representing the convergence and divergence of entities within a system. The process of meaning-making (signification) can be seen in how these objects and morphisms are combined. The Lagrangian, capturing the dynamics of signification processes, provides a mathematical lens to understand the equilibrium states that these combinations seek to achieve.
  * Transversality, on the other hand, is evident in the ways in which these combinations challenge established boundaries. The intertwining dynamics of the Braid Group can be seen in the overlapping and intertwining of objects and morphisms, representing the continuous boundary-crossing inherent in limits and colimits.

- **Entities' Role, Intentionality, and Participation in Signification and Transversality**: Entities within category theory diagrams actively participate, shaping and being shaped by the dynamics of the system. Their roles in signification processes are defined by their position within the GL(2,C) Group's vector space. Each entity engages in coding, decoding, and overcoding, assigning and interpreting meaning based on its relationships with other entities. Transversality is evident in how these entities cross boundaries within the diagram. The Braid Group dynamics manifest in the movement and interactions of entities, challenging established positions and forging new connections. Entities possess agency, actively shaping the diagrams and influencing the overall structure of the system. This agency is deeply intertwined with the processes of signification and transversality. The intentional actions of entities, driven by the Lagrangian's dynamics, shape the coding, decoding, and overcoding processes, influencing the equilibrium states of the system.

Organizational Effects: The diagram serves as a map, capturing the relationships, potentials, and forces within the system. The combined symmetry group GL(2,C)×B3 provides a comprehensive view of this landscape, integrating the dynamics of signification and transversality. This map offers a snapshot of the organization's structure, illuminating the concatenation of meaning-making and boundary-crossing. An organization that understands and can effectively utilize the gauge transformations of

> ### Toward a theory of organizational morphogenesis *(cont'd)*
>
> signification and transversality can actively shape the configuration of assemblages in the ecosystem. By strategically applying these transformations, the organization can influence the current and future states of the ecosystem, guiding it toward desired outcomes. For instance, an organization might employ a "countersignifying" regime to decentralize its operations, fostering innovation and agility.
>
> - **Regimes of Signs**
>   - **Signifying Regime:** This regime is primarily associated with the Coding State, which represents the process of organizing signs into structured codes. In an organizational context, this could represent a structured, hierarchical system where information flows in a top-down manner.
>   - **Countersignifying Regime:** This regime is closely related to the Decoding State, representing the breakdown of established codes. In an organizational context, this could signify a more decentralized structure, where individual units or members operate with a degree of autonomy.
>   - **Postsignifying Regime:** Associated with the Overcoding State, this regime represents the restructuring or domination of codes. As a superposition of the Coding and Decoding states, indicating a state of near dominance but with elements of both coding and decoding. This could represent an organization undergoing a transition, where old structures are being redefined or replaced.
>   - **Presignifying Regime:** Representing a dynamic interplay between coding and decoding, this regime might be visualized as a balanced superposition of the Coding and Decoding states. It signifies a flexible organization that can swiftly adapt to changes by alternating between structured and decentralized modes of operation.
> - **Specific Interactions and Transformations Governed by the Lagrangian**
>   - **Decoding Provoking Boundary Crossings:** A critique of prevailing norms can lead to interdisciplinary engagements, e.g., a tech company breaking traditional industry boundaries by venturing into healthcare or finance.
>   - **Overcoding Restraining Transversality:** A dominant narrative can limit boundary crossings, e.g., a centralized authority imposing regulations that limit interdisciplinary collaborations.
>   - **Transformation of Signification through Transversality:** Interdisciplinary engagements can lead to the restructuring of signifying regimes, e.g., a multidisciplinary research institute fostering innovations that redefine traditional academic disciplines.
>   - **Transversality Inducing Decoding:** Interdisciplinary critiques can challenge established paradigms, e.g., an art–science collaboration challenging traditional artistic mediums through technological integration.
>   - **Equilibrium and Phase Transitions:** The Lagrangian can depict stable states and potential shifts, e.g., a business model shift in a company, from product sales to subscription

Toward a theory of organizational morphogenesis *(cont'd)*

services, causing market realignments.[13]

In summary, the gauge transformations and Lagrangian of the combined symmetry group $GL(2,C) \times B3$ for Signification and Transversality provide a mathematical framework for understanding and influencing the dynamics of ecosystems, as represented by diagrams and limit/colimit objects in category theory. By leveraging this framework, organizations can actively shape the evolution of the ecosystems they are a part of.

[12] Polk et al.'s (2023) study provides an illustration of how MSK navigates the complexities of genomics and precision medicine through the interplay of signification and transversality.

- Signification gauge transformations are observed as shifts in the organization's approach to genomics and precision medicine, where new interpretations of genomic data lead to changes in treatment protocols and patient care strategies: (1) Coding State: Initially, genomic data might be structured into specific codes or patterns that suggest certain treatment pathways. (2) Decoding State: As new research and patient data become available, these initial structures or codes might be broken down or questioned, leading to a reevaluation of treatment protocols. (3) Overcoding State: Subsequently, a new dominant narrative or treatment protocol might emerge, restructuring or dominating the initial genomic codes based on the latest genomic research and clinical outcomes.
- Transversality gauge transformations are evident in the creation of interdisciplinary teams and the integration of diverse scientific and medical fields, fostering innovation and enabling the organization to adapt to the rapidly evolving field of genomics: (1) Boundary State: The traditional boundaries between different medical and research disciplines. (2) Crossing State: The formation of interdisciplinary teams represents the crossing of these boundaries, enabling a flow of ideas and innovations across previously distinct fields. (3) Interdisciplinary State: The result is an environment where collaborative, cross-functional teams drive forward the application of genomics in cancer treatment, continuously seeking to break down silos and foster innovation.
- Furthermore, the application of a gauge-invariant Lagrangian to this organizational context captures the dynamic equilibrium between the processes of coding, decoding, and overcoding of genomic information and the boundary-crossing activities that propel the organization toward innovative solutions in precision medicine. The Lagrangian would reflect the organization's capacity to navigate the tension between established treatment protocols (potential terms) and the innovative potential of genomics and precision medicine (kinetic terms). This equilibrium reflects the organization's capacity to maintain its foundational objectives while dynamically incorporating new genomic knowledge and technologies.

[13] MSK's approach to precision oncology, as reported by Polk et al. (2023), illustrates the dynamic interplay between coding, decoding, overcoding, and the transversal actions that lead to boundary crossings within the healthcare and research ecosystem.

- Various regimes of signs can be noted: (1) Establishment of new centers and the MSK-IMPACT platform aligns with the Signifying Regime, characterized by the structured, hierarchical organization of signs (in this case, genomic data) into codes or structured frameworks that guide clinical decisions and research directions. (2) Early Drug Development (EDD) service illustrates the Countersignifying Regime, where individual units or members (e.g., molecular champions) operate with a degree of autonomy, challenging established codes (traditional treatment protocols) by focusing on molecular pathways rather than cancer types. (3) MSK's transition to a tissue-agnostic approach and the testing of mechanism-based therapies represent the Postsignifying Regime, characterized by a restructuring or domination of codes, where old structures (cancer treatment based on organ location) are being redefined or replaced with new codes (treatment based on genetic mutations, regardless of cancer type). (4) Initiatives like the Ecosystems Project aligns with the Presignifying Regime, which. Signifies

> ## Toward a theory of organizational morphogenesis *(cont'd)*
>
> an organization's ability to swiftly adapt to changes by alternating between structured codes and the breakdown of these codes, allowing for rapid integration of new research findings and therapeutic approaches.
>
> - Several specific interactions and transformations governed by the Lagrangian are exemplified in this case: (1) Decoding Provoking Boundary Crossings: MSK's approach to precision oncology provokes boundary crossings by critiquing prevailing norms (traditional cancer treatment paradigms) and venturing into interdisciplinary engagements (combining genomics with traditional oncology). (2) Overcoding Restraining Transversality: The establishment of structured platforms and protocols, while promoting innovation, also imposes a framework that could potentially limit boundary crossings by enforcing a new dominant narrative around precision oncology. (3) Transformation of Signification through Transversality: The interdisciplinary engagements at MSK, particularly through the EDD service, foster innovations that redefine traditional academic and clinical disciplines, leading to the restructuring of signifying regimes within the field of oncology. (4) Transversality Inducing Decoding: The collaborative efforts across different departments and with external pharmaceutical companies challenge established paradigms, promoting the decoding of traditional treatment modalities through technological and methodological innovation.

Case: Morphology and morphogenesis in healthcare

The previous chapter ended with a case that explores the digital health ecosystem for breast cancer care through category theory, illustrating how entities like patients, healthcare providers, and technology form dynamic assemblages and undergo transformation via functors. Here, we will continue with that context to illustrate the ideas introduced in this chapter.

Organizational Morphology

- Dimensions of Morphogenesis
  - Stability of the Functor Set:
    Stable Functor Set: Traditional medical practices that have been established over decades. For instance, mammograms as a standard diagnostic tool for breast cancer detection. The process of using mammograms has been consistent and is widely accepted.
    Variable Functor Set: Innovative medical practices that are adaptable. For instance, the introduction of new diagnostic tools or techniques, like advanced genetic testing or AI-driven image analysis for detecting breast cancer.
  - Focus on Limit or Colimit Objects:
    Limit Object Focus: An "ideal" state of breast cancer care. This could be a future where breast cancer is detected at the earliest stage in every patient, ensuring the highest survival rates. All processes, whether traditional or innovative, aim to achieve this state.

Colimit Object Focus: Expansion and diversification of breast cancer care. This could involve exploring a variety of treatments, from surgery to radiation, chemotherapy, and newer methods like targeted therapies or immunotherapies.

- Typology of Organizations
  - **Stable Functor Set + Limit Object Focus (Structur*ed* Corporations):** Traditional hospitals that rely on mammograms for detection and have a clear protocol for treatment, aiming for the best patient outcomes. Their goal is early detection and efficient treatment to achieve the "ideal" state of patient health.
  - **Variable Functor Set + Limit Object Focus (Agenc*ed* Corporations):** Innovative medical research institutions that constantly explore new diagnostic tools, like AI-driven analyses, but with a clear goal: early and accurate detection of breast cancer. They might also research new treatments but always with the aim of achieving the best possible patient outcomes.
  - **Stable Functor Set + Colimit Object Focus (Structur*ing* Collectives):** Community health clinics that use mammograms but also incorporate traditional or alternative medicine practices. They aim to provide a range of treatments tailored to individual patient needs, emphasizing both modern and traditional care.
  - **Variable Functor Set + Colimit Object Focus (Agenc*ing* Collectives):** Biotech startups or research labs that not only develop new diagnostic tools but also explore a wide range of potential treatments, from gene therapies to personalized medicine. Their approach is dynamic, aiming to expand the horizons of breast cancer care without being tied to a single "ideal" outcome.

Organizational Morphogenesis

- **Signification Gauge Transformations:** These transformations adjust the meaning or interpretation of objects and morphisms in the diagram without changing their fundamental structure. In the context of a breast cancer care ecosystem, this could correspond to a shift in the roles or functions of certain entities (like treatment modalities or diagnostic tools) without altering the overall network of interactions.
  - **Coding State**: Represents the process of organizing signs into structured codes. In the context of the breast cancer care ecosystem, this could represent structured treatment protocols.
  - **Decoding State**: Represents the breakdown or deconstruction of these codes. This could signify alternative or experimental treatments that deviate from standard protocols.
  - **Overcoding State**: Represents the restructuring or domination of codes. In the ecosystem, this might represent authoritative guidelines or best practices that override previous treatment modalities.
- **Transversality Gauge Transformations**: These transformations introduce or remove connections (morphisms) between objects in the diagram. This could represent the emergence or dissolution of interactions between entities in the breast cancer care ecosystem, such as the introduction of a new treatment approach or the phasing out of an outdated diagnostic method.
  - **Boundary State**: Represents established boundaries or disciplines. In the breast cancer care ecosystem, this could signify traditional treatment modalities.

- **Crossing State**: Represents the act of crossing or transgressing boundaries. This could signify interdisciplinary treatments that combine oncology with other medical disciplines.
- **Interdisciplinary State**: Represents the dynamic interplay of disciplines. This could signify holistic treatment approaches that integrate medical treatment with psychological and social care.
- The **Lagrangian** captures the dynamics of signification, subjectification, and boundary transgression. In the breast cancer care ecosystem, this could represent the tension between traditional treatment modalities and innovative approaches. The kinetic terms might represent the active processes of treatment, while the potential terms capture the inherent tension or potential for innovative treatments.
- **Regimes of Signs**
  - **Signifying Regime**: A hospital following established breast cancer treatment protocols.
  - **Countersignifying Regime**: Patient advocacy groups that challenge established treatments and seek alternative therapies.
  - **Postsignifying Regime**: A research institution that challenges established protocols and introduces new treatment modalities.
  - **Presignifying Regime**: Collaborative efforts between hospitals, research institutions, and patient advocacy groups to integrate traditional and alternative treatments.
- **Specific Interactions and Transformations Governed by the Lagrangian**:
  - **Decoding Provoking Boundary Crossings**: A radical new research finding that challenges established breast cancer treatments, leading to interdisciplinary collaborations.
  - **Overcoding Restraining Transversality**: Regulatory bodies imposing strict guidelines that limit the exploration of alternative treatments.
  - **Transformation of Signification through Transversality**: Interdisciplinary collaborations leading to the integration of oncology with nutrition, psychology, and alternative medicine.
  - **Transversality Inducing Decoding**: A breakthrough in genomics leading to a paradigm shift in breast cancer treatment.
  - **Equilibrium and Phase Transitions**: The breast cancer care ecosystem reaching a balance between traditional treatments and innovative approaches, with the potential for sudden shifts due to breakthroughs.

CHAPTER 10

Architectural transformation in an ecosystem

In the previous chapter, we saw how Diagrams offer a visual representation of complex Machinic Life-Experience Ecosystems (MLXEs), with "limits" and "colimits" merging structures and underscoring assemblage—organization interactions. Diagrams represent the present organizational state, with limit objects indicating ideal configurations and colimit objects showcasing growth potential. We saw how category theory introduces a typology for organizational analysis, distinguishing between types like Structured Corporations and Agencing Collectives based on functor stability and limit/colimit focus. The ecosystem's macro-architecture, crafted through diagrams, emerges from foundational category interactions. Higher categories, such as "2-categories," capture intricate intercategory dynamics. The ecosystem's architectural diagram is anchored in the platform category, subtly influencing entities. Addressing misalignments demands the design of co-innovation platforms, as entities actively mold these structures, with diagrams evolving through organizational morphogenesis, interwoven with meaning-making and boundary-crossing. We saw how grasping these transformations empower organizations to steer the ecosystem in which it functions toward envisioned goals.

Organizations are dynamic constructs with entities forming multiple assemblages, each with unique dynamics. Complex organizations operate as intricate ecosystems, with diverse interactions. In this chapter, we discuss in more detail how Sheaf/Stack Analysis provides insights into these complex interactions of organizational morphogenesis, especially in multifaceted sectors like healthcare MLXEs. We discuss the duality of potentiality (Body without Organs, BwO) and actualization (Organized Body, OB). Diagrams visually trace this dynamic, with nodes symbolizing entities and edges denoting relationships. As organizations evolve, new potentials (BwO elements) integrate into the established structure, transitioning to OB. The BwO and OB dynamics parallel the Molecular (adaptability-focused) and Molar (stability-focused) lines of organizations. Complex organizations operate as intricate ecosystems, with diverse interactions. Energy dynamics between BwO and OB are encapsulated by kinetic and potential terms, governed by the Lagrangian. Shifts between adaptability and stability are influenced by these dynamics. In enterprise modeling, misalignments in diagrams indicate inconsistencies. Addressing these challenges can involve adjusting diagrams, using functors and natural transformations, or employing higher abstraction levels like 2-categories.

Extending sheaves to stacks for multilayered global to local analysis

Since various healthcare professionals, patients, clinical procedures, equipment and devices, data, and computing services, etc., within and between healthcare enterprises enroll in multiplex ways as assemblages in local communities of practice and global fields of practice, enterprises themselves act like small ecosystems of their own with all kinds of intersecting local communities of practice across global fields of practice. More generally, in any multifaceted system, be it an organization, a technological process, or a natural ecosystem, integrating various components or stages is not always a linear endeavor. Each component has its own set of protocols, data, and objectives. When these components intersect, the overlaps might not always be straightforward. Instead of a direct merger, there can be "twisted" conditions where the integration involves additional layers, transformations, or decision points. These conditions underscore the need for adaptability and a holistic understanding. It is not merely about progressing from one stage to the next; it is about weaving together diverse threads of information, understanding the nuances at each overlap, and ensuring that the integrated system is both comprehensive and tailored to its specific context and objectives. Thus, we need to use a more sophisticated approach to capture the global-to-local relationships.[1]

As we saw earlier in Chapter 7, a sheaf is a tool for tracking local data and how it patches together to form global data. For example, if you have a topological space (like a manifold), you can assign data to each open set in a way that respects the topology. If you have two overlapping open sets, the data on the overlap should be compatible. This is the idea of a sheaf. A stack takes this idea a step further. It is still about tracking local data and how it patches together, but now the data are more complex. Instead of just assigning sets to open sets, we might assign categories or even higher structures. A stack can provide a mathematical framework for studying complex systems with overlapping entities. In the context of our healthcare ecosystem, a stack would provide a way to model the complex, multilayered relationships between different entities (professionals, patients, procedures, equipment, data, etc.) and organizations (enterprises, communities, etc.) across multiple categories. A professional, patient, clinical procedure, equipment, data, etc. could be part of multiple assemblages across categories and also belong to the same or multiple enterprises, organizations, or communities. The stack would also allow us to study the healthcare ecosystem as a whole while still keeping track of the local details. For example, we could use the stack to study how changes in one part of the ecosystem (e.g., the introduction of a new clinical procedure) affect

---

[1] In their study of how an emergency department (ED) functions as a microcosm of a broader healthcare ecosystem, characterized by dynamic, nonlinear interactions among various healthcare professionals, patients, clinical procedures, and support systems, Nugus et al. (2010) underscore the challenges and necessities of managing patient trajectories in an environment where protocols, data, and objectives are continuously intersecting and evolving. By highlighting the ED's role as a nexus of local and global practice communities, they demonstrate the limitations of linear models in capturing the full spectrum of healthcare integration and the importance of adaptability and holistic understanding. Their findings resonate with the concept that integration in healthcare and other multifaceted systems requires a sophisticated approach to account for the "twisted" conditions of overlapping components, advocating for a model that appreciates the dynamic, relational nature of such systems, and the need for a comprehensive strategy to navigate these complexities effectively.

other parts of the ecosystem (e.g., the work of healthcare professionals, the experiences of patients, etc.).[2]

The gluing process can also be more complex. For example, we might have "twisted" gluing conditions where the data on an overlap is not just the intersection of the data on the two sets but involves some additional structure or transformation. DRT could inform the definition of the gluing conditions, which would need to capture the dynamic, relational nature of the interactions between entities. For example, the gluing conditions could be defined in such a way that they reflect the ongoing exchanges and transformations that occur within and between entities. Stack is a generalization of a sheaf that allows for "twisted" or "nontrivial" gluing conditions. Stacks can be used to model situations where the local-to-global principle of sheaves is too restrictive, such as when the local data cannot be simply "glued together" to form the global data. In the context of a healthcare ecosystem, a stack could be used to model the complex interactions and overlaps between different healthcare enterprises, organizations, communities, and collectivities. Each of these entities could be seen as a local *section* of the stack, and the gluing conditions could capture the ways in which these entities interact and overlap with each other.[3]

Complex Gluing in Breast Cancer Care: Breast cancer care is multifaceted, involving various stages from prevention to diagnosis and treatment. Each stage can be seen as a local unit with its own set of diagrams and limit/colimit objects, representing specific protocols, interventions, and patient data. When integrating these stages to form a comprehensive care pathway, overlaps and intersections arise. These overlaps might not always be straightforward and can involve intricate "twisted" conditions.

- **Prevention and Diagnosis Overlap:**
    * **Standard Intersection:** A woman undergoes regular mammograms as a preventive measure. An anomaly is detected, leading directly to a diagnostic procedure.
    * **Twisted Gluing Condition:** Consider a scenario where a woman undergoes genetic testing as part of a preventive measure and discovers she has a BRCA gene mutation, increasing her risk for breast cancer. This information does not directly diagnose cancer but transforms her preventive pathway. She might opt for more frequent screenings or even prophylactic surgery. The overlap here is not a direct transition from prevention to diagnosis but involves an added layer of risk assessment and decision-making.

---

[2] As detailed by Polk et al. (2023), MSK Cancer Center's work in developing new programs, services, and an infrastructure for genomic practices point up the necessity of a sophisticated mathematical framework, like stacks, to capture the intricate, multilayered relationships between different entities within a healthcare ecosystem. Through its efforts to make genomic results actionable and to access targeted drugs within a precision medicine ecosystem, MSK exemplifies the dynamic, relational approach to organizing and integrating complex biomedical and organizational data, mirroring the stack's capacity to model complex interactions and overlaps between different healthcare enterprises, organizations, and communities.

[3] Trnka's (2021) exploration of how care and therapeutic practices are constituted through a network of interconnected sites, and the need for healthcare providers to navigate the fluid, often complex relationships between different care settings, suggests a role for "twisted" gluing conditions within the stack framework. This approach offers a nuanced methodology for capturing the ongoing exchanges and transformations within and between entities in a healthcare ecosystem, reinforcing the value of stacks in analyzing complex, interconnected systems where simple mergers are insufficient to convey the full scope of relational dynamics.

- **Diagnosis and Treatment Overlap:**
  - **Standard Intersection:** A biopsy confirms the presence of a malignant tumor, leading directly to a treatment plan, be it surgery, chemotherapy, or radiation.
  - **Twisted Gluing Condition:** Imagine a patient diagnosed with a particularly aggressive form of breast cancer. Instead of moving directly to a standard treatment, the oncologist might recommend enroling in a clinical trial for an experimental drug. Here, the overlap between diagnosis and treatment is transformed by the introduction of experimental therapy, adding complexity to the patient's care pathway.
- **Prevention, Diagnosis, and Treatment Triad:**
  - **Standard Intersection:** Regular screenings (prevention) lead to early detection (diagnosis) and prompt treatment, following established protocols.
  - **Twisted Gluing Condition:** A woman with a family history of breast cancer (raising preventive concerns) discovers through advanced diagnostic tools that she has precancerous cells. Instead of standard treatments, she has offered a preventive mastectomy or a novel immunotherapy treatment to halt the progression. The data here does not just merge prevention, diagnosis, and treatment but introduce a transformative decision point based on risk, advanced diagnostics, and innovative treatments.

Such "twisted" gluing conditions underscore the need for healthcare providers to be adaptable and consider multiple data points, patient histories, and emerging medical advancements when integrating care pathways. It is not always about moving linearly from one stage to the next; sometimes, it is about weaving together diverse threads of information to create a tailored, comprehensive care plan. In essence, the integration of breast cancer care, from prevention to treatment, is not just about piecing together consecutive stages. It involves understanding the nuances, complexities, and transformative decisions at each overlap, ensuring that the care provided is both comprehensive and personalized.

In the context of a stack (or sheaf) analysis, the key idea is to understand how local data (in this case, the diagrams and limit/colimit objects associated with different members or parts of the enterprise) can be "glued" together to form a global picture. If different members or parts of the enterprise are working with different diagrams and limit/colimit objects, this could be seen as different local sections of the stack. Each local section represents a particular perspective or understanding of the enterprise, based on the specific roles, responsibilities, and interactions of the members or parts in question. The challenge in a stack analysis is to find a way to "glue" these local sections together to form a coherent global picture of the enterprise. This involves identifying the overlaps or intersections between the local sections, while ensuring that the data in these overlaps are consistent across all the sections. In the context of an enterprise, this could involve ensuring that the different diagrams and limit/colimit objects are compatible with each other and can be integrated into a single, unified model of the enterprise. If the local sections cannot be glued together in a consistent way, this could indicate a problem or conflict within the enterprise, such as a misalignment of goals or a lack of coordination between different parts of the enterprise. In this case, the stack analysis can help to identify and diagnose these issues, providing valuable insights for improving the operation and management of the enterprise. In this way, a stack (or sheaf) analysis can provide a powerful tool for understanding and managing the complex, dynamic relationships and interactions within an enterprise, in line with the principles of DRT.[4]

Complex dynamics of order and scale in organizations

In organizational dynamics, the interplay between potentiality and actualization, between fluidity and stability, is of critical significance. The BwO serves as a metaphorical plane of pure potential, a space brimming with what could be. In contrast, the OB represents the crystallization of these potentials into tangible, identifiable forms. This duality, further nuanced by the Molecular and Molar lines of segmentation, paints a picture of organizations as ever-evolving entities. They oscillate between the micro and macro, the innovative and the established. These concepts shape the very fabric of organizations, influencing their strategies, operations, and growth trajectories (for details, please see "Unified Gauge Theory of Organization" in the **Appendix**).

Order as Body without Organs (BwO) vs. Organized Body (OB). Organizations are intricate configurations intersecting countless lines of relationships, strategies, and goals. Within this complex web, entities often find themselves part of multiple assemblages, each with its unique dynamics and aspirations. The interplay between the Body without Organs (BwO) and the Organized Body (OB) captures the essence of this multifaceted nature. Here, we will review how units within organizations navigate this dynamic, how they integrate potentialities into their structured operations, and the challenges and opportunities that arise from these interactions.

In any complex ecosystem, entities do not exist in isolation. They enroll in multiple assemblages, forming intricate networks of relationships. A single entity might be part of an assemblage in a local community of practice and simultaneously be part of a broader field of practice. These assemblages are not static. They evolve, intersect, and sometimes diverge. An entity might be part of multiple assemblages, each with different goals and operational dynamics.

- This fluidity is a hallmark of the BwO dynamic. BwO represents the realm of potentialities, innovations, and uncharted territories within an organization. It is the space of pure possibilities that have not been actualized or standardized. This includes potential strategies, methods, and collaborations that are still in the exploratory phase. OB symbolizes the structured, established, and standardized

Continued

---

[4] Spicer et al.'s investigation (2014) into the multifaceted process of scaling up health innovations in Ethiopia, India, and Nigeria demonstrates the essence of "gluing" together disparate parts of a health initiative—ranging from government advocacy, community engagement, to aligning with health systems and priorities—to achieve a coherent, unified health strategy. This methodology resonates with the principles of stack analysis, where different local sections (in this context, components of health innovation scale-up) must be harmonized to form a comprehensive global framework. By highlighting the necessity for diverse strategies, substantial support from various stakeholders, and the alignment of innovations with broader health systems, Spicer et al. provide a real-world example of how stack or sheaf analysis can be instrumental in navigating the complexities of integrating different diagrams and limit/colimit objects within an enterprise.

Complex dynamics of order and scale in organizations *(cont'd)*

protocols and practices within an organization. It is the embodiment of known pathways, established relationships, and formalized collaborations.

- Each unit or department within an organization can be seen as navigating through the BwO-OB dynamic. Their unique approach to tasks, collaborations, and goals can be mapped using transformations, reflecting how they integrate BwO elements into their OB.
- Diagrams capture the structure and dynamics of organizational pathways. Nodes in these diagrams could represent various entities (e.g., team members, tools, strategies) while edges signify interactions and relationships. Incompatibilities, representing BwO elements, introduce new nodes that need integration into the established diagram, transitioning them from BwO to OB.[5]

Different units within an organization operate based on distinct diagrams, reflecting their unique priorities. One unit's diagram, emphasizing one aspect, is distinct from another unit's diagram, which might prioritize a different outcome.

- Each unit's diagram points toward a limit object, representing its "ideal" state. This "ideal" represents the convergence of the unit's goals and aspirations. While the limit object represents a convergence toward an "ideal," the colimit object symbolizes potential diversification and expansion. It represents the continuous push toward new frontiers without a fixed endpoint in sight.
- The processes or transformations that an organization (or its units) uses to navigate its ecosystem can either be stable or variable. Organizations (or their units) with stable functors use established, consistent processes, often resulting in a more traditional and hierarchical structure. This stability can influence how they interpret and act upon their diagrams and limit/colimit objects. Organizations (or their units) with variable functors are adaptable, changing their processes in response to environmental shifts. Their adaptability can lead to a more fluid interpretation of their diagrams and a dynamic approach toward achieving their limit or colimit objects.

The diverse nature of diagrams across units means they might not always be directly compatible. The goals of one unit might sometimes clash with the goals of another due to their distinct priorities.[6]

- The challenge for the overarching organization is to navigate these incompatibilities, ensuring that each unit can operate optimally while also facilitating interunit collaboration. This might involve creating mechanisms that can bridge one unit's operations with another's, facilitating collaboration without compromising each unit's unique goals.
- While the limit objects provide a vision for each unit to strive toward, the colimit objects offer opportunities for growth and diversification. By recognizing and harnessing these opportunities, organizations can adapt to emerging challenges and capitalize on new possibilities.

> ### Complex dynamics of order and scale in organizations *(cont'd)*
>
> - The choice between a stable and variable functor set is not merely operational; it has strategic implications. Stability can lead to efficiency and reliability, while variability can lead to innovation and adaptability. Organizations need to understand their strategic priorities and choose their approach accordingly.
>
> The challenge of integrating processes or units with different diagrams is analogous to the transition between the BwO and OB. The BwO, with its potentialities, can be seen as a space of diagrams that have not been actualized or standardized. These are the myriad ways an organization could evolve, innovate, or diversify. The OB, on the other hand, symbolizes the structured, established diagrams. These are the tried-and-tested pathways, the established strategies, and the formalized collaborations.
>
> - The limit object in the context of organizational units represents an "ideal" state toward which an organizational unit aspires. In the context of the BwO and OB, the limit object corresponds to the OB, which is a structured, actualized state. It is the culmination of processes and strategies that an organization aims for. The colimit object, representing potential diversification, aligns with the BwO's realm of potentialities. It captures the myriad ways an organization could evolve, innovate, or diversify, embodying the fluid and chaotic nature of the BwO.
> - A stable functor set, with its established processes, corresponds to the OB. It represents processes that have been standardized and are resistant to change. In the context of diagrams, these are the established pathways and strategies.[7] A variable functor set, adaptable to changes, aligns with the BwO's fluid nature. It represents processes that are flexible, innovative, and can adapt to the ever-changing dynamics of an organization. In terms of diagrams, these are the evolving strategies and methods that have not been fully actualized.[8]
>
> Transitions and dynamics between BwO and OB:
>
> - **Inversion operator** toggles between the BwO and the OB. When applied to a state, it can transition an entity from a state of pure potential (BwO) to one of actualization (OB) or vice versa. In the context of diagrams, this transformation can be seen as a shift between potential diagrams (unactualized strategies or methods) to established diagrams (standardized protocols).
> - Introducing a phase shift, **latent operator** captures the nonlinear dynamics of the transition between chaos (BwO) and organization. This phase shift can be likened to the unpredictable and complex nature of organizational dynamics, where potentialities undergo various phases before crystallizing into tangible forms.
> - Differentiating between the BwO and OB states, **dominance/bifurcation operator** represents the tension between potential and actualization. In terms of limit/colimit objects, the tension between what an organizational unit aspires to become (limit object) and the myriad ways it could diversify or expand (colimit object) is evident.[9]

Continued

> ### Complex dynamics of order and scale in organizations *(cont'd)*
>
> **Scale as Molecular vs. Molar.** Scaling from the granular to the overarching, organizations exhibit two contrasting yet intertwined dynamics: the Molecular and the Molar. The Molecular Line, with its emphasis on adaptability and specificity, contrasts sharply with the Molar Line's focus on standardized protocols and overarching structures. Through the lens of Sheaf/Stack Analysis, we have seen how these lines influence the data, strategies, and operations of organizations, and how they come together to form a cohesive, holistic system view.
>
> The Molecular Line delves into the granular, emphasizing microlevel, fluid processes tailored to specific situations. It is about the details, from individual interactions to the use of specialized tools. This line is characterized by its dynamic, transformative nature, constantly adapting to new challenges.
>
> - In terms of Sheaf/Stack Analysis, the Molecular Line underscores the granularity of local data, often visualized as diagrams and represented by limit/colimit objects. These objects capture the unique dynamics of individual system units, which are often less standardized due to their adaptive nature.
>
> On the other hand, the Molar Line represents the system's macro-level, focusing on standardized, large-scale protocols and best practices. It is about the overarching structures that remain stable over time.
>
> - In the realm of Sheaf/Stack Analysis, the Molar Line pertains to the more established sections of a system, operating based on well-defined protocols. Their representations, in contrast to the Molecular Line, are more fixed and standardized.
>
> Central to our understanding of these systems is the concept of Local Data, which comprises diagrams and limit/colimit objects associated with different system parts. These diagrams and objects encapsulate the specific roles, responsibilities, and interactions of system entities. The Limit represents the convergence of strategies, the ideal state, while the Colimit showcases the spectrum of potential strategies. Integrating these diverse representations cohesively to form a global picture is a challenge, often referred to as the "gluing" process.
>
> - Sheaves and Stacks further elucidate this process. Sheaves track how local data amalgamates to form a global perspective, ensuring data consistency. Stacks, conversely, model the overlaps and interactions between system entities, with "gluing conditions" capturing these intricate collaborations. Occasionally, this "gluing" is not straightforward and might involve "twisted" conditions, necessitating adaptability and a comprehensive understanding. It is a process of weaving diverse information threads, understanding overlaps, and ensuring a holistic system view.
>
> Transitions and dynamics between Molecular and Molar line:
>
> - **Inversion operator** toggles between the Molecular and the Molar lines. When applied, it transitions an entity from a micro-level, fluid process (Molecular) to a macro-level, stable structure (Molar) or vice versa. This transformation captures the fluidity of organizational units as

Complex dynamics of order and scale in organizations *(cont'd)*

they navigate between adaptability and stability.

- Introducing a phase shift, **latent operator** captures the nonlinear dynamics between the Molecular and Molar lines. This phase shift represents the complex interlock between adaptability and overarching structures within organizations and the challenges and complexities in the gluing process, especially when dealing with overlaps or twisted conditions.
- Differentiating between the Molecular and Molar states, **dominance/bifurcation operator** represents the tension between micro and macro scales. In the context of "gluing" across sheaves and stacks, this tension is evident as organizations attempt to weave together diverse information threads, ensuring a holistic system view.[10]

Dynamics of Order and Scale. With respect to the energy dynamics and interactions between the BwO and the OB, as well as the Molecular and Molar lines, the kinetic and potential terms capture the active and stored energies, respectively, governing the interplay between organization and scale, and the dynamics of chaos and structure.

The kinetic term represents the "energy" associated with the active processes of organization and scaling.

- **BwO/OB**: Represents the active processes of transitioning from potentiality (BwO) to actualization (OB). It is the energy associated with the dynamic shifts and transformations within an organization.
- **Molecular/Molar**: Captures the energy of scaling, transitioning from micro (Molecular) to macro (Molar) scales. It embodies the active processes of navigating between adaptability and stability.

The potential term captures the "energy" or tension associated with the potential of the BwO and its process of organization, as well as the energy associated with the scaling process.

- **BwO/OB**: Captures the tension or "energy" associated with the potential of the BwO and its process of organization. It is the latent energy waiting to be actualized.
- **Molecular/Molar**: Represents the energy or tension associated with the scaling process. It is the potential energy of an organization as it contemplates scaling up or down.[11]

Mechanics of Interactions Governed by the Lagrangian:

- **Organized Body Stabilizing Molar Structures**: The formation of an OB can reinforce and stabilize Molar structures. This is represented by the dominance/bifurcation transformation stabilizing both states.[12]
    * The OB, represented by the limit object and stable functor set, reinforces Molar structures. It is the process of solidifying diagrams and ensuring that the overarching structures (Molar) are stable. The gluing process ensures that these structures are cohesive and integrated. Sheaves are apt for this as they glue data that can be consistently patched together.
    * The OB's inherent structure and stability mean that it has a lot of stored energy. This energy is used to

Continued

> ### Complex dynamics of order and scale in organizations *(cont'd)*
>
> stabilize and bring order to the Molar structures. As the Molar structures actively stabilize under the influence of the OB, there is a dynamic process at play, making the kinetic energy associated with Molar structure significant.[13]
>
> - **Organized Body Negotiating with Molecular Forces**: The Organized Body continuously negotiates with Molecular forces for new formations. This negotiation is represented by the latent transformation on both states.[14]
>   * The OB continuously negotiates with Molecular forces for new formations. This is the process of integrating new, variable functors into the established framework (limit object). Sheaves ensure that these new formations are cohesively integrated. The gluing process ensures that these new formations are integrated cohesively.
>   * The OB's inherent structure and stability mean it has a lot of stored potential energy. This energy is used in the negotiation process. The Molecular forces are dynamic and active, and as they negotiate with the OB, the kinetic energy associated with them becomes significant.[15]
> - **Molar Suppression of BwO**: Strong Molar structures can suppress the BwO's potentials. This is represented by the inversion transformation acting on the BwO and the dominance/bifurcation transformation stabilizing the Molar state.[16]
>   * Strong Molar structures, represented by stable functors, can suppress the BwO's potentials (colimit objects). The suppression process is about integrating the potentialities of the BwO in a consistent manner. Sheaves ensure that these integrations are consistent with the overarching Molar structures. The gluing process ensures that the potentialities of the BwO are integrated but not dominant.
>   * The Molar structures are actively suppressing the BwO, making this a dynamic process. The tension in the BwO as it resists this suppression represents stored energy, which gets released or transformed during the suppression.[17]
> - **BwO Inducing Molecular Transformations**: A surge in the BwO's intensity can provoke a shift toward more Molecular processes, challenging established Molar structures. This is represented by the inversion transformations on both states.[18]
>   * The BwO, with its realm of potential diagrams (colimit objects), can induce transformations at the Molecular level. This is akin to introducing new, variable functors that challenge established Molar structures. This can introduce inconsistencies in the established framework. Stacks are needed to account for these inconsistencies and ensure a smooth transition. The gluing process here involves integrating these new Molecular processes into the existing framework.
>   * BwO's active push toward transformations is a dynamic process. The BwO's inherent chaotic nature is driving the system out of equilibrium, making its kinetic energy crucial. The Molecular line's potential to transform

> **Complex dynamics of order and scale in organizations** *(cont'd)*
>
> is being activated by the BwO. This stored energy or tension in the Molecular line is released during the transformation.[19]
>
> - **Molecular Disruptions Leading to BwO Emergence**: Intense Molecular activity can destabilize the OB, leading to the BwO's emergence. This is represented by the inversion transformation on both states.[20]
>   * Intense Molecular activity, represented by variable functors, can destabilize the OB, leading to the BwO's emergence. This is the process of challenging established diagrams and pathways. The disruptions at the Molecular level can introduce nontrivial inconsistencies in the established structures. Stacks help in reconciling these disruptions. The gluing process here involves reconciling these disruptions with the existing framework.
>   * The Molecular disruptions are active, dynamic processes. These disruptions are causing the system to move, making the kinetic energy associated with Molecular activity dominant. The potential energy stored in the BwO gets activated by these Molecular disruptions, leading to its emergence.[21]
> - **BwO Dissolving Molar Structures through Deterritorialization**: The BwO can dissolve rigid Molar structures. This is represented by the latent transformation on both states.[22]
>   * The BwO can challenge and dissolve rigid Molar structures. This is represented by the colimit object's potential diversifications challenging the established diagrams. The dissolution and reformation process can introduce complexities that are not consistently patchable. Stacks are needed to account for these complexities. The gluing process here is more about dissolution and reformation.
>   * The BwO's active process of dissolution is dynamic, making its kinetic energy dominant. The potential energy stored in the Molar structures gets released during this dissolution, as the structures are broken down.[23]
>
> [5] In Grudniewicz et al.'s (2018) study on Ontario's Health Links as a "complexity-compatible" policy for integrated care, the initiative's "low rules" approach, aimed at fostering voluntary networks for better care coordination, allowed for the emergence of new strategies, collaborations, and operational methods, resonating with the BwO dynamic where potentialities are explored and actualized. The process of forming Health Links to coordinate care involves the integration of innovative approaches and potential collaborations into structured and identifiable forms, embodying the OB's characteristics by formalizing partnerships and operational practices. Moreover, the preference for flexibility reflects the Molecular aspect, emphasizing innovation and adaptation at a granular level, whereas the need for standardization speaks to the Molar aspect, underscoring the importance of stability and uniformity across the healthcare system. The evolving nature of diagrams, with new nodes (BwO elements) being integrated into the established structure (OB), illustrates the ongoing negotiation between potentiality and actualization within the healthcare ecosystem.
>
> [6] Our typology of organizational morphology from Chapter 9 with Structured Corporations and Agenced Corporations, along with Structuring and Agencing Collectives, can also provide a nuanced framework for analyzing organizational units based on their stability, adaptability, and focus on achieving an "ideal" state or expansion. This differentiation helps in comprehending how each unit's unique priorities, encapsulated in their distinct diagrams and aimed toward limit or colimit objects, contribute to the broader organizational ecosystem's complexity. The challenge of aligning these diverse

Continued

> # Complex dynamics of order and scale in organizations *(cont'd)*
>
> units, each with its aspirations and operational methodologies, underlines the importance of a strategic approach that accommodates both stability and variability. It involves crafting mechanisms that not only respect each unit's uniqueness but also foster collaboration toward the overarching organization's collective vision. This balancing act between maintaining unit-specific efficacies and promoting interunit synergy is pivotal for the organization's holistic growth and adaptability in a dynamic environment.
>
> [7] The key nuance here is that while both the limit object and the stable functor set correspond to the OB, they do so in different ways. The limit object is about the "end goal" or "ideal state," representing the culmination or target of organizational processes. In contrast, the stable functor set is about the processes that have been standardized, signifying the established and consistent methods within the organization.
>
> [8] Similarly, while both the colimit object and the variable functor set align with the BwO, their representations diverge. The colimit object embodies the myriad potential pathways and diversifications that the organization could pursue, capturing the BwO's inherent fluidity and potentiality. On the other hand, the variable functor set represents the adaptable and evolving processes within the organization, highlighting the BwO's dynamic and transformative nature.
>
> [9] In Grudniewicz et al.'s (2018) study, each unit within an integrated care framework might operate based on distinct diagrams, reflecting their unique priorities and objectives about different aspects of patient care, research, or community engagement. Integrated care inherently involves creating mechanisms for collaboration across different units within healthcare organizations, despite their distinct priorities and diagrams. For integrated care to be effective, healthcare organizations need to adopt a more fluid interpretation of their diagrams and a dynamic approach towards achieving their goals, where policies allow for both convergence towards shared goals (limit objects) and the embrace of diversification and expansion (colimit objects). The policy framework's adaptability and emphasis on integrated care reflect a strategic choice for variable over stable functor sets to foster innovation and responsiveness. Concepts of the inversion operator, latent operator, and dominance/bifurcation operator describe the processes and challenges involved in navigating between potentiality and actualization, innovation and standardization, and adaptability and stability:
>
> - In the context of integrated care, the inversion operator could represent the ability of healthcare systems to move between standardized care protocols (OB) and innovative, patient-centered care models (BwO) as needed.
> - In the process of implementing new healthcare technologies or practices, latent operator captures the non-linear dynamics of transitioning from the initial chaotic state, where the potential benefits and applications of the innovation are not fully understood, to a more organized state, where the innovation is integrated into standard care practices.
> - Dominance/bifurcation operator represents the tension that exists between the aspiration to provide holistic, patient-centered care (a limit object or ideal state) and the practical challenges of expanding services or integrating new care models (colimit objects representing diversification and expansion).
>
> [10] Grudniewicz et al.'s (2018) study highlights the need for healthcare policies and practices to be adaptable and tailored to specific contexts, embodying the Molecular line's emphasis on the granular, micro-level fluid processes, from individual patient interactions to the use of specialized tools and approaches that are tailored to specific situations. Integration and coordination across different levels of healthcare provision requires a "gluing" process ensuring that the diverse needs and practices at the micro-level are cohesively integrated into the overarching structure of healthcare system, i.e., an application of Sheaf/Stack Analysis, where sheaves track local data integration to form a global perspective, and stacks model the overlaps and interactions between system entities. In this context, limit objects represent the 'ideal' state of fully integrated care tailored to patient needs, while colimit objects symbolize the expansion and diversification of care models and practices. We can further elaborate on this interplay between adaptability and structure within organizations through the gauge transformation operators:
>
> - The inversion operator represents healthcare entities' ability to oscillate between Molecular adaptability—tailoring care to individual needs—and Molar stability—adhering to evidence-based guidelines and protocols.
> - The latent operator captures the policy's capacity to manage the non-linear dynamics between these scales, facilitating a phase shift

Complex dynamics of order and scale in organizations *(cont'd)*

towards integrated care models that blend personalized care with standardized practices.
- The dominance/bifurcation operator signifies the policy's ability to navigate the tension between micro-level care customization and macro-level standardization, ensuring that healthcare systems can evolve and adapt without losing sight of their core objectives.

[11] In Grudniewicz et al.'s (2018) study about "complexity-compatible" policy, where different organizational units must adapt and scale their processes to meet patient needs effectively, kinetic and potential terms represent the active and stored energies governing the interplay between organization and scale, and the dynamics of chaos and structure. Kinetic energy is associated with the effort and dynamism required to navigate the complexities of healthcare integration in transitioning from potentiality (BwO) to actualization (OB) and from micro (Molecular) to macro (Molar) scales in integrated care settings. Energy associated with the potentialities within the healthcare system (BwO) and the process of organizing these potentials into structured, integrated care pathways (OB) reflects the tension between the desire to innovate and adapt healthcare practices at a Molecular level and the necessity to maintain stable, overarching structures at a Molar level.

[12] The inherent structure and stability of the OB, coupled with its capacity to utilize stored energy for order and stabilization, highlights a dynamic toward maintaining and enhancing Structured Corporations, aiming for an 'ideal' state through strategic planning and optimization of activities.

[13] Returning to Grudniewicz et al. (2018), the Health Links initiative functions as an OB that encourages the formation of voluntary networks, or Health Links, to improve care coordination for individuals with complex healthcare needs, thus stabilizing Molar structures through improved protocols and procedures. This policy framework, acting as a limit object, ensures that the overarching Molar structures are cohesive and integrated, using stored energy to bring order and stability to these structures by formalizing connections between different healthcare providers and services.

[14] A characteristic of successful Agenced Corporations is the incessant negotiation between the OB's inherent stability and the dynamic nature of Molecular forces facilitated by sheaves, which enable the seamless incorporation of innovative approaches and models in a strategic pivot toward a more structured, yet still adaptable and future-oriented organizational configuration.

[15] In Grudniewicz et al. (2018), implementation of Health Links illustrates how integrated care policies negotiate with Molecular forces as new variable functors (adaptive policies and practices) are integrated into the established Molar framework, ensuring cohesive integration. In this process, OB leverages stored potential energy to embody the dynamic negotiation between structured protocols of integrated care and the adaptive, molecular forces necessitated by patient-centered approaches and technological advancements.

[16] In this dynamic characteristic of Structuring Collectives, innovative and expansive potentials are harmonized with the organization's stable, foundational processes, not only maintaining coherence within the overarching structure but also fostering a controlled environment for expansion and diversification.

[17] In Health Links study (Grudniewicz et al., 2018), the policy framework, while promoting integration and coordination, can limit the full realization of innovative potentials by imposing certain structures and protocols. However, the initiative's 'low rules' approach and emphasis on local flexibility show an attempt to balance innovation with structure, allowing for some degree of BwO emergence within the confines of an organized system.

[18] The active induction of Molecular transformations by the BwO suggests an organizational landscape that values and encourages adaptability, innovation, and the exploration of new frontiers. Such an environment is conducive to the growth of Agencing Collectives, which are pivotal in navigating the ecosystem effectively while continuously evolving and innovating.

[19] Grudniewicz et al.'s (2018) study also exemplifies how the intensity of BwO provokes a shift toward more Molecular processes. Health Links, by fostering local innovation and adapting to patient needs, embody the policy's flexibility to accommodate innovative practices within the integrated care framework, thus facilitating Molecular transformations within the healthcare system.

[20] Such disruptions, indicative of the organization's adaptability and innovation, signify a departure from established structures and protocols toward more fluid, expansive, and innovative approaches. This might effectively mean a transformative dynamic within an overarching organization, moving from an Agenced Corporation toward an Agencing Collective.

[21] We can also see in Grudniewicz et al. (2018) how Health Links have developed novel interconnections and practices (intense Molecular activity) beyond traditional care models, thereby disrupting

Continued

III. Organizations and creative transformation

> **Complex dynamics of order and scale in organizations** *(cont'd)*
>
> traditional care structures (OB) to embrace more holistic and patient-centered approaches (emergence of BwO).
>
> [22] By embracing the colimit object's potential for diversification, the organization undergoes a dynamic dissolution and reformation from Structuring Collective toward Agencing Collective, necessitating the use of stacks to manage the complexities introduced by this transformation. This signals a shift toward an organization that is not only adaptive and expansive but also one that actively navigates its ecosystem through continuous evolution and innovation.
>
> [23] By embracing innovation and patient-centricity (dissolution of Molar structures), facilitated by the initiative's flexibility and the encouragement of local solutions (BwO) to systemic healthcare challenges, Health Links policy allows for the exploration of new care models and technologies (deterritorialization), thus fostering a more adaptable and responsive healthcare ecosystem (Grudniewicz et al., 2018).

Dynamic relationality #9: Strategic architecturing as transformations of diagrams

When different diagrams and limit/colimit objects are not compatible with each other, and cannot be integrated into a single, unified model of the enterprise, that is, if the local sections cannot be glued together in a consistent way: this suggests that there are inconsistencies or conflicts within the enterprise that need to be addressed. From a category theoretical perspective, there are several ways to address this problem:

- Adjustment of Diagrams or Limit/Colimit Objects: One approach could be to adjust the diagrams or limit/colimit objects themselves to make them compatible. This might involve changing the categories, objects, morphisms, or functors in the diagrams, or adjusting the properties of the limit/colimit objects. This would essentially involve a process of negotiation and compromise among the different parts of the enterprise to agree on a common model.
- Use of Functors and Natural Transformations: Another approach could be to use functors to map between the different diagrams or limit/colimit objects. A functor can transform one category into another in a way that preserves the structure of the category. If a suitable functor can be found, it could be used to transform the diagrams or limit/colimit objects into a common form that can be integrated into a single model. If the diagrams or limit/colimit objects can be viewed as functors, then a natural transformation could be used to provide a 'bridge' between them. A natural transformation is a way of transforming one functor into another while preserving the structure of the categories. This could provide a way to reconcile the differences between the diagrams or limit/colimit objects.

- Use of 2-Categories or Higher Categories: If the inconsistencies cannot be resolved within the framework of ordinary category theory, it might be necessary to move to a higher level of abstraction, such as 2-categories or higher categories. These provide a way to deal with situations where the standard concepts of category theory (categories, objects, morphisms, functors) are not sufficient.

In all cases, the goal would be to find a way to reconcile the differences and inconsistencies, and to create a unified, coherent model of the enterprise that accurately reflects its complex, dynamic nature. This is in line with the principles of Dynamic Relationality Theory, which emphasizes the importance of understanding and managing the complex, dynamic relationships and interactions within an enterprise.

Adjustment of diagrams or limit/colimit objects

In a healthcare organization comprising a hospital, a network of clinics, and a research institute, each entity might operate with different diagrams and limit/colimit objects, reflecting their unique roles, responsibilities, and goals. The hospital's diagram might include categories like patients, healthcare providers, and treatments, with the limit object representing an 'ideal' state of optimal patient care and the colimit object capturing the diverse range of patient care scenarios. The clinics might have a diagram with categories like patients, healthcare providers, and community resources, aiming for an 'ideal' state where all patients have access to necessary resources and a colimit that represents the variety of community interactions. The research institute might have a diagram with categories like researchers, research projects, and funding sources, with the limit object representing an 'ideal' state of fully funded, valuable research and the colimit object showcasing the spectrum of research endeavors.

These diagrams and limit/colimit objects, while individually effective, are not directly compatible due to their distinct perspectives and goals. However, they can be adjusted to create a common model that integrates all parts of the organization. This could involve adding new categories, objects, morphisms, or functors to the diagrams to represent interactions between the hospital, clinics, and research institute. The limit/colimit objects could also be adjusted to reflect a shared 'ideal' state for the entire organization, incorporating elements from each of the original limit/colimit objects. For example, the new limit object might represent an 'ideal' state where all patients receive optimal care, have access to necessary resources, and benefit from research projects, while the colimit object could capture the broad range of healthcare and research scenarios across the organization.

This process of adjusting the diagrams and limit objects would require negotiation and compromise among the different parts of the organization, a shared understanding and commitment to the organization's overall goals, and a willingness to adapt individual perspectives and goals to achieve these shared goals. This approach is generally suitable for most of the interactions, especially when there's a need to integrate new formations or transformations into an established framework. However, its suitability might vary depending on the dynamic nature of the interactions and the complexities introduced.

Regarding interactions in a Structured Corporation, where the OB reinforces and stabilizes Molar structures, there is a clear emphasis on ensuring that the overarching structures are stable and cohesive. The challenge is to ensure that these structures are integrated in a way that reflects the enterprise's complex, dynamic nature. The use of sheaves is generally apt for this interaction as they glue data that can be consistently patched together, further emphasizing the role of the OB in stabilization. Since the OB is already represented by the limit object and stable functor set, adjusting the diagrams or limit/colimit objects might not necessarily be the primary approach. However, minor adjustments to the diagrams or limit/colimit objects could ensure that the Molar structures are fully stabilized and integrated, reflecting the enterprise's complex, dynamic nature.[24]

- In the healthcare system's new breast cancer program, the central health authority, i.e., OB, aims to integrate prevention, diagnosis, and treatment strategies. Recognizing the diverse protocols across departments, the OB seeks to unify these approaches. The prevention department focuses on community outreach and genetic screening, the diagnostic department emphasizes various imaging techniques, and the treatment department offers a range of therapeutic options. The OB identifies potential integrations: linking genetic screening with MRI protocols, aligning post-treatment follow-ups with mammography schedules, and expanding community outreach to cover treatment education. Adjustments are made to each department's protocols, creating a cohesive patient journey. Training sessions and resource allocations ensure these changes are effectively implemented, resulting in a comprehensive, patient-centric breast cancer program.

Regarding interactions in an Agenced Corporation, where the OB is in a continuous negotiation with Molecular forces, the challenge is to ensure that this process results in cohesive and consistent integrations, without leading to inconsistencies or conflicts within the enterprise. There's a need to adjust the established diagrams or limit/colimit objects to accommodate the new formations. This might involve changing the categories, objects, morphisms, or functors in the diagrams to ensure compatibility. By making necessary adjustments, the OB can ensure

---

[24] In the context of Nugus et al.'s (2010) study on integrated care in the emergency department (ED), the ED functions as a pivotal interface within the healthcare system, coordinating a multitude of services and patient care trajectories that span across various departments and external healthcare entities. This coordination is challenged by the diverse objectives, practices, and operational logics (diagrams) of these interconnected units, each with its 'ideal' state of care delivery (limit object) or potential diversification and expansion (colimit object). Addressing these challenges necessitates aligning the diverse operational models, goals, and practices of the ED with those of other departments and external healthcare providers to achieve a cohesive and integrated model of patient care. This might include aligning treatment protocols, patient transfer policies, and communication mechanisms to ensure that patient care is continuous, coherent, and responsive to the dynamic needs of patients across the healthcare continuum.

that the new formations introduced by the Molecular forces are integrated cohesively into the established framework. This approach involves a process of negotiation and compromise to agree on a common model.[25]

- In the realm of breast cancer research, new diagnostic tools and treatments—representing "Molecular forces"—continually emerge. The OB, a central health authority, faces the challenge of integrating these innovations into existing breast cancer management protocols. A recent discovery of a new genetic marker associated with high breast cancer risk exemplifies this. Previously, the diagnostic department primarily relied on mammography, ultrasound, MRI, and biopsies. Recognizing the marker's significance, the OB adjusted the diagnostic pathway: individuals testing positive for this marker now receive more frequent MRI screenings. Furthermore, the treatment protocol was revised to offer these patients a tailored chemotherapy regimen. To ensure a smooth transition, the OB initiated training sessions for healthcare professionals about the new marker and its implications. Feedback mechanisms were also established to monitor patient outcomes closely. Through this adaptive approach, the OB successfully negotiated with the Molecular forces, ensuring the healthcare system remains responsive and patient-centric.

Use of functors and natural transformations

In a healthcare organization, let us consider two departments: nursing and pharmacy. The nursing department, favoring a stable functor set and limit object focus, prefers established care processes and aims for a specific 'ideal' state of patient care. They resist changes and focus on optimizing current practices. Conversely, the pharmacy department, favoring a variable functor set and colimit object focus, advocates for adaptability and expansion. They are open to process changes based on new medical research or patient needs and aim to expand services or collaborations. This discrepancy in perceptions can lead to conflicts, with the nursing department viewing the pharmacy's proposed changes as disruptive, and the pharmacy department seeing the nursing department as rigid and resistant to innovation.

To resolve this, the organization can facilitate dialogs to understand these different perceptions and work toward a shared understanding. This might involve negotiating a balance between maintaining established processes and being open to change, and between aiming for a specific 'ideal' state and being open to new possibilities. This could result in a shared vision that incorporates both limit and colimit object focuses, and a shared strategy that allows for both stability and variability in the functor set. Recall from the discussion from the end of Chapter 8, where we posited the Platform as the higher order category that discursively

---

[25] As detailed in Polk et al.'s (2023) case study of Memorial Sloan Kettering (MSK) Cancer Center's shift toward precision oncology, compatibility issues addressed at MSK involved reconciling the traditional anatomic site-based organization of oncology with the novel, tissue-agnostic approach necessitated by precision medicine. This entailed a significant adjustment of the organizational and epistemic frameworks (diagrams) and the goals and practices (limit/colimit objects) to facilitate the integration of genomic and bioinformatic platforms, testing mechanisms, and patient cohort management systems. Negotiation and compromise among various stakeholders at MSK were pivotal in agreeing on a common model that could integrate the traditional and genomic-based approaches to cancer treatment. This process involved redefining the roles and affiliations of oncologists (as molecular champions), fostering interdepartmental collaborations, and establishing working groups to ensure that all necessary expertise was involved in research decision-making.

represents the diagrams of categories and functors as well as affords a way to align and steer the limit/colimit objects of different entities and departments across organizations and ecosystem. The examples above then fit into that context where the functor (e.g., dialogs etc.) would be the action of the territorializing functor Resourced Capabilities from Co-Innovation category to the Platform category.

In category theory, a transformation of diagrams can be seen as a natural transformation of functors. Consider a diagram as a functor D from an indexing category $\mathcal{I}$ into a category C. The objects and morphisms in the category $\mathcal{I}$ index the objects and morphisms of the diagram D in C. Therefore, to transform one diagram into another is equivalent to applying a natural transformation between two such functors. Here's a more detailed explanation:

1. **Diagrams as Functors:** A diagram in a category C can be thought of as a functor from an indexing category $\mathcal{I}$ to C. The objects of $\mathcal{I}$ index the objects in the diagram, and the morphisms of $\mathcal{I}$ index the morphisms in the diagram.
2. **Transforming Diagrams:** A transformation of one diagram into another can be seen as a natural transformation between the corresponding functors. Given two diagrams D: $\mathcal{I} \to C$ and D': $\mathcal{I} \to C$ (represented by two functors from the same indexing category $\mathcal{I}$ into the category C), a natural transformation η: D → D' gives a way to 'map' or 'transform' one diagram into the other. For each object i in $\mathcal{I}$, there is a morphism η_i: D(i) → D'(i) in C. These morphisms make the diagrams 'commute', meaning that for every morphism f: $i \to j$ in $\mathcal{I}$, the square in Fig. 10.1 commutes in C:

$$
\begin{array}{ccc}
D(i) & \xrightarrow{\eta\_i} & D'(i) \\
D(f) \downarrow & & \downarrow D'(f) \\
D(j) & \xrightarrow{\eta\_j} & D'(j)
\end{array}
$$

FIGURE 10.1 Transformation of diagrams as a natural transformation.

This expresses that transforming along the diagram using D and then applying η at the target is the same as applying η at the source and then transforming along the diagram using D'. In conclusion, transformations of diagrams in a category theoretical context correspond to natural transformations of functors that represent these diagrams.

The use of functors and natural transformations is highly suitable for most of the interactions. It provides a robust framework to bridge the gap between different structures, ensuring a unified, coherent model. The dynamic nature of the interactions, especially those involving the BwO, further underscores the need for such tools to reconcile differences and inconsistencies.

Regarding interactions in an Agencing Collective, where the BwO's surge in intensity challenges established Molar structures by introducing new Molecular processes, there is a clear introduction of potential inconsistencies in the established framework. The challenge is to integrate these new, variable processes into the existing Molar framework without causing disruptions. Adjusting the diagrams or limit/colimit objects might be a way to accommodate the new Molecular processes. However, this approach might not fully capture the dynamic nature of the BwO's surge in intensity and its effects on the Molar structures. Given that the BwO introduces new, variable functors that challenge the Molar structures, using functors and natural transformations seems highly relevant. A functor can map the new Molecular processes into the existing Molar framework, while a natural transformation can provide a bridge between the old and new processes. This approach provides a structured way to integrate the new Molecular processes introduced by the BwO into the existing Molar framework, ensuring a smooth transition and maintaining the integrity of the enterprise's model.[26]

- It's plausible to imagine that a groundbreaking "Nano-Imaging" technique emerges, promising early tumor detection at a cellular level, far surpassing traditional mammography. This innovation, representing a surge in the BwO intensity, challenges the established diagnostic pathway. Traditionally, breast cancer diagnosis starts with mammography, followed by supplementary tests like ultrasound or MRI, culminating in a biopsy for confirmation. To integrate Nano-Imaging seamlessly into this established pathway, we employ category theory tools: functors and natural transformations. We use two functors: one mapping the traditional pathway and another representing the new process with Nano-Imaging. A natural transformation then bridges these two functors, suggesting potential replacements or sequence alterations in the diagnostic steps. For instance, Nano-Imaging might precede or even replace the mammography step. Through this structured approach, the healthcare system can smoothly incorporate Nano-Imaging, ensuring patients benefit from its early detection capabilities without disrupting the existing diagnostic framework. This integration exemplifies how advanced mathematical concepts can guide the assimilation of innovative medical techniques into established protocols.

Regarding interactions in a shift from an Agenced Corporation to an Agencing Collective, where intense Molecular activity destabilizes the Organized Body (OB) leading to the emergence of the BwO, there's a clear emphasis on addressing the disruptions at the Molecular level. These disruptions can introduce nontrivial inconsistencies in the established structures, and the challenge is to reconcile these disruptions with the existing framework. The use of stacks helps in reconciling these disruptions, but there might be a need to adjust the diagrams or limit/colimit objects to accommodate the emerging BwO. Intense Molecular activity that challenges established diagrams and pathways can benefit from the use of functors and natural transformations. While the Molecular activity is represented by variable functors, using additional functors or natural transformations might be necessary to bridge the gap between the disrupted OB and the emerging BwO. Functors and natural transformations can provide

---

[26] In the context of scaling-up health innovations, as studied by Spicer et al. (2014), the diverse operational models (diagrams) and objectives (limit/colimit objects) of various stakeholders (government, development partners, civil society organizations, etc.) must be aligned, leveraging the theoretical framework of functors and natural transformations to achieve a cohesive and integrated approach, to ensure effective implementation and integration at scale. Capacity building among implementers to catalyze scale-up involves developing a shared understanding of the innovation and its potential impact, that is, finding a suitable functor that can transform disparate diagrams into a common form. Effective advocacy requires presenting strong evidence and engaging policy champions, that is, using natural transformations to bridge the gap between different stakeholders' objectives and the innovation's goals.

a 'bridge' between the disrupted OB and the emerging BwO. They can transform one category into another, preserving the structure, and thus help in reconciling the differences.[27]

- In breast cancer research, the established treatment protocols, symbolized as the OB, face disruption with the discovery of a new Molecular marker indicating potential resistance to certain chemotherapy drugs. This marker challenges the OB, hinting at a new treatment paradigm, the BwO. To reconcile this, we employ category theory tools. The current protocol, viewed as a limit object, represents standardized treatments. The new marker suggests an alternative pathway, seen as a colimit object, diversifying treatments based on the marker's presence. Using functors, we map both the established protocol and the new paradigm to a shared category. A natural transformation then bridges these functors, indicating adjustments in the treatment protocol for patients with the marker. Through this structured approach, the healthcare system can seamlessly integrate the new Molecular discovery into the existing treatment framework, ensuring personalized care and potentially improved patient outcomes.

Use of 2-categories or higher categories

In category theory, when you consider transformations of diagrams that involve categories, functors, sheaves, and natural transformations, you are moving to a higher-level categorical structure called 2-categories. In a 2-category:

- The objects are categories.
- The 1-morphisms (or simply morphisms) between two objects (categories) are functors.
- The 2-morphisms are natural transformations between functors.

The notion of transformation between such diagrams corresponds to modifications, which are essentially "natural transformations between natural transformations." Specifically, they are called 2-natural transformations or modifications. So, if one has two functors $F, G: C \to D$ and two natural transformations between them, $\alpha, \beta: F \Rightarrow G$, then a modification from α to β is a family of 2-morphisms in D such that certain coherence conditions (analogous to naturality squares) are satisfied. If we include sheaves in the picture, we are essentially talking about categories enriched over a base category, often the category of sets (Set), but potentially other categories like topological spaces (Top) or categories of sheaves. This takes us into the realm of enriched category theory and sheaf theory, which are sophisticated generalizations of the basic 2-categorical structures. The transformation of such enriched diagrams would then involve higher categorical notions, potentially including things like enriched functors, enriched natural transformations, and so on.

---

[27] In Thorndike's (2020) study on choice architecture in nudging supermarket customers toward healthier choices can be seen as an effort to apply a 'functor' aimed to map the domain of supermarket shelf arrangements to the codomain of consumer purchase behavior, with the hope of preserving the structural properties of healthier eating choices among the public. The lack of significant change, despite the intervention, underscore the need for a more complex functor or a series of functors, for example, integrating additional strategies such as price incentives, simplified health labels, and public health collaborations, which could act as natural transformations between functors, providing a 'bridge' that better aligns with consumer preferences and behaviors.

The use of 2-categories or higher categories seems to be quite suitable for most of the interactions, especially when there is a need for a more detailed and nuanced representation of the processes. The complex nature of the interactions, with multiple layers of dependencies and structures, underscores the potential benefits of moving to a higher level of abstraction.

Regarding interactions in a Structuring Collective, where strong Molar structures suppress the potentials of the BwO, the challenge is to ensure that the potentialities of the BwO are integrated into the overarching Molar structures without allowing them to dominate. Given that the Molar structures are represented by stable functors and they are suppressing the BwO's potentials (colimit objects), functors and natural transformations can provide a 'bridge' that ensures the BwO's potentials are integrated into the Molar structures without allowing them to dominate. However, if the suppression dynamics involves multiple layers of interactions and dependencies that introduce profound inconsistencies that cannot be addressed using functors and natural transformations, then moving to a higher level of abstraction like 2-categories might be necessary. This approach would provide a way to deal with the complex suppression dynamics, ensuring that the BwO's potentials are integrated into the Molar structures without dominating.[28]

- In the context of breast cancer treatment, established protocols (Molar structures) have been standardized based on extensive research. These protocols represent the OB. However, the rise of personalized medicine, tailored to individual genetic profiles, introduces a new realm of potential treatments, symbolizing the BwO. As personalized medicine gains traction, there is a challenge: how to integrate these new potentials without overshadowing the established protocols. To address this, we can turn to higher mathematical abstractions, specifically 2-categories. Within this framework, established treatments and personalized plans are viewed as objects. The transitions between them are 1-morphisms. The influence of established protocols on personalized treatments, ensuring they do not dominate, is represented as 2-morphisms. This 2-category approach models the intricate dynamics between traditional and personalized treatments, ensuring a balanced approach to patient care. In essence, while embracing the potentials of the BwO (personalized medicine), the integrity of the Molar structures (established protocols) is maintained, achieving an optimal balance in breast cancer treatment.

Regarding interactions in a shift from a Structuring Collective to an Agencing Collective, where the BwO challenges and dissolves rigid Molar structures, the main requirement is to ensure that the dissolution process does not lead to inconsistencies or conflicts within the

---

[28] In Trnka's (2021) exploration of multisited therapeutic assemblages in the context of youth mental health support, therapeutic practices and their efficacy are not merely anchored to single sites but are constituted through a network of shifting relations among multiple sites and elements. This complexity can lead to inconsistencies within therapeutic assemblages, such as conflicting practices or outcomes across different sites. The use of 2-categories or higher categories offers a methodological approach to reconcile these inconsistencies by viewing the assemblages as higher-dimensional structures where the inconsistencies are not merely problems to be solved but are integral to understanding the dynamic nature of therapeutic practices.

enterprise, and to find a way to reintegrate the dissolved structures into a new, unified model. Given that the BwO's potentials (colimit objects) are challenging and dissolving the established Molar structures, functors and natural transformations can provide a 'bridge' that ensures the dissolved structures are reintegrated in a way that aligns with the BwO's potentials. However, given the dissolution dynamics, where the BwO is actively dissolving the Molar structures, there might be complexities that cannot be addressed within the standard framework of category theory. Moving to a higher level of abstraction like 2-categories would provide a way to deal with the complex dissolution dynamics, ensuring that the dissolved Molar structures are reintegrated into a new, unified model that aligns with the BwO's potentials.[29]

- In the realm of breast cancer research, traditional treatments like mastectomy and chemotherapy (Molar structures) are being challenged by innovative therapies such as CRISPR gene editing (BwO). As these groundbreaking treatments show promise, they begin to dissolve long-standing medical practices. This dissolution poses a challenge: how to ensure consistent treatment delivery while integrating these new methods. To address this, we turn to the mathematical concept of 2-categories. In this framework, individual treatments are objects. The shift from a traditional to an innovative treatment is a 1-morphism, and the dynamics of how new treatments challenge established ones are 2-morphisms. For instance, the transition from mastectomy to CRISPR is a 1-morphism, with the challenge CRISPR poses to surgery being the 2-morphism. Using 2-categories, the medical community can model these complex interactions, ensuring that as traditional methods are modified or replaced, they are integrated into a holistic, advanced approach to breast cancer care.

Designing life-experience ecosystems via pragmatic semiotics of diagrams

Pragmatic Semiotics of Diagrams. In category theory, the process of interpreting a diagram and bringing forth a new set of objects and morphisms can be seen as the creation of a new category. The interpretation of a diagram can also lead to the identification or creation

---

[29] In Dixon et al.'s (2021) research study on the evolution and challenges of antibiotic use within Zimbabwe's healthcare system, the transition from Rational Drug Use (RDU) to Antimicrobial Stewardship (AMS) represents a shift from a Structuring Collective focus on individual patient health and system affordability to an Agencing Collective concern for the global threat of antimicrobial resistance (AMR), emphasizing the safeguarding of medication efficacy itself. The challenge of moving beyond a narrow focus on 'rational' versus 'irrational' use of antibiotics toward a reconfiguration of the global health architecture of healthcare, pharmaceutical distribution, and usage patterns, necessitates adopting 2-categories or higher categories.

of new objects and morphisms within an existing category.[30] This process can be understood in terms of the following concepts:

- Functors: Functors can be seen as the actions of the interpreter on the diagram. When the interpreter interacts with the diagram, they are effectively applying a functor that maps the objects and morphisms of the diagram to a new set of objects and morphisms. This can be seen as a transformation of the diagram, which corresponds to a change in the system of categories, functors, and natural transformations, or the creation of a new category.
- Natural Transformations: Natural transformations can be seen as the changes in the interpretation of the diagram by the interpreter. When the interpreter changes their interpretation of the diagram, they are effectively applying a natural transformation that maps one functor to another. This can be seen as a change in the way the diagram is understood, which corresponds to a change in the system of categories, functors, and natural transformations.
- Universal Constructions: Universal constructions, such as limits and colimits, can be seen as the outcomes of the interpreter's interaction with the diagram. When the interpreter interacts with the diagram, they are effectively determining the limit or colimit of a certain diagram in the category. This can be seen as the result of the interaction, which corresponds to a change in the system of categories, functors, and natural transformations or the creation of a new object or morphism in the new category.
- Cone and Cocone: In the context of a diagram, a cone is a way of "completing" the diagram by adding a new object and morphisms that make the diagram commute. Similarly, a cocone is a way of "capping" the diagram with a new object and morphisms. Both cones and cocones can be seen as ways of creating new objects and morphisms through the interpretation of a diagram.
- Adjunctions: An adjunction is a pair of functors that have a certain relationship. An adjunction can be seen as a way of creating new objects and morphisms through the interpretation of a diagram. The left adjoint functor can be seen as a way of "extending" the diagram by adding new objects and morphisms, while the right adjoint functor can be seen as a way of "restricting" the diagram by identifying certain objects and morphisms as equivalent.
- Yoneda Lemma: The Yoneda Lemma, a fundamental result in category theory, states that every category is fully and faithfully represented by its presheaf category. This can be interpreted as saying that every interpretation of a diagram (i.e., every functor from the diagram to a set of objects and morphisms) corresponds to a unique presheaf on the

---

[30] Cambrosio et al.'s (2022) study on cancer clinical research as 'oncopolicy' suggests that oncology clinical trials, by engaging in molecular profiling and the off-label use of drugs, create a new operational category within clinical research that transcends traditional trial objectives. These trials are not just about assessing the efficacy of treatments; they are about redefining the healthcare ecosystem, including regulatory practices, professional norms, and the evidential basis for treatment protocols. They pragmatically interpret and utilize the existing structures (the diagram of current oncology practice) to generate new objects (trial designs, policy initiatives) and morphisms (relations between different stakeholders) within the healthcare category.

diagram. This suggests that different interpretations of the diagram can lead to the creation of different presheaves, and hence different categories.[31]

Designing Life-Experience Ecosystems with Diagrams. Consider the limit/colimit object in C_9 that represents the CARE architecture by combining the objects and relationships from categories C_1 to C_8 through the diagram D and functors L_n. This construction captures the essence of the CARE architecture and its constituents as a single object in category C_9.

We can define natural transformations between the functors L_n to represent the interactions between the constituents of the CARE architecture. The limit/colimit construction of the CARE architecture, along with the natural transformations between the functors L_n, provides a more comprehensive understanding of the relationships between the categories and the constituents of the CARE architecture. This enables researchers and practitioners to study the interplay of events, actors, and the environment within the MLXE and design more engaging and immersive experiences.

By leveraging the universal properties of the limit and colimit objects, we can also gain insights into the most general and least constrained aspects of the CARE architecture. These insights can guide researchers in identifying areas where the architecture can be improved or refined, and help practitioners develop more effective strategies for managing and adapting to the rapidly evolving landscape of the MLXE.

In category theory, limits and colimits are used to describe universal properties in a category. They provide a way to talk about how objects and morphisms in a category relate to each other in a very general way. Here are a few ways they can be used:

- Describing Structures: Limits and colimits can describe various types of structures in a category. For example, the product of two objects in a category is a type of limit, and the coproduct (or sum) of two objects is a type of colimit. More complex structures can be described using more complex limits or colimits.
- Solving Diagrams: Limits and colimits can be used to "solve" diagrams in a category. A diagram in a category is a way of describing a particular pattern of objects and morphisms. The limit or colimit of a diagram is an object that represents the "solution" to that pattern, in the sense that it has a universal property with respect to the diagram.
- Functorial Properties: Limits and colimits are used to describe properties of functors. For example, a functor is said to preserve limits if it maps limits in one category to

---

[31] Cambrosio et al.'s (2022) study exemplifies how oncologists and trial designers, acting as interpreters, apply specific actions (functors) to the existing healthcare and research system (the diagram), transforming it into a new set of objects (trials, policies, practices) and morphisms (relations between trials and healthcare outcomes). As the understanding and interpretation of what constitutes effective clinical research evolve, so does the approach to designing and implementing these trials, moving toward a more integrated approach that blends research and policy-making, i.e., a natural transformation. Through the interaction with the healthcare and research system, these trials determine new "limits" (e.g., the limitations and possibilities of off-label drug use) or "colimits" (e.g., the expansion of drug indications based on molecular profiling), as universal constructions. The trials also serve to "complete" the healthcare policy diagram by introducing new objects (policies, guidelines) and morphisms (policy actions, healthcare practices) to reframe healthcare as a learning system, connecting epistemic, organizational, and economic issues seamlessly. Ultimately, every interpretation of a clinical trial and its policy implications (every functor) corresponds to a unique representation within the healthcare system (a presheaf), leading to the creation of diverse healthcare practices and policies (categories), showcasing the dynamic and multifaceted nature of oncology healthcare systems.

limits in another category. This is a useful property that can help us understand how functors behave.
- Constructing New Categories: Limits and colimits can be used to construct new categories from existing ones. For example, the category of all diagrams of a certain shape in a given category can be constructed using limits and colimits.
- Abstract Problem Solving: Limits and colimits provide a level of abstraction that can simplify problem solving. By working with these universal properties, category theorists can often prove results in great generality, which can then be applied to many specific cases.

In the context of DRT, limits and colimits could be used to describe the structure and dynamics of digital ecosystems in a very general and abstract way. They could provide a way to talk about how different parts of the ecosystem relate to each other, and how these relationships change over time.[32]

Case: Healthcare application of strategic architecturing as transformation of diagrams

Recall that, in Chapter 9, we had introduced a typology of organizational morphology and development, and illustrated the concepts using the context of breast cancer prevention, diagnosis, and treatment. Now, consider again New York City with its diverse population. The healthcare ecosystem in this city is vast, comprising multiple hospitals, clinics, research institutions, pharmaceutical companies, and tech startups focusing on health innovations. Each of these entities has its own set of practices, protocols, and priorities. The city's healthcare landscape is a complex web of interactions, collaborations, and sometimes, conflicts.

Let us consider a large hospital, "MetroHealth," as our primary focus. In NYC's healthcare landscape, strategic architecturing using concepts like stacks, diagrams, and limit/colimit objects becomes essential. It ensures that large entities like MetroHealth can provide seamless care despite their internal complexities. By understanding how each department's local practices can be glued into a global, coherent model, healthcare providers can ensure optimal patient care and institutional success.

The stack perspective: MetroHealth is not just a single entity but a conglomerate of various departments, ranging from cardiology to oncology, from outpatient care to intensive

---

[32] In Cambrosio et al. (2022), the integration of cancer clinical research and policy-making within the CARE architecture can be understood through the construction of limit and colimit objects that represent the synthesis of various elements from categories C_1 to C_8. Here, clinical research findings (emanating from one or multiple categories) serve as inputs that, through the diagram D and functors L_n, contribute to shaping the overarching healthcare ecosystem's policy and practice landscape. The limit object might represent an "ideal" state of cancer care derived from the synthesis of current knowledge and policy frameworks, while the colimit object could embody the potential expansion and diversification of cancer care practices and policies. Natural transformations between the functors L_n within the CARE architecture could represent the fluid exchange of insights, evidence, and imperatives between cancer research and policy domains. "Oncopolicy" substantiates the importance of these transformations by showcasing real-world instances where research and policy feedback into each other, driving the evolution of cancer care practices. Moreover, universal properties inherent in limit and colimit constructions facilitate a holistic understanding of how discrete elements within the cancer care ecosystem—ranging from individual patient experiences to broader policy initiatives—coalesce to inform and reshape the ecosystem.

research wings. Each department functions as its own mini-ecosystem, with specific professionals, equipment, and protocols. Using the concept of a stack, each department at MetroHealth can be viewed as a local section. The cardiology department might be using a specific ECG machine and following a set protocol for heart patients. In contrast, the oncology department might be collaborating with a pharmaceutical company for experimental cancer drugs. The "gluing" in this stack framework happens when patients require interdisciplinary care. For instance, a heart patient diagnosed with a tumor would need both the cardiology and oncology departments to collaborate. The stack's gluing conditions ensure that the patient receives seamless care, even though two different departments with potentially different protocols are involved.

Diagrammatic alignment and strategic architecture of organizational forms:

Within MetroHealth, different departments might operate based on different diagrams. The pediatric wing might prioritize patient-family centered care, focusing on both the child's health and the family's well-being. Their diagram might prioritize emotional well-being as much as physical health. In contrast, the intensive research wing, working on cutting-edge treatments, might have a diagram that prioritizes rapid data collection, experimentation, and publication.

- **The Challenge of Limit/Colimit Objects:** The pediatric wing's 'ideal' state (limit object focus) might be a fully holistic care model where every child patient and their family feel fully supported. On the other hand, the research wing might not have a fixed 'ideal' but is continuously expanding its understanding (colimit object focus), always looking for the next breakthrough.
- **Strategic Architecturing through Stack Analysis:** To ensure MetroHealth functions optimally, a stack analysis becomes crucial. This analysis would help identify if the diverse diagrams and limit/colimit objects across departments can be glued together coherently.
  - For instance, if the research wing develops a new drug beneficial for pediatric care, can it be integrated seamlessly into the pediatric wing's care model? If the pediatric wing's diagram prioritizes emotional well-being, how does the introduction of a new drug fit in? Does it cause anxiety or hope? The stack analysis would help in understanding these overlaps and potential points of friction.

Addressing Incompatibilities in MetroHealth's Organizational Framework: MetroHealth, as previously described, is a conglomerate of various departments, each operating with its unique diagrams and limit/colimit objects. As the organization grows and evolves, it is inevitable that some of these diagrams and objects might not be directly compatible, leading to potential conflicts or inefficiencies.

- **Adjustment of Diagrams or Limit Objects:**
  - **The Challenge:** The cardiology department at MetroHealth might have a diagram that prioritizes rapid patient diagnosis and treatment, with the 'ideal' state being immediate care for heart attack patients. In contrast, the administrative department might prioritize cost-effective care, with the 'ideal' state being the most efficient allocation of resources.

- **The Solution:** To create a unified model, MetroHealth could adjust these diagrams to include categories that represent both rapid care and cost-effectiveness. The new 'ideal' state might represent a balance between immediate patient care and resource allocation, ensuring that patients receive timely treatment without straining the hospital's resources.
- **Use of Functors:**
  - **The Challenge:** The radiology department, always keen on adopting the latest imaging technologies, might operate with a variable functor set and a colimit object focus, always looking for the next innovation. Meanwhile, the hospital's IT department, responsible for maintaining and securing patient data, might prefer a stable functor set and a limit object focus, emphasizing data security and stability.
  - **The Solution:** To bridge this gap, MetroHealth could introduce a functor that maps the radiology department's innovative drive to the IT department's focus on stability. This functor could represent a protocol where any new technology adoption in radiology undergoes a rigorous IT security assessment before implementation.
- **Use of Natural Transformations:**
  - **The Challenge:** The outpatient care department, focusing on patient follow-ups and long-term care, might have a diagram that emphasizes continuous patient monitoring. In contrast, the in-patient department might focus on acute care, emphasizing immediate treatment and rapid recovery.
  - **The Solution:** A natural transformation could serve as a bridge between these two departments. This transformation might be a protocol ensuring that once a patient moves from in-patient to outpatient care, there is a seamless transfer of medical data, ensuring continuous care without any gaps.
- **Use of 2-Categories or Higher Categories:**
  - **The Challenge:** MetroHealth's research wing, always collaborating with external institutions and pharmaceutical companies, might have complex interactions that can't be captured using standard category theory.
  - **The Solution:** By moving to a higher level of abstraction, such as 2-categories, MetroHealth can capture the intricate relationships between its research wing, external institutions, and pharmaceutical companies. This would allow for a more nuanced understanding of collaborations, intellectual property rights, and shared research goals.

Integrating the platform as the higher order category: In the context of MetroHealth, the Platform can be envisioned as a digital infrastructure that integrates all departments, ensuring seamless data flow, communication, and collaboration. This Platform would represent the diagrams of categories and functors and provide a way to align and steer the limit/colimit objects of different departments.

PART IV

Complex transformative emergence and evolution

CHAPTER 11

Emergent transformation in a machinic life-experience ecosystem

In **Part 3** (Chapters 7–10), we discussed in detail creative transformations and organizational configurations along a molar/molecular stack in an MLXE and within and across sections of MLXEs. We saw how the interplay between local (molecular) and global (molar) dynamics is emphasized, with sheaf theory analyses contextualizing how local cognitive processes intertwine with global dynamics. We discussed the concept and application of 'strategic architecturing' as transformation of diagrams. We saw how organizations, as assemblage configurations, navigate between their current and aspirational states. Diagrams represent the present organizational state, underscoring assemblage—organization interactions with limit objects indicating ideal configurations and colimit objects showcasing growth potential. The MLXE's architectural diagram is anchored in the platform category, subtly influencing entities. Addressing misalignments demands the design of co-innovation platforms, as entities actively mold these structures, with diagrams evolving through organizational morphogenesis, interwoven with meaning-making and boundary-crossing. We also saw how complex organizations operate as dynamic constructs within ecosystems, as intricate MLXEs organizationally, balancing adaptability (BwO) and stability (OB) with sheaf/stack analyses providing insights into assemblage—organization interactions.

In the preceding chapter, we ventured into the intricate world of organizational enterprises, portraying them not as monolithic entities but as vibrant, living MLXEs themselves. An enterprise MLXE is composed of a myriad of components—professionals, practices, technologies, and data—that come together in complex and often unpredictable ways. Within and between these enterprises, we find a rich network of intersecting communities of practice, each contributing its unique patterns to the overall fabric of the organization. These communities, whether local or global, are not isolated; they are in constant interaction, shaping and being shaped by the broader fields of practice in which they are situated.

However, traditional models for understanding these enterprise ecosystems often fall short. They tend to impose rigid structures and linear pathways onto systems that are inherently fluid and dynamic. These models, while useful in certain contexts, struggle to capture the nuanced interactions and overlaps between different entities within the ecosystem. They often simplify the rich, multifaceted relationships into binary connections, losing the depth and complexity that are essential to understanding the true nature of these systems.

Enter the mathematical concepts of stacks and sheaves, tools that offer a fresh perspective on these complex, overlapping entities. Unlike traditional models, stacks and sheaves embrace the inherent complexity of ecosystems. A sheaf, in this context, serves as a tool for tracking local data and how it patches together to form a more comprehensive, global picture. It respects the topology of the system, ensuring that data assigned to overlapping regions are compatible and coherent. A stack takes this idea further. It allows for more complex data and more intricate gluing conditions, accommodating situations where local data cannot be simply pieced together to form the global data. In essence, sheaves and stacks provide a robust framework to model the intricate interactions and overlaps between different entities within an ecosystem, offering a more nuanced and flexible approach than traditional models.

But as we delve deeper into the nature of these ecosystems, a compelling question arises: How do we understand the emergence and co-evolution within (sectors of) these complex systems? This question prompts us toward a need for a theoretical framework—one that can elucidate the dynamic processes that give rise to new forms and structures within these ecosystems, as shown in Fig. 11.1.

Immanence within a machinic life-experience ecosystem

The inevitability and persistence of creative emergence and evolution in MLXEs are underscored through the concept of 'immanence.' This principle dives deep into the very fabric of dynamism, spotlighting how potentialities and actualities are not mere possibilities but are immanent forces driving transformation in MLXEs. On the one hand, the monadic structure serves as a foundational tool in elucidating this perspective. Its unit transformation component highlights subjective experiences of emerging entities via the endofunctor's continuous transformational potential. The multiplication transformation captures the interconnectedness of entities. Incorporeal universes act as abstract transformational vector fields. Taken together, the monadic structure illustrates the inherent, ongoing metamorphosis present within entities of the ecosystem. Their intricate connections and influence on each other draw a map of the immanent forces at play. On the other hand, the gauge theory framework offers structured insights into the complex interplay of Existential Life Territories, Energetic-Signaletic Ecosystem Flows, Incorporeal Experience Universes, and Abstract Machinic Phyla, where the incorporation of Lagrangian dynamics emphasizes the kinetic and potential energies guiding the system's evolution, capturing both the movement and tension within the an MLXE's transformations delineating the perpetual processes of internalization and

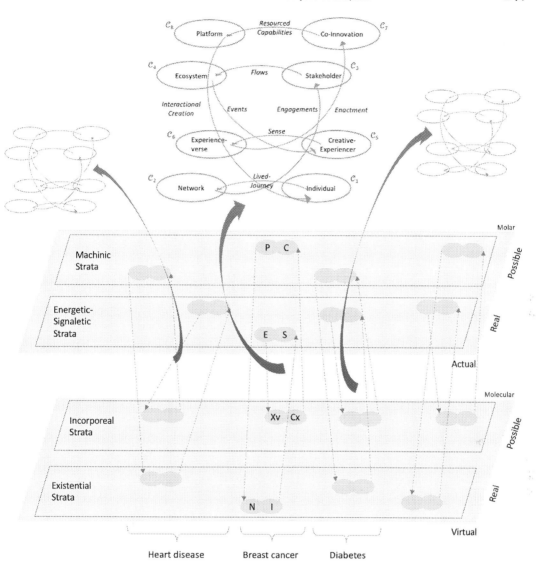

FIGURE 11.1 Creative emergence and co-evolution within MLXEs.

externalization and portraying the incessant folding and unfolding of entities in ecosystemic contexts. The contribution of gauge theory to evolutionary aspect of immanence will be the subject matter of Chapter 12. In this chapter, we will expand on the role of monadic structure for understanding immanence of emergence in MLXEs.

Imagine a primordial, virtual space—a space teeming with potential, where entities exist not as fixed identities but as dynamic processes of transformation. Within this space, we can envision a process of differentiation, a fracturing into distinct yet interconnected domains. This is not a mere unfolding of pre-existing possibilities but a creative emergence, a birth of new forms that spring from the virtual into actualized, concrete entities.[1] This process, intricate and multifaceted, calls for tools with the finesse to capture its complexity. Here, again, category theory and differential topology emerge as powerful lenses through which to view this fracturing and the subsequent evolution of the system.

In this light, the philosophy of Deleuze and Guattari offers invaluable insights. Their concept of 'lines of flight' represents movements of deterritorialization and transformation that disrupt and destabilize existing structures and systems, leading to the creation of new assemblages and territories. These lines of flight are not anomalies; they are integral to the very nature of ecosystems, driving change and evolution. They embody the vibrant, restless energy that propels systems toward new possibilities, shaping the trajectories of entities within the ecosystem.

As we navigate through this theoretical landscape, we discern a structure—a monadic structure—that provides a systematic approach for strategic analysis and development within ecosystems. This structure, informed by the analysis of the differentiated domains, serves as a guide to understanding the transformations that occur as a result of interactions within the system. It offers a blueprint for mapping the broader ecosystem of practices and experiences, identifying specific contexts or 'territories' within which entities operate, and analyzing the dynamic processes that respond to changes in these territories. Furthermore, this structure illuminates the guiding principles, goals, and values that shape responses to changes in the existential territories and energetic flows within the ecosystem. It allows us to examine the organization of the system and how it adapts to these changes, providing a comprehensive, integrated view of the ecosystem as a whole.

In summary, as we journey from the concept of enterprises as MLXEs, through the limitations of traditional models, and into the rich, complex world of sheaves and stacks, we are led to a significant realization. To truly understand the creative emergence and evolution within these ecosystems—to grasp the creative forces that drive their constant transformation—we require a theoretical framework that is as dynamic and multifaceted as the ecosystems themselves. In this chapter, we will start exploring such an extended framework, delving into the nuanced coordination of differentiation and emergence, and unveiling the profound insights it offers for the strategic analysis and development of vibrant, living MLXEs—whether as enterprises, offerings, or the broader landscape of interconnected ecosystems in which enterprises operate.

Monadic structure of differentiation into four domains

As we have discussed in Chapter 1, Deleuze's concept of **differentiation**, as articulated in "Difference and Repetition," refers to the process by which a virtual multiplicity actualizes

---

[1] Guattari (2013, pp. 52–53) describes this immanent process of creativity thus: "our project of a mapping of Effects and Affects ... have as its object ... Assemblages that can be subject to radical transformations, to schizzes or relinkages that change their configurations, to re-orderings through fluctuation, irrevocable implosions, etc ... They arise from a general Plane of immanence that implies them all in relations of presupposition that will be considered as so many levels of consistency of energy ... it is the fracturing of the Plane of Consistency that each one of these entities makes happen which manifests specific levels of energy ... it will only be possible to 'discernibilize these intensive entities and the quanta of energy relative to the consistency of their (actual and virtual) inter-relations through the complex Assemblages that semiotize them ..."

into diverse, distinct entities. This process is not a simple unfolding of pre-existing possibilities but rather a creative and dynamic emergence. In the context of DRT, we interpreted the emergence of the four domains (**L, E, X, M**) from the Plane of Consistency/Immanence (**P**) as a process of differentiation. Each domain actualizes a different aspect of the virtual multiplicity contained within **P**, becoming a distinct subcategory with its own objects and morphisms.

The monadic structure can be seen as the mechanism of this differentiation process. The endofunctor represents the internal dynamics of the primordial category, driving the process of actualization. The unit transformation represents the emergence of new entities, and the multiplication transformation represents the interaction and merging of these entities. We can use a monad to capture the idea that each entity, including Plane of Consistency/Immanence (**P**), is a world unto itself, with its own internal logic and potential for transformation.[2]

---

[2] In Part 3, we saw how the monad in category theory, consisting of the endofunctor, unit transformation, and multiplication transformation, serves as a powerful tool that encapsulates the intricate dynamics *within* categories. There, drawing parallels from differential topology, we established that: (a) the endofunctor mirrors the transformative flow of vector fields on pre-existing smooth manifolds and captures global structures akin to differential forms and de Rham cohomology, (b) the unit transformation provides foundational structure, reminiscent of the local Euclidean descriptions on manifolds, and (c) the multiplication transformation ensures the coherence of transformations, analogous to how Lie derivatives measure changes along flows. What we are about to develop here aims to provide a geometric and topological intuition for the abstract constructs of category theory in emergence and co-evolution in complex systems, particularly the monad and its components. Here is a brief reconciliation:

- There (in Part 3), the endofunctor was likened to a vector field on a pre-existing manifold, representing the dynamics of transformations *within* a category. Here (in Chapter 11), the endofunctor represents the internal dynamics of the Plane of Consistency/Immanence (**P**) as primordial category itself and gives birth to smooth manifolds through Energetic-Signaletic Flows (**E**). Both descriptions emphasize the continuous transformational nature of the endofunctor. The manifold structure captures this continuous change, and the vector field provides the directionality of these transformations.
- There, the unit transformation was related to the local Euclidean descriptions of points on a manifold, providing a foundational structure for transformations. Here, the unit transformation represents the creation of new entities within **P**, implemented by the Existential Territories (**L**) as topological spaces. Both descriptions focus on the foundational or "starting" nature of the unit transformation. The topological spaces capture the relationships and "closeness" of entities, aligning with the foundational role of the unit transformation.
- There, the multiplication transformation was likened to Lie derivatives, capturing the consistency and structure of transformations. Here, the multiplication transformation represents the merging of entities within **P**, implemented by the Abstract Machinic Phyla (**M**) as fiber bundles. Both descriptions emphasize the interaction and merging nature of the multiplication transformation. Fiber bundles, with their complex interconnections, provide a geometric representation of this merging.
- There, vector fields guide the transformations of the endofunctor and multiplication transformation. Here, each vector in the field represents a "force" or "direction" influencing transformations. Both descriptions align in their portrayal of the vector fields as guiding or influencing forces in the transformational dynamics.
- To recap, in Part 3, monadic structure accounts for Deleuze's differenciation within categories, highlighting evolution via endofunctor dynamics, unit transformation foundations, and coherent multiplication transformations. Chapter 11 shifts focus, using these components to explain foundational differentiation from the Plane of Consistency/Immanence (**P**) into domains (**L, E, X, M**), marking creative emergence.

- **Endofunctor**: The endofunctor represents the process of transformation within the Plane of Consistency/Immanence (**P**). In our model, this could be implemented by the Energetic-Signaletic Flows (**E**), represented as **smooth manifolds.** The manifold structure allows us to model the continuous transformations that entities in **P** can undergo, with each point on the manifold representing a possible state of an entity, and the paths on the manifold representing possible transformations.
    * The endofunctor of the monad could be seen as the transformation of these processes as the system evolves. The endofunctor T: **P** → **P** represents the process of transformation within the Plane of Consistency/Immanence (**P**). We can think of the objects of **P** (the entities) as being mapped to their transformed states within **P**. This transformation process can be understood in terms of the Energetic-Signaletic Flows (**E**), where each point on the smooth manifold represents a state of an entity (a MAB-CAE nexus), and the paths on the manifold represent the transformations of these states. When we apply the endofunctor T to an object in **P**, it maps the object to a new object in **P**, representing a transformed state of the entity. Similarly, when we apply T to a morphism in **P**, it maps the morphism to a new morphism in **P**, representing a transformed transformation process. This can be understood as moving from one point to another on the smooth manifold (for objects) and altering a path on the smooth manifold (for morphisms). This movement and alteration represent changes in the state and transformation process of the entity, respectively.
- **Unit Transformation**: The unit transformation represents the creation of a new entity within **P**. This could be implemented by the Existential Territories (**L**), represented as **topological spaces.** The topological structure allows us to model the "nearness" or "closeness" of entities in **P**, with each point in the space representing an entity, and the topology representing the relationships between entities.
    * The unit transformations of the monad could be seen as the creation of new narratives as entities navigate their experiences. The unit transformation υ: Id → T represents the initial or "default" transformation associated with each entity within **P**. This transformation can be metaphorically linked to the Existential Territories (**L**), which are represented as topological spaces. In this context, the Existential Territories (**L**) provide the 'space' where entities exist and undergo transformations. However, it is important to note that **L** is a category and not a functor, so the unit transformation doesn't directly map to **L** in a strict category theoretical sense.
- **Multiplication Transformation**: The multiplication transformation represents the merging of entities within **P**. This could be implemented by the Abstract Machinic Phyla (**M**), represented as **fiber bundles.** The fiber bundle structure allows us to model the complex interconnections between entities in **P**, with each fiber representing a possible set of connections between entities.
    * The multiplication transformation of the monad could be seen as the coordination and integration of these structures and organizations as they respond to changes. The multiplication transformation μ: T^2 → T represents the merging of entities within the Plane of Consistency/Immanence (**P**). This can be metaphorically linked to the Abstract Machinic Phyla (**M**), which are represented as fiber bundles. In the context of the multiplication transformation μ: T^2 → T, we can think of the objects of **P** (the entities) as being mapped to their merged states within **P**. This merging process can

be understood in terms of the Abstract Machinic Phyla (**M**), where each fiber in the fiber bundle represents a possible set of connections between entities. The merging of entities in **P** can then be seen as the formation of new fibers in the fiber bundle, representing new sets of connections.

- The Incorporeal Universes (**X**), represented as **vector fields**, while technically not a part of a monad, could play a role in guiding the transformations represented by the endofunctor and the multiplication transformation. Each vector in the field represents a "force" or "direction" that influences these transformations.
  - The vectors in the vector field could be seen as the "forces" or "directions" that these principles, goals, and values provide in guiding the transformations.

Assemblages can also be seen to act as functors between the subcategories of **P**, representing the transversal connections that Guattari emphasizes. An assemblage functor would take an object in one domain and map it to an object in another domain, and similarly for morphisms. In the context of our model, we can interpret these functors as processes of translation or transformation between the different domains. Each functor would take an entity in one domain and transform it into an entity in another domain, according to the specific rules and dynamics of the two domains. In the context of a monad, the functors we have described could be seen as additional structure that complements and interacts with the monadic operations. Here's a possible interpretation.

- **Functor from L to E (Engagement)**: This functor could be interpreted as the process by which an existential territory (a point in a topological space) gives rise to an energetic-signaletic flow (a path on a smooth manifold). This could represent, for example, the way in which a personal identity (**L**) can generate a process of change or transformation (T). This functor could be seen as a way of "exporting" the entities created by the unit transformation in **L** (which creates new entities within **P**) into the domain of T. This would allow the transformations represented by the endofunctor in T to act on these entities.
- **Functor from E to X (Event)**: This functor could be interpreted as the process by which an energetic-signaletic flow (a path on a smooth manifold) gives rise to an incorporeal universe (a vector in a vector field). This could represent, for example, the way in which a process of change or transformation (T) can generate a new idea or concept (**X**). This functor could be seen as a way of "exporting" the transformed entities from **E** (which are transformed by the endofunctor) into the domain of **X**. This would allow the vectors in **X** to guide further transformations of these entities.
- **Functor from X to M (Enactment)**: This functor could be interpreted as the process by which an incorporeal universe (a vector in a vector field) gives rise to an abstract machinic phylum (a fiber in a fiber bundle). This could represent, for example, the way in which an idea or concept (**X**) can generate a complex system or structure (**M**). This functor could be seen as a way of "exporting" the entities guided by the vectors in **X** into the domain of **M**. This would allow the fiber bundles in **M** to create complex interconnections between these entities.
- **Functor from M to L (Interactional Creation)**: This functor could be interpreted as the process by which an abstract machinic phylum (a fiber in a fiber bundle) gives rise to an existential territory (a point in a topological space). This could represent, for example, the way in which a complex system or structure (**M**) can generate a new

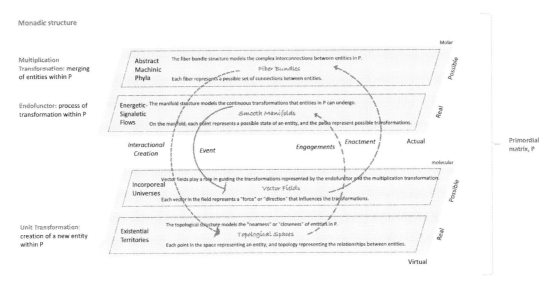

FIGURE 11.2 Monadic structure of differentiation into L, E, X, and M.

personal identity (**L**). This functor could be seen as a way of "importing" the interconnected entities from **M** back into the domain of **L**. This would allow the topological spaces in **L** to create new entities based on these interconnections, completing the cycle and setting up the next round of transformations.

This model captures the key elements of Guattari's philosophy: the emphasis on transformation and change, the multiplicity of domains of existence, and the transversal connections between these domains. It also makes use of the mathematical structures of category theory and differential topology to provide a rigorous and precise representation of these ideas. Through the interplay of these four domains, the virtual multiplicity of **P** fractures into a rich array of concrete entities, each with its own identity and relationships, yet all part of the same dynamic process of becoming.

Following Fig. 11.2, let's construct a theoretical narrative that captures the fracturing of the Plane of Consistency/Immanence (**P**) into the four subcategories (**L**, **E**, **X**, **M**) using the tools of category theory and differential topology:

- The Plane of Consistency/Immanence (**P**) is a primordial category, a virtual multiplicity teeming with potential. It is a space of pure becoming, where entities exist not as fixed identities but as dynamic processes of transformation. The objects in this category are the entities of the world, and the morphisms are the transformations that these entities can undergo.
- The fracturing of **P** into the four subcategories (**L**, **E**, **X**, **M**) can be seen as a process of differentiation, where the virtual multiplicity of **P** actualizes into distinct, concrete entities. This process is not a simple unfolding of pre-existing possibilities (what Deleuze called 'differenciation'), but a creative emergence, a birth of new forms from the womb of the virtual.

- The Existential Territories (**L**), represented as topological spaces, are where new entities emerge within **P**. As the unit transformation of our monad, **L** gives birth to new entities, embedding them within the fabric of **P**. Each point in the space represents an entity, and the topology represents the relationships between entities. **L** is the realm of identity, the space where entities carve out their own territories of existence.
- The Energetic-Signaletic Flows (**E**), represented as smooth manifolds, play a crucial role in this process. As the endofunctor of our monadic structure, **E** drives the internal dynamics of **P**, guiding the continuous transformations of entities. Each point on the manifold represents a possible state of an entity, and the paths on the manifold represent possible transformations. **E** is the engine of change, the force that propels the entities of **P** along their paths of becoming.
- The Incorporeal Universes (**X**), represented as vector fields, guide the transformations represented by the endofunctor and the multiplication transformation. Each vector in the field represents a "force" or "direction" that influences the transformations. **X** is the realm of the abstract, the space of ideas and concepts that shape the trajectories of entities within **P**.
- The Abstract Machinic Phyla (**M**), represented as fiber bundles, are where entities within **P** merge and interact. As the multiplication transformation of our monad, **M** weaves together the entities of **P**, creating complex interconnections. Each fiber represents a possible set of connections between entities. **M** is the realm of platformized interaction, the space where entities come together to form complex systems and structures.[3]

---

[3] Rhodes' (2018) study on methadone treatment's implementation and its effects on individuals' recovery processes in Kenya illustrate the process of differentiation:

- Recovery Potential as unit transformation in Existential Territories (**L**): The narratives of recovery potential shared by participants, where methadone clears the mind and promises a future beyond addiction signal the birth of new identities and trajectories out of the virtual potential of recovery.
- Methadone's Implementation as endofunctor in Energetic-Signaletic Flows (**E**): The dynamic implementation of methadone treatment, including the material aspects of its delivery and the social interactions it fosters, guide the continuous change and actualization of recovery potential into observable and lived experiences.
- Guiding Forces as vector fields in Incorporeal Universes (**X**): The desire for normalcy, social acceptance, and a drug-free life shapes the trajectory of individuals' recovery, influencing the paths of transformation within the broader assemblage of methadone treatment.
- Interconnected Changes as fibers in Abstract Machinic Phyla (**M**): The complex interconnections formed through recovery processes, such as restored familial relationships and social re-acceptance, exemplify the creation of complex systems and structures out of the interactions between distinct entities.
- Methadone treatment assemblages as functors of transversality: The transformation of individuals' lives through methadone treatment, from their existential territories to their social relations and self-perceptions, exemplifies the translation and transformation processes across different domains.

Case: Healthcare system and COVID-19

Let us consider the healthcare system as a virtual multiplicity within the Plane of Consistency/Immanence (**P**). In this context, the entities could be patients, healthcare providers, medical facilities, diseases, treatments, and so on. The transformations could be the processes of diagnosis, treatment, recovery, and so on.

- Existential Territories (**L**): In the healthcare context, these could be the personal health experiences of individuals. Each individual's health status, medical history, and personal experiences with the healthcare system could be seen as a point in a topological space. The "closeness" or "nearness" in this space could represent similarities in health status or experiences. For example, two individuals with similar health conditions might be "close" in this space. The functor from **L** to **E** could represent the process by which personal health experiences influence the dynamics of the healthcare system, such as how individual health outcomes contribute to public health data.
- Energetic-Signaletic Flows (**E**): These could represent the dynamic processes within the healthcare system, such as the flow of patients through the system, the administration of treatments, the spread of diseases, and so on. These processes could be modeled as transformations on a smooth manifold, with each point on the manifold representing a state of the healthcare system, and the paths on the manifold representing possible transformations. The functor from **E** to **X** could represent the influence of these dynamic processes on the guiding principles and goals of the healthcare system.
- Incorporeal Universes (**X**): These could represent the guiding principles, goals, and values of the healthcare system. For example, principles of medical ethics, goals for public health, and societal values about healthcare could be seen as vectors in a vector field, guiding the transformations within the healthcare system. The functor from **X** to **M** could represent the influence of these guiding principles on the structure and organization of the healthcare system.
- Abstract Machinic Phyla (**M**): These could represent the structure and organization of the healthcare system. For example, the network of healthcare providers, the organization of medical facilities, the structure of health insurance systems, and so on could be seen as fibers in a fiber bundle. Each fiber represents a possible set of connections within the healthcare system. The functor from **M** to **L** could represent the influence of the system's structure on individual health experiences.

Through the interplay of these four domains and the action of the functors between them, the virtual multiplicity of the healthcare system fractures into a rich tangle of concrete entities, each with its own identity and relationships, yet all part of the same dynamic process of becoming. This is the healthcare system as seen through the lens of Deleuze and Guattari, a system of constant transformation and creative emergence, a system where everything is interconnected and interdependent.

In the context of the healthcare system, the monadic structure can be seen as the mechanism that drives the differentiation of the Plane of Consistency/Immanence (**P**) into the four subcategories (**L, E, X, M**), and guides the transformations within and between these categories.

- Endofunctor: The endofunctor in our monadic structure could be seen as representing the internal dynamics of the healthcare system. It captures how the system changes and evolves over time, driven by the processes of diagnosis, treatment, disease spread, and so on. This aligns with our interpretation of the Energetic-Signaletic Flows (**E**) as representing the dynamic processes within the healthcare system.
- Unit Transformation: The unit transformation could be seen as representing the creation of new entities within the healthcare system. This could include the emergence of new patients, new diseases, new treatments, and so on. This aligns with our interpretation of the Existential Territories (**L**) as representing the personal health experiences of individuals.
- Multiplication Transformation: The multiplication transformation could be seen as representing the merging or interaction of entities within the healthcare system. This could include the interaction between patients and healthcare providers, the combination of different treatments, the spread of a disease within a population, and so on. This aligns with our interpretation of the Abstract Machinic Phyla (**M**) as representing the structure and organization of the healthcare system.

In this way, the monadic structure provides a mathematical framework for understanding the complex dynamics of the healthcare system. It captures the continuous transformations that occur within the system, the emergence of new entities, and the interactions between entities. By mapping these elements of the monadic structure to the four subcategories (**L**, **E**, **X**, **M**), we can gain a deeper understanding of how the healthcare system evolves and changes over time, and how it can fracture into a rich ensemble of concrete entities, each with its own identity and relationships.

Let us consider the emergence and spread of COVID-19 as a process of differentiation within the healthcare system, fracturing from a virtual multiplicity in the Plane of Consistency/Immanence (**P**) into a rich array of concrete entities. We will iterate through the four domains (**L**, **E**, **X**, **M**) three times, representing three distinct phases of the pandemic. This process of differentiation was driven by the monadic structure, with the endofunctor, unit transformation, and multiplication transformation guiding the transformations within and between the four domains (**L**, **E**, **X**, **M**).

Phase 1: Emergence of the virus

- *Existential Territories* (**L**): In late 2019, the first cases of a novel coronavirus were reported in Wuhan, China. These initial patients, their symptoms, and their experiences with the healthcare system became new points in the topological space of **L**, creating new existential territories.
  - The unit transformation, represented by the Existential Territories (**L**), was the process by which new entities emerged within the healthcare system. This included the initial patients in Wuhan, and the new hospital wards and protocols created to treat them.
- *Energetic-Signaletic Flows* (**E**): The virus quickly spread within Wuhan and to other parts of China. This rapid spread of the virus could be modeled as a transformation on a smooth manifold, with each point on the manifold representing a state of the healthcare system, and the paths on the manifold representing possible transformations. The manifold of **E** was reshaped by this new energetic-signaletic flow.

- The endofunctor, represented by the Energetic-Signaletic Flows (**E**), guided the internal dynamics of the healthcare system as it responded to the emergence of the virus. This included the rapid spread of the virus within Wuhan and to other parts of China, and the transformations within the healthcare system to manage this spread.
- *Incorporeal Universes* (**X**): As the severity of the virus became apparent, new guiding principles and goals were established for the healthcare system. These included principles for managing the spread of the virus, such as quarantine measures and contact tracing, and goals for treating infected individuals. These principles and goals created new vectors in the field of **X**, guiding the transformations within the healthcare system.
- *Abstract Machinic Phyla* (**M**): The healthcare system in Wuhan and other affected areas had to rapidly adapt to the emergence of the virus, creating new structures and organizations. These included new hospital wards for treating COVID-19 patients, and new protocols for testing and quarantine. These new structures and organizations could be seen as new fibers in the fiber bundle of **M**, representing new connections within the healthcare system.
  - The multiplication transformation, represented by the Abstract Machinic Phyla (**M**), was the process by which these new entities interacted and merged within the healthcare system. This included the interactions between different parts of the healthcare system in Wuhan, and the merging of resources to create new structures and organizations.

Phase 2: Global spread of the virus

- *Existential Territories* (**L**): By March 2020, the virus had spread globally, leading the World Health Organization to declare a pandemic. Millions of people around the world became infected, creating new existential territories. Each of these individuals, their experiences with the virus, and their interactions with the healthcare system became new points in the topological space of **L**.
  - The unit transformation included the millions of people around the world who became infected, and the new testing centers and protocols created to manage the pandemic.
- *Energetic-Signaletic Flows* (**E**): The global spread of the virus represented a new dynamic process within the healthcare system. This process could be modeled as a transformation on a smooth manifold, with each point on the manifold representing a state of the healthcare system, and the paths on the manifold representing possible transformations. The manifold of **E** was reshaped by this new global energetic-signaletic flow.
  - The endofunctor included the spread of the virus to new countries and regions, and the transformations within healthcare systems around the world to manage this spread.
- *Incorporeal Universes* (**X**): The global spread of the virus created new vectors in the field of **X**, representing new guiding principles and goals for the healthcare system. These included principles for managing the global spread of the virus, such as international

travel restrictions, and goals for treating infected individuals on a global scale. These principles and goals guided the transformations within the global healthcare system.
- *Abstract Machinic Phyla* (**M**): Healthcare systems around the world had to adapt to the global spread of the virus, creating new structures and organizations. These included new testing centers, new protocols for social distancing and mask wearing, and new systems for tracking and reporting cases. These new structures and organizations could be seen as new fibers in the fiber bundle of **M**, representing new global connections within the healthcare system.
  - The multiplication transformation included the interactions between different healthcare systems around the world, and the merging of resources to create new structures and organizations.

Phase 3: Vaccination

- *Existential Territories* (**L**): By the end of 2020, several vaccines for COVID-19 had been developed and approved for use. As individuals received vaccinations, they became new points in the topological space of **L**, creating new existential territories. Their postvaccination health status and experiences with the vaccination process added to the complexity of **L**.
  - The unit transformation included the individuals who received vaccinations, and the new vaccination centers and protocols created to administer the vaccines.
- *Energetic-Signaletic Flows* (**E**): The vaccination process represented a new dynamic process within the healthcare system. This process could be modeled as a transformation on a smooth manifold, with each point on the manifold representing a state of the healthcare system, and the paths on the manifold representing possible transformations. The manifold of **E** was reshaped by this new energetic-signaletic flow of vaccination.
  - The endofunctor included the distribution of vaccines to different countries and regions, and the transformations within healthcare systems to manage this process.
- *Incorporeal Universes* (**X**): The vaccination process created new vectors in the field of **X**, representing once again new guiding principles and goals for the healthcare system. These included principles for managing the distribution of vaccines, such as prioritizing healthcare workers and vulnerable populations, and goals for achieving herd immunity. These principles and goals guided the transformations within the healthcare system during the vaccination process.
- *Abstract Machinic Phyla* (**M**): The healthcare system had to adapt to the vaccination process, creating new structures and organizations. These included new vaccination centers, new systems for tracking vaccination status, and new protocols for vaccinating large populations. These new structures and organizations could be seen as new fibers in the fiber bundle of **M**, representing new connections within the healthcare system related to vaccine distribution.
  - The multiplication transformation included the interactions between different parts of the healthcare system to manage the vaccination process, and the merging of resources to create new structures and organizations.

In each phase, the COVID-19 pandemic fractured from a virtual multiplicity in **P** into a rich amalgam of concrete entities, each with its own identity and relationships, yet all part of the same dynamic process of becoming.

Differentiation of ecosystem sectors via double articulation

The process of differentiation, as described by Deleuze, can be seen as the process by which these domains are articulated or brought into being. This involves a "double articulation"—the formation of machinic assemblages of bodies (the physical, material aspects of the system) and collective assemblages of enunciation (the social, communicative aspects of the system). Each domain is thus characterized by a unique combination of these two types of assemblages, reflecting the specific dynamics and processes associated with that domain.

Recall that the Plane of Consistency/Immanence, that is, category **P**, is the overarching category that encompasses all entities and their potential transformations. The objects in this category, denoted as **Obj(P)**, are the abstract entities (MAB-CAE nexuses) of the world. These entities are not static but are dynamic processes of transformation. The morphisms in this category, denoted as **Mor(P)**, are the potential transformations that these entities can undergo. These transformations are not just physical changes but also include changes in the way these entities are understood, represented, and articulated, reflecting the double articulation of substance and form of content and expression. Also remember that the monad is a triad (T, υ, μ), where T is the endofunctor, υ is the unit transformation, and μ is the multiplication transformation.

The fracturing of **P** into the four subcategories (**L, E, X, M**) can be seen as a process of differentiation, where the virtual multiplicity of **P** differentiates into distinct, yet still virtual, domains. As we have seen, this process is not a simple unfolding of pre-existing possibilities, but a creative emergence, a birth of new forms from the womb of the virtual. This emergence is guided by the double articulation of substance and form of content and expression, which shapes the way these entities are formed and understood (see Fig. 11.3).

In the language of category theory and differential topology, we can describe the differentiation of the Plane of Consistency/Immanence (**P**) into the four subcategories (**L, E, X, M**) as follows:

1. Existential Territories (**L**): Let us denote this category as **L**. The objects in this category, denoted as **Obj(L)**, are the states or conditions of entities from **P**.[4] These can be represented as points in a topological space. The morphisms in this category, denoted as **Mor(L)**, are the changes in these states or conditions over time or due to specific events. These can be represented as paths in the topological space. **L** is the realm of identity, the space where entities carve out their own territories of existence (or 'cutouts,' as introduced in Chapter 1).

---

[4] The entities of the world (Obj(**P**)) have specific states or conditions that are represented as objects in the Existential Territories (Obj(**L**)).

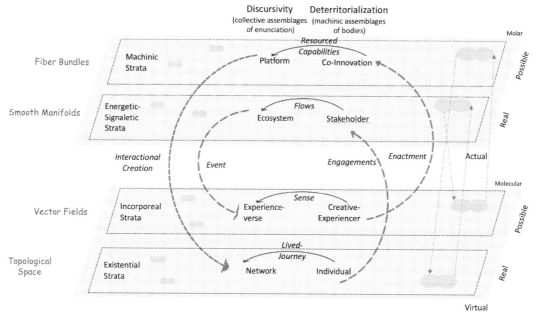

FIGURE 11.3 Differentiation of ecosystem sectors via double articulation.

- Here, the double articulation of substance and form of content and expression shapes the way these entities are formed and their relationships are structured. The substance of content could be the physical and emotional experiences of individuals, while the form of content could be the specific ways these experiences are structured. The substance of expression could be the personal narratives and discourses that individuals create to make sense of their experiences, and the form of expression could be the specific ways these narratives are articulated. The MABs here could be the entities themselves, and the CAEs could be the narratives and discourses they create. We can think of the substance and form of content as the actual state of the Machinic Assemblages of Bodies (MABs) and the substance and form of expression as the actual state of the Collective Assemblages of Enunciation (CAEs) within the Existential Territories (**L**). Both states together form a point in the topological space, representing the entire nexus of MAB and CAE at a specific moment in time within **L**. The paths in the topological space then represent the transformations of this nexus over time, reflecting the dynamic interplay and transformations of both MAB and CAE within **L**. A functor from **L** to **E** exports these new entities into the domain of **E**, where they can undergo transformation.
2. Energetic-Signaletic Flows (**E**): Let us denote this category as **E**. The objects in this category, denoted as **Obj(E)**, are the states of a system, that is, dynamic processes

within the entities from **P**.[5] These can be represented as points on a smooth manifold. The morphisms in this category, denoted as **Mor(E)**, are the transformations of these states due to various processes. These can be represented as paths on the smooth manifold. **E** is the engine of change, the force that propels the entities of **P** along their paths of becoming (or 'complexions' in the language of Chapter 1).

- Here, the double articulation of substance and form of content and expression shapes the transformations that these entities undergo and the paths they follow. The substance of content could be the actual processes, while the form of content could be the specific ways these processes are structured. The substance of expression could be the policies, guidelines, and protocols that guide these processes, and the form of expression could be the specific ways these policies and guidelines are articulated. The MABs here could be the entities involved in these processes, and the CAEs could be the policies, guidelines, and protocols that guide them. We can think of the substance and form of content as the actual state of the Machinic Assemblages of Bodies (MABs) and the substance and form of expression as the actual state of the Collective Assemblages of Enunciation (CAEs) within the Energetic-Signaletic Flows (**E**). Both states together form a point on the smooth manifold, representing the entire nexus of MAB and CAE at a specific moment in time within F. The paths on the smooth manifold then represent the transformations of this nexus over time, reflecting the dynamic interplay and transformations of both MAB and CAE within **E**. A functor from **E** to **X** exports these transformed entities into the domain of **X**, where they can be guided by abstract concepts and ideas.

3. Incorporeal Universes (**X**): Let us denote this category as **X**. The objects in this category, denoted as **Obj(X)**, are guiding principles, goals, and values that influence the transformations within the entities from **P**. These can be represented as vectors in a vector field. The morphisms in this category, denoted as **Mor(X)**, are the influences these exert on the transformations within a system, or the entities from **P**. These can be represented as transformations of the vectors in the vector field. **X** is the realm of the abstract, the space of ideas and concepts (or 'constellations,' as we called them in Chapter 1) that shape the trajectories of entities within **P**.

- Here, the double articulation of substance and form of content and expression is crucial, shaping the way these abstract concepts and ideas guide the transformations. The substance of content could be the actual principles, goals, and values, while the form of content could be the specific ways these principles, goals, and values are prioritized and balanced. The substance of expression could be the public discourses, debates, and narratives that articulate these principles, goals, and values, and the form of expression could be the specific ways these discourses, debates, and narratives are conducted. The MABs here could be the entities who articulate these

---

[5] The term "system" in the context of Energetic-Signaletic Flows (**E**), Abstract Machinic Phyla (**M**), and Incorporeal Universes (**X**) corresponds to the Plane of Consistency/Immanence (**P**) and the dynamic nature of the entities and transformations within it. The "system" refers to the dynamic processes, connections, and influences within the entities of the world (Obj(**P**)) and their transformations (Mor(**P**)). These aspects of the system are represented as objects and morphisms in the categories **E**, **M**, and **X**. Note that "system" refers to all MAB-CAE nexuses and their interactions and transformations within the Plane of Consistency/Immanence (**P**), not just a single MAB–CAE nexus.

principles, goals, and values, and the CAEs could be the public discourses, debates, and narratives that shape them. A functor from **X** to **M** exports these guided entities into the domain of **M**, where they can form complex interconnections.

4. Abstract Machinic Phyla (**M**): Let us denote this category as **M**. The objects in this category, denoted as **Obj(M)**, are specific sets of connections within a system, or the entities from **P**. These can be represented as fibers in a fiber bundle. The morphisms in this category, denoted as **Mor(M)**, are the creation, modification, or dissolution of these connections. These can be represented as transformations of the fibers in the fiber bundle. **M** is the realm of interaction, the space where entities come together to form complex systems and structures (see Chapter 1 for the concept of 'rhizomes').

   * Here, the double articulation of substance and form of content and expression is at work, shaping the way these entities interact and the connections they form. The substance of content could be the actual structures and organizations, while the form of content could be the specific ways these structures and organizations are arranged and coordinated. The substance of expression could be the administrative policies, procedures, and protocols that structure and organize the healthcare system, and the form of expression could be the specific ways these policies, procedures, and protocols are articulated. The MABs here could be the entities that structure and organize the system, and the CAEs could be the policies, procedures, and protocols that guide them. We can think of the substance and form of content as the actual state of the Machinic Assemblages of Bodies (MABs) and the substance and form of expression as the actual state of the Collective Assemblages of Enunciation (CAEs). The merging of these states (MAB-CAE nexuses) can be seen as the formation of new fibers in the fiber bundle, representing new sets of connections. The paths on the fiber bundle then represent the transformations of these connections over time, reflecting the dynamic interplay and transformations of both MAB and CAE. A functor from **M** to **L** imports these interconnected entities back into the domain of **L**, where they can form new identities, completing the cycle and setting up the next round of transformations.

In this framework, **P** is the overarching category that encompasses all entities and their potential transformations. **L, E, X,** and **M** are subcategories of **P** that focus on specific aspects of these entities and their transformations. The objects and morphisms in **L, E, X,** and **M** are derived from the objects and morphisms in **P**, reflecting the specific focus of each subcategory. In each category, the double articulation process involves the interplay between the substance and form of content (the objects and their transformations) and the substance and form of expression (the understanding, representation, and articulation of these objects and transformations) within each subcategory. This interplay can be represented mathematically as a functor from the category to itself, which maps each object to its image under the double articulation process and each morphism to its image under the double articulation process.[6]

In Deleuze and Guattari's philosophy, lines of flight represent movements of deterritorialization and transformation that disrupt and destabilize existing structures and systems, leading to the creation of new assemblages and territories (see Fig. 11.4). Using this concept, we can narrate the functors between the different strata and assemblages as follows:

- From Existential Territories to Economy of Flows: The functor maps collective assemblages of enunciation on the existential territories strata (unary, discontinuous, mixture processes within the virtual real, or the potentialities that are realized in subjective, lived experiences) to machinic assemblages of bodies on the economy of flows strata (finite, reversible, close-to-equilibrium processes within the actual real, or the concrete, material processes of economic production and exchange). This represents a shift from the existential, subjective, and lived experiences of individuals and communities to the impersonal, objective, and systemic processes of economic production and exchange. This transition can be seen as a line of flight that deterritorializes the existential territories and reterritorializes them within the economy of flows, i.e., actualizes the unary, discontinuous, mixture processes of the virtual real within the finite, reversible, close-to-equilibrium processes of the actual real.
  - This could represent the transition from individual lived experiences and personal narratives of breast cancer (existential territories) to the broader healthcare system

---

[6] Rhodes' (2018) study on methadone treatment in Kenya demonstrates the complex interplay of narrative and material elements in the recovery process from addiction:

- **Topological Spaces in Existential Territories (L)**: The interplay between physical/emotional realities of addiction and treatment (substance of content—SC) and the structured treatment protocols (form of content—FC) alongside the individual narratives of recovery (substance of expression—SE) and the collective discourse on treatment (form of expression—FE) creates a dynamic, topological space. This space maps the diverse, continuous recovery journeys, emphasizing the adaptability of individual paths within a collectively structured treatment framework.
- **Smooth Manifolds of Energetic-Signaletic Flows (E)**: The interplay between the operational dynamics (SC) and the organization of treatment delivery (FC), alongside policies and professional discourses (SE) against public health efforts (FE), creates a manifold reflecting the healthcare system's structure. This framework emphasizes how localized operational changes and interactions within methadone distribution centers contribute to the broader treatment ecosystem's coherence, demonstrating the seamless integration of individual experiences with systemic objectives.
- **Vector Fields in Incorporeal Universes (X)**: The interplay between guiding principles (SC), strategic planning (FC), collective aspirations (SE), and narrative constructions (FE) forms a vector field within the methadone treatment ecosystem. These elements collectively direct recovery trajectories, influencing both the pace and direction of individual journeys towards normalization and social reintegration, thereby manifesting a structured yet dynamic pathway towards achieving desired health outcomes and societal acceptance.
- **Fiber Bundles of Abstract Machinic Phyla (M)**: The dynamic interplay between personal transformation (SC), evolving support networks (FC), advocacy efforts (SE), and collaborative societal responses (FE) forms a complex structure. Each individual's recovery journey (a point in the base space) connects to a multitude of potential pathways and outcomes (fibers), illustrating the interconnected and multi-dimensional nature of recovery within the methadone treatment ecosystem.

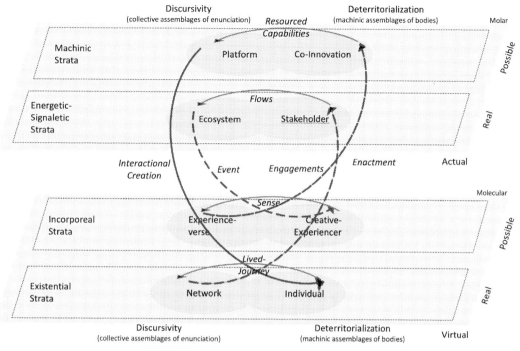

FIGURE 11.4 Lines of flight between strata.

and economic structures that govern access to prevention, diagnosis, and treatment (economy of flows). For example, a woman's personal experience with a breast cancer scare could lead to a greater understanding of the need for regular screenings, which could influence healthcare policies and funding decisions.
- From Economy of Flows to Incorporeal Universes: The functor maps collective assemblages of enunciation on the economy of flows strata (plural, continuous, fusional processes within the actual real) to machinic assemblages of bodies on the incorporeal universes strata (infinite, irreversible, far-from-equilibrium processes within the virtual possibility, or the potentialities that exist in the realm of symbolic, immaterial processes of meaning-making and significant). This represents a shift from the material, economic processes of production and exchange to the immaterial, symbolic processes of meaning-making and significant. This transition can be seen as a line of flight that deterritorializes the economy of flows and reterritorializes them within the incorporeal universes, that is, deterritorializes the plural, continuous, fusional processes of the actual real and reterritorializes them within the infinite, irreversible, far-from-equilibrium processes of the virtual possibility.
  - This could represent the transition from the economic and systemic processes of the healthcare system (economy of flows) to the symbolic and cultural meanings of breast cancer (incorporeal universes). For example, the allocation of resources for breast cancer research and treatment could influence societal perceptions and

understandings of the disease, shaping narratives around survivorship, femininity, and health.

- From Incorporeal Universes to Machinic Phyla: The functor maps collective assemblages of enunciation on the incorporeal universes strata (unary, discontinuous, of mixture processes within the virtual possibility) to machinic assemblages of bodies on the machinic phyla strata (infinite, irreversible, far-from-equilibrium processes within the actual possibility, or the potentialities that are realized in the technical, material processes of machine production and operation). This represents a shift from the symbolic, immaterial processes of meaning-making and significant to the technical, material processes of machine production and operation. This transition can be seen as a line of flight that deterritorializes the incorporeal universes and reterritorializes them within the machinic phyla, i.e., actualizes the unary, discontinuous, of mixture processes of the virtual possibility within the infinite, irreversible, far-from-equilibrium processes of the actual possibility.
  * This could represent the transition from the symbolic and cultural meanings of breast cancer (incorporeal universes) to the technical and material processes of medical research and treatment (machinic phyla). For example, societal narratives around breast cancer could drive innovation and development in medical technologies and treatment strategies.
- From Machinic Phyla to Existential Territories: The functor maps collective assemblages of enunciation on the machinic phyla strata (plural, continuous, fusional processes within the actual possibility) to machinic assemblages of bodies on the existential territories strata (finite, reversible, close-to-equilibrium processes within the virtual real). This represents a shift from the technical, material processes of machine production and operation to the existential, subjective, and lived experiences of individuals and communities. This transition can be seen as a line of flight that deterritorializes the machinic phyla and reterritorializes them within the existential territories, i.e., deterritorializes the plural, continuous, fusional processes of the actual possibility and reterritorializes them within the finite, reversible, close-to-equilibrium processes of the virtual real.
  * This could represent the transition from the technical and material processes of medical research and treatment (machinic phyla) back to the individual lived experiences and personal narratives of breast cancer (existential territories). For example, advancements in medical technology could lead to more effective treatments and improved quality of life for individuals living with breast cancer.

In each case, the functor represents a line of flight that moves between different types of processes (unary, discontinuous, mixture vs. plural, continuous, fusional; and finite, reversible, close-to-equilibrium vs. infinite, irreversible, far-from-equilibrium) and between the virtual and the actual, the possible and the real. This sequence of functors captures the dynamic interplay between the existential, economic, symbolic, and technical dimensions of reality, and the constant flux and transformation between different levels and dimensions of reality. In each case, the functor represents a line of flight that moves between different dimensions of the breast cancer experience, from the personal to the systemic, the symbolic to the technical. This sequence captures the complex interplay between individual experiences, societal structures, cultural narratives, and medical technologies in the context of breast cancer prevention, diagnosis, and treatment.[7]

Case: Healthcare system and COVID-19 (Continued)

Let us delve into how the double articulation of substance and form of content and expression between machinic assemblages of bodies (MABs) and collective assemblages of enunciation (CAEs) is implicated in each of the subcategories (L, E, X, M) in the **context of the COVID-19 pandemic**:

- **Existential Territories (L)**: In the context of COVID-19, the Existential Territories (L) could be seen as the personal experiences of individuals as they navigate the pandemic. In this category, objects could be individuals' health statuses, and morphisms could be the changes in these statuses over time or due to specific events (like receiving a vaccine). The category could entail a multiplicity of topological spaces, each representing a different aspect of personal health experiences. A typical point in the topological space could be an individual's health status at a specific point in time.
  * The MABs here could be the individuals themselves, and the CAEs could be the personal narratives and discourses they create. The substance of content here could be the physical and emotional experiences of individuals - getting tested, falling ill, recovering, getting vaccinated, etc. The form of content could be the specific ways these experiences are structured - the sequence of events, the progression of the disease, the process of vaccination, etc. The substance of expression could be the personal narratives and discourses that individuals create to make sense of their experiences. The form of expression could be the specific ways these narratives are articulated - through social media posts, conversations with others, personal diaries, etc. The double articulation process here could involve the interplay between the physical health status of individuals (substance and form of content) and their personal experiences and narratives of their health (substance and form of expression). For instance, a person's health status (an object) changes from being unvaccinated to vaccinated (a morphism). This change is a form of content that is then expressed in the person's narrative of their health experience. This narrative, in turn, becomes part of the CAE around vaccination, influencing societal understanding and discourse.

---

[7] Rhodes (2018) reveals how methadone, as an "evidence-making intervention," deterritorializes individuals from their entanglement with drug addiction (Existential Territories) and reterritorializes them within new assemblages of recovery hope and social inclusion (Economy of Flows and Incorporeal Universes). Methadone's "becoming" in the social fabric of Kenya serves as a line of flight, disrupting previous structures of addiction and enabling new territories of recovery potential, showcasing the transition from the virtual (potentialities of recovery) to the actual (realized changes in individuals and their social worlds). Moreover, in a movement from incorporeal universes to machinic phyla, symbolic and cultural understandings of addiction and recovery influence the material practices of methadone treatment delivery, e.g., the methadone queue, as a material dynamic of treatment delivery, embodies a space where potentiality (hope for recovery) and actuality (the everyday realities of accessing treatment) coexist and interact.

- The unit transformations of the monad could be seen as the creation of new personal narratives as individuals navigate the pandemic. The double articulation of substance and form of content and expression between MABs and CAEs could entail the unit transformations of the monad by creating new points in the topological space. For example, when a person gets vaccinated, a new point representing their postvaccination health status is created in the topological space.
- **Energetic-Signaletic Flows (E)**: In the context of COVID-19, the Energetic-Signaletic Flows (**E**) could be seen as the dynamic processes within the healthcare system as it responds to the pandemic. Here, objects could be states of the healthcare system, and morphisms could be the transformations of these states due to various processes (like the spread of a disease or the implementation of a new policy). The category could entail a multiplicity of smooth manifolds, each representing a different process within the healthcare system. A typical point on the smooth manifold could be a state of the healthcare system at a specific point in time.
  - The MABs here could be the healthcare workers, hospitals, and other entities involved in these processes, and the CAEs could be the policies, guidelines, and protocols that guide them. The substance of content here could be the actual processes—testing, treating, vaccinating, etc. The form of content could be the specific ways these processes are structured—the sequence of steps in testing or treatment, the distribution of vaccines, etc. The substance of expression could be the policies, guidelines, and protocols that guide these processes. The form of expression could be the specific ways these policies and guidelines are articulated - through official documents, press releases, public announcements, etc. The double articulation process could involve the interplay between the actual processes and transformations in the healthcare system (substance and form of content) and the interpretation, representation, and discourse around these processes (substance and form of expression). For example, the spread of COVID-19 (a morphism) changes the state of the healthcare system (an object). This change is a form of content that is then expressed in public health communications, policy decisions, and societal discourse, forming part of the CAE around the pandemic response.
  - The endofunctor of the monad could be seen as the transformation of these processes as the pandemic evolves. The double articulation of substance and form of content and expression between MABs and CAEs could entail the endofunctor of the monad by transforming points on the smooth manifold. For example, as the healthcare system adapts to the pandemic, the state of the system changes, moving along a path on the smooth manifold.
- **Incorporeal Universes (X)**: In the context of COVID-19, the Incorporeal Universes (**X**) could be seen as the guiding principles, goals, and values that shape the response to the pandemic. In this category, objects could be guiding principles, goals, and values, and morphisms could be the influences these exert on the transformations within the healthcare system. The category could entail a multiplicity of vector fields, each representing a different set of guiding principles, goals, and values. A typical vector in the vector field could be a specific principle, goal, or value that influences the response to the pandemic

- The MABs here could be the public health officials, scientists, and policymakers who articulate these principles, goals, and values, and the CAEs could be the public discourses, debates, and narratives that shape them. The substance of content here could be the actual principles, goals, and values - public health, scientific integrity, social responsibility, etc. The form of content could be the specific ways these principles, goals, and values are prioritized and balanced. The substance of expression could be the public discourses, debates, and narratives that articulate these principles, goals, and values. The form of expression could be the specific ways these discourses, debates, and narratives are conducted - through media coverage, public discussions, social media debates, etc. The double articulation process could involve the interplay between these abstract principles, goals, and values (substance and form of content) and their concrete impacts on the healthcare system and societal understanding of health (substance and form of expression). For instance, a principle of public health (an object) influences the healthcare system's response to the pandemic (a morphism). This influence is a form of content that is then expressed in the actions taken, policies implemented, and discourse around public health, contributing to the collective assemblage of enunciation (CAE) around public health.
- The vectors in the vector field could be seen as the "forces" or "directions" that these principles, goals, and values provide in guiding the response to the pandemic. The double articulation of substance and form of content and expression between MABs and CAEs could entail a "force" or "direction" that influences the endofunctor and the multiplication transformation of the monad by guiding the transformations within the healthcare system. For example, the principle of public health could be a vector that guides the healthcare system's response to the pandemic, influencing the transformations represented by the endofunctor and the multiplication transformation.

- **Abstract Machinic Phyla (M)**: In the context of COVID-19, the Abstract Machinic Phyla (M) could be seen as the structure and organization of the healthcare system as it responds to the pandemic. In this category, objects could be specific sets of connections within the healthcare system, and morphisms could be the creation, modification, or dissolution of these connections. The category could entail a multiplicity of fiber bundles, each representing a different aspect of the structure and organization of the healthcare system. A typical fiber in the fiber bundle could be a specific set of connections within the healthcare system, such as the network of healthcare providers involved in treating Covid-19 patients.
  - The MABs here could be the administrative bodies, organizations, and systems that structure and organize the healthcare system, and the CAEs could be the policies, procedures, and protocols that guide them. The substance of content here could be the actual structures and organizations - hospitals, testing centers, vaccination sites, etc. The form of content could be the specific ways these structures and organizations are arranged and coordinated. The substance of expression could be the administrative policies, procedures, and protocols that structure and organize the healthcare system. The form of expression could be the specific ways these policies, procedures, and protocols are articulated - through official documents, organizational charts, procedural manuals, etc. The double articulation process could involve the

interplay between the actual structure and organization of the healthcare system (substance and form of content) and the understanding, representation, and discourse around this structure and organization (substance and form of expression). For example, a network of healthcare providers treating COVID-19 patients (an object) is expanded with the addition of new providers (a morphism). This expansion is a form of content that is then expressed in the understanding and discourse around the healthcare response to the pandemic, forming part of the collective assemblage of enunciation (CAE) around healthcare provision.

- The multiplication transformation of the monad could be seen as the coordination and integration of these structures and organizations as they respond to the pandemic. The double articulation of substance and form of content and expression between MABs and CAEs could entail the multiplication transformation of the monad by creating new fibers in the fiber bundle. For example, as new healthcare providers join the network of those treating COVID-19 patients, new fibers representing these new connections are created in the fiber bundle.

CHAPTER 12

Evolutionary transformation in a machinic life-experience ecosystem

In the previous chapter, we saw how the inevitability and persistence of emergence and evolution in ecosystems dives deep into the very fabric of dynamism, spotlighting how potentialities and actualities are not mere possibilities but are immanent forces driving transformation in ecosystems. We saw how the monadic structure serves as a foundational tool in elucidating this perspective. Its unit transformation component highlights subjective experiences of emerging entities via the endofunctor's continuous transformational potential. The multiplication transformation captures the interconnectedness of entities. Incorporeal universes act as abstract transformational vector fields. Taken together, the monadic structure illustrates the inherent, ongoing metamorphosis present within entities of the ecosystem. Their intricate connections and influence on each other draw a map of the immanent forces at play. While gauge theory provides a mathematical framework for understanding the complex interplay of physical fields, which permeate the universe and govern the fundamental forces of nature, DRT brings gauge theory into the realm of social-behavioral phenomena, emphasizing the kinetic and potential energies guiding the evolution of MLXEs, capturing both the movement and tension within its transformations, delineating the perpetual processes of internalization and externalization, and portraying the incessant folding and unfolding of entities in dynamic relationalities.

Grounding social-behavioral phenomena with gauge theory

As we introduced at the outset in Chapter 1, **P** is the overarching total space that encompasses all entities and their potential transformations. **L**, **E**, **X**, and **M** are subspaces of **P** that focus on specific aspects of these entities and their transformations. The Plane of Consistency/Immanence (**P**) serves as a comprehensive category in the universe of social-behavioral dynamics, encompassing all entities and their potential transformations.

The entities in **L**, **E**, **X**, and **M** can be seen as "sections" of **P**, and the transformations in **L**, **E**, **X**, and **M** can be seen as "connections" on **P**. This mirrors the structure of gauge theory, where the total space is a fiber bundle over the base space, with the fibers representing the internal states of the fields at each point in space-time. The fields themselves are represented as sections of this fiber bundle, and the forces are represented as the curvature of a connection on the fiber bundle.

The Plane of Consistency/Immanence (**P**) serves as the total space, an overarching category that encompasses all healthcare entities - physicians, nurses, patients, nonhuman artifacts, technologies, and their potential transformations. This is akin to the total space in gauge theory, which represents the entire physical system under consideration, including all possible states of the fields.

The stage for this complex interplay of the MLXE is the Existential Territories (**L**), akin to the base space in gauge theory. Instead of a traditional spacetime, we propose the concept "L-Spacetime," a fusion of physical spacetime with the condition or situation of the system, providing a more comprehensive view of the socio-behavioral ecosystem. It effectively captures the multidimensional and dynamic nature of the system, where entities and their states are not just situated in physical space and time but also in a broader context that includes their specific conditions or situations. Each point in these territories represents a state or condition of an entity, and the structure of the territories allows us to describe changes and motions within this existential space.

- In the Existential Territories (**L**), the Machinic Assemblages of Bodies (MAB) and Collective Assemblages of Enunciation (CAE) play a crucial role in defining and shaping the states or conditions of entities. The MAB, representing the physical, biological, and material aspects of the healthcare system, contributes to the tangible, observable states of entities. For instance, the physical health of a patient, the skills and expertise of a healthcare provider, or the capabilities of a medical device would be aspects of the MAB. On the other hand, the CAE, representing the discursive, symbolic, and communicative aspects of the healthcare system, contributes to the intangible, conceptual states of entities. These could include the diagnosis of a disease, the treatment plan for a patient or the healthcare policies that guide practice.

The Energetic-Signaletic Flows (**E**), represented by dynamic processes on smooth manifolds, entail elements in this societal configuration defined by flows that represent the states of a system, that is, dynamic processes within the entities from **P**. They exist at every point in the existential territories, and their values determine the transformations experienced by entities. However, these flows are not static entities; they have internal degrees of freedom that can change without affecting the healthcare phenomena they govern. This is where the concept of double articulation comes into play. These articulations can change the internal states of the flows without changing the observable transformations, reflecting a fundamental symmetry of the healthcare dynamics, similar to the gauge transformations in gauge theory. Through the process of double articulation, the MAB and CAE interact and influence each other, shaping the trajectories of entities through the Existential Territories, the dynamics of the Energetic-Signaletic Flows, the guiding principles of the Incorporeal Universes, and the connections within the Abstract Machinic Phyla. This process allows the healthcare system to adapt and evolve in response to changes in both the physical, tangible aspects of the system and the discursive, symbolic aspects of the system, reflecting a fundamental symmetry of the healthcare dynamics.

- In the Energetic-Signaletic Flows (**E**), the MAB and CAE influence the dynamic processes within the entities. The MAB could influence these processes through physical changes, such as the progression of a disease or the application of a treatment. The CAE could influence these processes through changes in understanding, communication, or decision-making, such as a new interpretation of a patient's symptoms or a change in treatment guidelines.

The Abstract Machinic Phyla (**M**) captures this additional layer of structure. The base space of these phyla is the Existential Territories (**L**) and attached to each point in this base space is a fiber representing the internal states of the entity at that point. The entities themselves can be seen as sections of this fiber bundle, mathematical entities that assign a value in the fiber to each point in the base space. This mirrors the structure of gauge theory, where the fibers represent the internal "degrees of freedom" or "states" that particles can have at each point in space-time.

- In the Abstract Machinic Phyla (**M**), the MAB and CAE contribute to the specific sets of connections within the system. The MAB could influence these connections through physical, tangible interactions, such as the relationships between healthcare providers and patients or the protocols for using a medical device. The CAE could influence these connections through intangible, conceptual interactions, such as the communication between healthcare providers and patients or the guidelines for diagnosing and treating a disease.

Finally, to describe how these entities interact and evolve, the model introduces the Incorporeal Universes (**X**). These universes represent guiding principles, goals, and values that influence the transformations within the entities from **P**. These serve a role similar to the forces in gauge theory, which represent the influences or effects that the fields have on particles.[1]

- In the Incorporeal Universes (**X**), the MAB and CAE shape the guiding principles, goals, and values that influence the transformations within the entities. The MAB could influence these principles, goals, and values through tangible, observable changes in the healthcare system, such as the development of a new medical technology or the results of a clinical trial. The CAE could influence these principles, goals, and values through changes in discourse, policy, or understanding, such as a shift in healthcare policy or a new theory of disease.

---

[1] The relationship between the Incorporeal Universes (**X**) and the Energetic-Signaletic Flows (**E**) is that the Incorporeal Universes provide a measure of the "intensity" or "strength" of the Energetic-Signaletic Flows at each point in the Territorial Spacetime. They are computed from the values of the Energetic-Signaletic Flows and their derivatives. In this way, the Incorporeal Universes (**X**) would represent the "difference" in the Energetic-Signaletic Flows (**E**). However, in Deleuze's philosophy, difference is not something that exists between two identical entities, but rather something that is constitutive of the entities themselves (Deleuze, 1994). Thus, we should interpret the Energetic Signaletic Flows (**E**) not as static entities with fixed identities but as dynamic processes that are constantly becoming through their interactions with the Incorporeal Universes (**X**). The Incorporeal Universes (**X**), in turn, could be seen not as external forces that act upon the Energetic-Signaletic Flows, but as integral aspects of the flows themselves, shaping and being shaped by them in a continuous process of mutual becoming. In this way, the relationship between the Energetic-Signaletic Flows (**E**) and the Incorporeal Universes (**X**) would not be one of cause and effect, but one of mutual constitution and transformation, i.e., an event, or a "quasi-cause" (Deleuze, 1990), represented by F_4 in our MLXE framework.

In this way, the intricate meshing of the healthcare entities, their symmetries, and their interactions, is fully described within the mathematical framework of this model, just as the intricate meshing of the physical fields, their symmetries, and their interactions, is fully described within the mathematical framework of gauge theory. Each field is governed by a gauge-invariant Lagrangian, a function on the space of all possible field configurations that determines the equations of motion for the field. The gauge symmetry of the theory reflects the fundamental physical principle that the laws of physics should be the same at every point in spacetime. Similarly, the healthcare entities and their transformations are guided by the double articulation of substance and form of content and expression.

The healthcare system can be visualized as a fiber bundle, with the base space representing the Plane of Consistency/Immanence (**P**), the fibers representing the Abstract Machinic Phyla (**M**), and the sections representing the entities. The MAB-CAE nexuses are points in the fibers, and the Incorporeal Universes (**X**) are vectors in the base space. This visualization captures the complex and interconnected nature of the healthcare system, where multiple entities are interacting and evolving over time, and where a single moment or situation can involve many different entities and states.

- Fiber Bundle: Visualize the base space as the Plane of Consistency/Immanence (**P**), where each point corresponds to a distinct existential territory or condition, a 'cutout' within larger landscape of healthcare. Attached to each of these points is a fiber, which, while representing the Abstract Machinic Phyla (**M**), embodies the myriad of specific connections and potential interactions that emanate from or influence the corresponding existential condition. Thus, each fiber not only represents the abstract machinic connections but also encapsulates the dynamic potentialities and interactions associated with its point of origin in the existential territories. In essence, the fibers extend from the existential territories (**L**), through the intricate web of Energetic-Signaletic Ecosystem Flows (**E**), influenced by the guiding principles of Incorporeal Experience Universes (**X**), to express the complex relational dynamics of health states within the healthcare ecosystem.
- Section: A section of this fiber bundle is a curve that cuts through the cylinder, touching each fiber exactly once. It can be visualized as a line spiraling around the surface of the cylinder. Each section represents an entity such as physicians, nurses, patients, or technologies, selecting a specific state from the fiber at each point in the base space.
- MAB-CAE Nexus: Each point in the fiber represents a specific Machinic Assemblages of Bodies (MAB) and Collective Assemblages of Enunciation (CAE) nexus. This nexus represents a specific configuration or state of the healthcare system at a given moment.
- Multiple Sections: Since multiple entities can be involved in a single MAB-CAE nexus, and a single entity can be involved in multiple MAB-CAE nexuses over time, visualize this as multiple sections (lines) criss-crossing each point on a fiber. This captures the complex and interconnected nature of the healthcare system.

Returning to the visualization in Fig. 1.5 in Chapter 1, we see the health and mobility ecosystems as fiber bundles, with each fiber emanating from a different existential territory or condition within the respective ecosystem. In the healthcare ecosystem, we have fibers representing different health states: healthy, acute disease, and chronic disease. Each point in these fibers represents a unique Machinic Assemblage of Bodies (MAB) and Collective Assemblage of Enunciation (CAE) nexus, or in simpler terms, a unique combination of physical, biological, material, discursive, symbolic, and communicative aspects of the healthcare system.[2]

Let us take John, the physician, as an example. John exists in all three fibers of the healthcare ecosystem, representing his role as a healthcare provider across different health states. He is a constant presence in the healthcare field, regardless of the health state of his patients. This is represented by the line connecting the points representing John in each fiber, forming a section across the healthcare field. Now, let us consider Sara, the patient. Sara exists in two fibers of the healthcare ecosystem: healthy and chronic disease. This represents her journey as a patient, moving between different health states. Like John, Sara forms a section across the healthcare field, but her section is limited to the fibers representing the health states she experiences.

In the mobility ecosystem, we have a fiber representing new car sales. Both John and Sara exist in this fiber, representing their roles in this ecosystem. John, as a customer, and Sara, as a car sales agent, form sections across the (auto)mobility field. This figure illustrates the concept of gauge invariance within a fiber. Despite their different roles and experiences, both John and Sara maintain their identities across different fibers and fields. Their roles and experiences might change depending on the fiber or field they are in, but their fundamental identities remain the same. This is the essence of gauge invariance: the internal states of entities can change without affecting their observable transformations.

Moreover, this figure shows how entities can move between different fibers via connections, representing their interactions and transformations within and across different ecosystems. These connections form a complex network that spans across the entire Territorial Spacetime, reflecting the intricate web of relationships and interactions within and between different ecosystems.

In DRT, each entity is fully described by considering their involvement (sections) across all fibers and fields. This comprehensive description is made possible by gauge theory, which provides a mathematical framework for capturing the complex interplay of entities and their transformations within and across different ecosystems. The ultimate goal is to find the appropriate Gauge-invariant Lagrangian for the motion of each entity, which would provide a complete description of their dynamics within and across different ecosystems.

---

[2] In her work on social capital and its application to global public health initiatives, Campbell (2020) shows how Treatment Action Campaign (TAC) and movements for mental health are formed within specific socio-political contexts (**L**), where marginalized groups experience adverse health outcomes due to unequal social determinants of health. The creation and utilization of social capital networks by these movements effectively change the flows within the socio-behavioral system (**E**), challenging existing power structures and advocating for redistribution of health-enabling resources. These networks cut across various levels of social organization (**M**), from local communities to global alliances, where each "fiber" represents a pathway of potential transformation within the system, influenced by the collective action of these movements. The movements' guiding principles, objectives, and values, serve as the forces (**X**) that drive transformations within the socio-behavioral system, compelling government and industry stakeholders to acknowledge and act upon the health rights of marginalized populations. Here, TAC's successful advocacy for HIV/AIDS treatment access in South Africa can be interpreted as a section of the fiber bundle, where the collective action of individuals and communities selected specific states (improved healthcare policies and access to treatment) from the fiber at each point in the base space (the socio-political and healthcare landscape).

From global immanence to well-being, wealth, empowerment, and welfare

As discussed earlier, the Plane of Consistency/Immanence (**P**), conceptualized as a primordial, virtual landscape, is a space of pure potentiality, where entities exist not as fixed identities but as dynamic processes of transformation. This plane sets the stage for the unfolding of complex social-behavioral dynamics, serving as the foundational ontology of this theoretical framework. The fracturing of **P** into subcategories (**L, E, X, M**) is not a mere division but a creative emergence. This fracturing represents the actualization of the virtual multiplicity of **P** into distinct, concrete entities, each embodying unique dynamics and characteristics.

Central to this framework are the unified gauge theories (for technical details, see the section titled "Applying gauge theory to social-behavioral phenomena" in the **Appendix** and Figure 1 therein), which encompass Desire and Territory (see boxed discussions in Chapters 3 and 4), Identity and Structure (see boxed discussion in Chapter 5), Significant and Transversality (see boxed discussion in Chapter 9), Organization (see boxed discussion in Chapter 10), Folding and Ecosophy, and Differentiation (latter two are covered in this Chapter 12). These theories serve as metaphysical blueprints, offering a lens through which complex systems—ranging from human desires to societal structures—can be analyzed and engaged with. They elucidate how the fracturing of **P** can be understood as a structured, yet dynamic, process. For the purposes of our analysis in this chapter, first, the Unified Gauge Theory of Folding and Ecosophy is spotlighted. It is posited as a metaphysical blueprint that integrates the evolutionary dynamics that follow the fracturing of **P** with the key sectors (Nature, Economy, Society, and Technology). This theory, with its gauges of Fold and Extended Ecosophy, offers a profound lens through which the complex interrelations between entities in these sectors and their ecological contexts can be understood. It provides a way to conceptualize how entities within these complex "NEST-networked" systems are continuously folding and unfolding in relation to their environments, embodying the dynamic interplay between the virtual (**P**) and the actual (**L, E, X, M**).

The Fold, inspired by Deleuze's interpretation of Leibniz, symbolizes the intricate processes of internalization and externalization, where the outside is continuously drawn into the inside, creating a complex and intricate internal landscape.[3] Extended Ecosophy, building upon Guattari's four domains of existence, offers a multidimensional perspective that captures the multifaceted, layered nature of reality and existence.[4] These domains—Existential Territories, Energetic-Signaletic Flows, Incorporeal Universes, and Phyla or Abstract Machines—are not isolated; they are in constant dialog, and transformations in one domain invariably resonate through the others.

Gauge-invariant Lagrangian and the gauge transformations of the Fold and Extended Ecosophy in Unified Gauge Theory of Folding and Ecosophy, provide the principle and mechanics, respectively, for explaining the evolutionary dynamics following the fracturing of the Plane of Consistency/Immanence (**P**) into the four subcategories (**L, E, X, M**) that we analyzed in Chapter 11 using the tools of category theory and differential topology. More specifically:

---

[3] See Deleuze (1993).

[4] See Guattari (1995) for his initial exposition of these four domains. A deeper analytical elaboration can be found in Guattari (2013).

- The gauge-invariant Lagrangian represents the principle of dynamics. It is a mathematical function that encodes the 'energy' or 'action' of the system, which is used to derive the equations of motion for the fields involved. In this context, the fields represent the various aspects of Folding/Unfolding and Extended Ecosophy. This Lagrangian can be seen as a mathematical representation of the dynamics, i.e., the actualization of virtual potentials into concrete realities, that are described qualitatively earlier above. It provides a way to quantify and analyze how the fracturing of the Plane of Consistency/Immanence (**P**) into the four subcategories (**L, E, X, M**) evolves over time, i.e., the virtual multiplicity of **P** differentiates into distinct, yet still virtual, domains through a creative emergence, a birth of new forms from the womb of the virtual.
- The gauge transformations represent the mechanics of the transformations between different states or configurations of the system. They are symmetries of the Lagrangian, meaning that they are transformations under which the 'action' of the system remains invariant. The combined symmetry group captures the intertwined dynamics of interiority/exteriority (Folding/Unfolding) and the four functorial domains following Guattari's Extended Ecosophy (**L, E, X, M**). These gauge transformations can be interpreted as the mathematical counterparts to the transformations and differentiations described in the previous section. They provide a precise way to describe how entities and their relations transform as they move through the various subcategories (**L, E, X, M**).[5]

To operationalize the monadic structure across the subcategories (**L, E, X, M**) using the gauge transformations and Lagrangian of the combined symmetry group, we can draw parallels between the mathematical structures in gauge theory and the conceptual structures in the monadic framework:

- **Unit Transformation and Existential Territories (L):**
    - In a monad, the unit transformation is a function that takes a value and puts it into a monadic context. In our context, this can be interpreted as the process of embedding entities within the fabric of **P** (Plane of Consistency/Immanence).

---

[5] In Campbell's (2020) exploration of Treatment Action Campaign's (TAC) efforts in mobilizing community support for access to HIV/AIDS treatment in South Africa, the internalization of global health knowledge (Folding) within local communities transforms into actionable strategies (Unfolding) that challenge existing healthcare paradigms. The campaign's strategies span across community identities and experiences (**L**), information dissemination and mobilization (**E**), shifting public opinions and policy discourse (**X**), and the creation of new organizational structures and alliances (**M**). The gauge-invariant Lagrangian can be seen as encapsulating the 'energy' or 'action' of the social movement system, defining the principles of dynamics that drive change within the health ecosystem. In contrast, gauge transformations illustrate the mechanics of these changes, depicting how shifts in community mobilization, policy advocacy, and healthcare access (states or configurations of the system) maintain the system's coherence through transformations that preserve the underlying principles of equity and access to care.

- In the gauge theory, this could be represented by a specific gauge transformation that maps the 'unfolded' state (Exteriority) to the 'folded' state (Interiority). This transformation can be seen as the process where new entities emerge within **P**, carving out their own territories of existence (**L**).
- **Endofunctor and Energetic-Signaletic Flows (E):**
  - In a monad, the endofunctor represents a way to apply a function within the monadic context, transforming the value inside the monad without changing the structure of the monad itself.
  - In the gauge theory, this could be represented by a set of gauge transformations that act on the energetic-signaletic flows (**E**). These transformations would correspond to the internal dynamics of **P**, guiding the continuous transformations of entities without changing the overall structure of the ecological interrelations.
- **Multiplication Transformation and Abstract Machinic Phyla (M):**
  - In a monad, the multiplication transformation is a function that takes a monad within a monad and flattens it to a single monadic layer. In our context, this can be interpreted as the process where entities within **P** merge and interact, creating complex interconnections (**M**).
  - In the gauge theory, this could be represented by certain nonabelian gauge transformations that intertwine different components of the system, effectively 'merging' or 'binding' them into more complex structures. This mirrors the way in which **M** weaves together the entities of **P**.
- **Incorporeal Universes (X) as Vector Fields Guiding Transformations:**
  - In our theoretical narrative, the Incorporeal Universes (**X**) are represented as vector fields that guide the transformations represented by the endofunctor and the multiplication transformation.
  - In the gauge theory, this could be represented by the potentials associated with the gauge fields. These potentials (akin to vector fields in classical physics) dictate how entities (fields) in the theory evolve and transform under the gauge transformations. They are the 'forces' or 'directions' that influence the transformations, analogous to the role of X in guiding the trajectories of entities within **P**.
- **Lagrangian and Dynamics of the Monad:** The Lagrangian in the gauge theory encodes the dynamics of the system, describing how the fields (representing entities and their relations) evolve over time. This can be seen as a mathematical representation of the dynamics of the monadic structure itself. The kinetic terms in the Lagrangian could represent the active, unfolding processes of the monad (how entities transform and evolve within the monadic context), while the potential terms could represent the inherent 'tensions' or 'energies' that are associated with the

various configurations of the monad (how entities are related and interact within the monadic context).[6]

In Chapter 6, we introduced differential forms as mathematical constructs that measure the "rate of change" or "flow" of transformations within and between categories. These forms are integral to the gauge-invariant Lagrangians, mathematical expressions that capture the dynamics of these transformations. At the global level of the ecological dynamics implied by the actualization of the virtual multiplicity of **P** into distinct, concrete entities, differential forms relevant at this scale can articulate how the energy of an ecological system changes as entities within that system (e.g., natural, economic, social, or technological entities) transform and evolve, providing a dynamic, quantifiable perspective on ecological interrelations.

Well-being, wealth, empowerment, and welfare are clear and intuitive meta-level outcomes/impacts of changes in existential life territories, energetic-signaletic ecosystem flows, incorporeal experience universes, and abstract machinic phyla, respectively. Well-being is an individual's holistic measure of health and happiness. Wealth is not just about financial resources; it also includes resources like good health. Empowerment involves gaining the knowledge, skills, and confidence to make decisions and take actions that improve one's life. Welfare refers to having access to resources, opportunities, and services as well as the ability of an individual or family to move up or down the social and economic ladder within a society. They provide a clear measure of the "intensity" or "rate" of transformations within each domain and can be easily understood and measured. For example, sustainable agri-food practices (**L**) that respect ecological dynamics can lead to widespread well-being by fostering food security and environmental health. Likewise, improvements in health (**E**) can be interpreted as shifts in the differential forms that measure the flow of transformations in health-related entities, and these shifts are closely tied to increases in wealth.

These outcomes affect the interactions between folding/unfolding and the Extended Ecological Interrelations (the four functorial domains of Guattari) by acting as constraints or targets in the Lagrangian, thereby shaping the dynamics of the entire system. Well-being might be associated with the harmonious Folding and Unfolding processes, reflecting a balanced internal and external complexity. Wealth might be interpreted as the abundance

---

[6] In Campbell's (2020) context of the Treatment Action Campaign (TAC) in South Africa.

• When gauge transformation takes a system from an "unfolded" state to a "folded" one, as seen in how TAC mobilized community members, the unit transformation "embeds" individuals within a broader socio-political and health-oriented framework, creating new territories of existence and action (**L**) for HIV/AIDS activism.

• When gauge transformations act on Energetic-Signaletic Flows (**E**), as TAC's information dissemination and mobilization efforts transform community knowledge and action, the endofunctor transforms values inside the monad without altering the inherent structure of the monad, i.e., community's social capital, itself.

• When gauge transformations intertwine different components of the system, e.g., TAC weaves together diverse stakeholders into a unified force for change, multiplication transformation creates complex interconnections (**M**) within the health advocacy ecosystem.

• Strategic directions of TAC serve as 'forces' (**X**) guiding the collective actions of individuals and groups within the health advocacy ecosystem, illustrating how potentials associated with gauge fields dictate the evolution and transformation of entities under gauge transformations.

and effective flow of Energetic-Signaletic Flows. Empowerment could be related to the capacity for an entity to navigate and influence its Incorporeal Universes. Welfare might be seen as the overall health of the Abstract Machines, indicating a system's resilience and adaptability.[7]

- In the context of Global Immanence, a surge in the intensity of Folding (Interiority) provokes a reconfiguration of Existential Territories within the primordial matrix. This signifies the emergence of interconnected and immanent subjective spaces of human experience. A surge in Folding, leading to new subjective spaces, may be associated with an increase in well-being. As Folding harmonizes internal and external complexities, individuals may experience a heightened sense of well-being due to this reconfiguration of Existential Territories.
- When a significant crisis arises in these Existential Territories, it may trigger a process of Unfolding. This Unfolding acts as a coping or adaptive strategy, where internal complexities are externalized within the matrix. An existential crisis that triggers Unfolding might be seen as a response to a decrease in well-being. This Unfolding, as a coping strategy, aims to restore well-being by externalizing internal complexities and reconfiguring Existential Territories.
- This act of Unfolding (Exteriority) may also stimulate the generation of new Energetic-Signaletic Flows. These flows encompass novel informational and communicative exchanges within and between the categories **L**, **E**, **X**, and **M**. The act of Unfolding that stimulates new Energetic-Signaletic Flows can be associated with the generation of wealth. As internal complexities are externalized, new informational and communicative exchanges (wealth in terms of information and relationships) are created.
- When these Energetic-Signaletic Flows intensify within the primordial matrix, they can induce a reciprocal process of Folding. This Folding process internalizes and integrates external complexities into a renewed subjective landscape. Intense Energetic-Signaletic Flows that induce Folding may be seen as a mechanism for accumulating wealth within an individual or system, as external complexities (resources, information) are internalized and integrated.
- Furthermore, Folding (Interiority) serves as a mechanism to integrate new Incorporeal Universes—collective assemblages of enunciation—into the subjective experience. This integration enriches the internal landscape with fresh senses and meanings. The process of Folding that integrates new Incorporeal Universes into subjective experience can be seen as an act of empowerment. This integration enriches the internal landscape with new senses and meanings, enabling an entity to navigate and influence its subjective and objective worlds more effectively.

[7] In Campbell's (2020) study, differential forms can be understood as tools that measure the rate of change or flow of transformations within and between social capital categories, integral to the dynamics of health activism. For instance, TAC not only engenders a significant transformation within the health ecosystem, which can be quantified by changes in well-being (a direct measure of health and happiness) and wealth (inclusive of resources like access to life-saving medication) but also influences the broader ecological system of public health, where the empowerment of communities to demand their rights impacts the welfare of individuals within these communities. The gauge-invariant Lagrangians represent underlying mechanisms of social capital dynamics—such as trust, cooperation, and shared norms—which act as forces that drive the success of public health campaigns and can be quantified through outcomes like increased access to treatment, improved health literacy, and stronger community support systems.

- These Incorporeal Universes, in turn, shape the dynamics of Folding and Unfolding within the primordial matrix, guiding how entities navigate internal and external complexities. The collective assemblages of enunciation (Incorporeal Universes) that shape the dynamics of Folding and Unfolding play a critical role in empowerment. They guide how internal and external complexities are navigated, thereby empowering entities to act with agency within their existential territories.
- Deep structures, known as Abstract Machines or Phyla, reside within the primordial matrix and steer the process of Unfolding. They determine the paths along which entities externalize their internal complexities. The deep structures (Abstract Machines or Phyla) that guide Unfolding can be associated with welfare. They determine the paths along which internal complexities are externalized in a manner that is conducive to the overall health and adaptability of the system, reflecting a system's resilience.
- As Unfolding (Exteriority) progresses within this matrix, it can reveal or activate new Abstract Machines. These deep structures subsequently guide the evolution of new existential territories within and between the categories **L**, **E**, **X**, and **M**. The process of Unfolding that reveals or activates new Abstract Machines can be seen as a mechanism for enhancing welfare. As new deep structures are revealed, they can guide the evolution of new existential territories in a way that promotes the system's overall health and adaptability.[8]

MLXE evolution as differentiation from virtual realities to actual possibilities

The approach discussed in the previous section is particularly effective when dealing with situations that require an understanding of the dynamic interplay between different

---

[8] In Campbell's (2020) examination of social capital in global public health, we can identify the following recurrent dynamic relational processes.
- Surge in the intensity of Folding, as TAC created new subjective spaces of human experience through the mobilization of local communities and the establishment of solidarity networks, that contributed to an increased sense of well-being among the marginalized communities.
- Unfolding process, as the existential crisis triggered by the AIDS epidemic and the South African government's initial denial of HIV/AIDS was externalized through public protests and legal actions, creating novel informational and communicative exchanges that contributed to the accumulation of wealth.
- Folding process, as TAC's dialogue and critical thinking programs enriched the internal landscape with fresh senses and meanings, empowering individuals to navigate and influence their worlds more effectively, through TAC's strategic use of both bonding, bridging, and linking social capital to confront and negotiate with power holders.
- Unfolding process, as new global networks and advocacy strategies revealed or activated deep structures that determine the paths along which entities externalize their complexities, enhancing welfare by promoting the system's overall health and adaptability.

sectors of an ecosystem (such as Nature, Economy, Society, and Technology) and their collective evolution, and examining how systemic changes, ecological pressures, and societal shifts influence the overall ecosystem's transformation. One would apply this perspective to map out the interactions between different sectors and how changes in one sector ripple through and impact others. This could involve assessing the ecological impact of technological innovations or the societal ramifications of economic shifts. The approach we will introduce in this section is ideal for delving into the micro-level transformations within each sector of the ecosystem, understanding how potentialities within these sectors are actualized and how they transition between different states or phases. We apply this perspective to explore how new concepts, technologies, or policies within a sector evolve from abstract potentials into tangible realities and how this process influences the sector's developmental trajectory.

Deleuze distinguishes between the Virtual, a dimension of unmanifested potentialities, and the Actual, the tangible realization of these potentials. This actualization is not mere replication but a dynamic process generating differences with each iteration, termed repetition. Each repetition introduces variations, ensuring that the Virtual, when actualized, brings forth something new and unexpected. While Deleuze focuses on the process of actualization, Guattari delves deeper, exploring potential future trajectories, transitioning from the Real (the present) to the Possible (potential futures). This shift is not predetermined but is a dynamic journey that can lead to unforeseen outcomes. Deleuze's concept of repetition, generating difference, mirrors Guattari's transition from the Real to the Possible, broadening the horizon of potentialities.[9]

As we have discussed, the fourfold schema of Guattari helps to understand the intricate dynamics shaping reality and subjectivity. Existential Territories (**L**), as personal, subjective realms, constantly evolving due to memories, desires, and external influences, represent the *virtual real*, the unmanifested potentials of existence. Energetic-Signaletic Flows (**E**), as the tangible currents, both material and semiotic, that shape our reality, symbolize the *actual real*, the processes that mold our existential territories. Incorporeal Universes (**X**), as abstract realms that influence our subjective territories, signify the *virtual possible*, the intangible forces guiding our aspirations. Abstract Machinic Phyla (**M**) symbolize concrete evolutionary paths, representing potential futures based on our current reality, the *actual possible*. The dynamics between these dimensions can be understood using Deleuze's concepts. For **L** to **E**, subjective experiences (**L**) actualize into tangible flows (**E**), similar to the Virtual becoming Actual. For **E** to **X**, material and semiotic flows (**E**) influence our understanding, leading to the creation of abstract meanings and potentials (**X**). This mirrors Deleuze's idea of repetition producing variations. For **X** to **M**, abstract potentials (**X**) actualize into potential evolutionary trajectories (**M**). For **M** to

---

[9] Remember that we first encountered and provided a tentative description of differentiation between Virtual and Actual on the one hand, and Real and Possible on the other hand, via Difference and Repetition in Chapter 1. Also, see Chapter 11 for an elaboration of differentiation from virtual realities to actual possibilities.

L, potential futures (**M**) interact with and reshape the existential territories (**L**), introducing new potentialities.[10]

Fig. 12.1 highlights the distinctions between virtual/actual and real/possible. The vertical and horizontal dynamics correspond to Deleuzean difference and repetition, respectively. Engagements and enactments produce difference, while events and interactional creation are related to repetition.[11]

Let us see how we can develop further insights about the differentiation of life-experience ecosystems from virtual realities to actual possibilities with the tools of differential geometry (for technical details, please see "Unified Gauge Theory of Differentiation" from the **Appendix**).

- **Representing Guattari's Domains in Transformations of Difference and Repetition**:
    - **Existential Territories (L)**: These are realms of unactualized potentialities, "Virtual" in its pure form, akin to a world of possibilities that have not yet come to fruition. While the potentialities within the Existential Territories are vast, they exist in a state of equilibrium or stasis, that underlies all transformations and actualizations. Imagine a canvas of life where various paths and outcomes are possible, but none have been actualized yet. This is where ideas, dreams, and potential scenarios exist before they take concrete form.
    - **Energetic-Signaletic Flows (E)**: This concept represents the tangible processes and activities, the "Actual," that shape and influence our day-to-day experiences in existential territories. As a consistent and stable pattern of actualization, the processes and flows in this category are regular and predictable, forming a structured and reliable framework within which existential territories are influenced and transformed. It is like the flow of energy and signals that drive the events and interactions in our lives, turning potentialities into actual, lived experiences.
    - **Incorporeal Universes (X)**: Here, we are looking at a domain constantly in flux, ever-shifting potentialities of the "Virtual," influenced by a variety of factors. It is a space of potential meanings and values that are always changing and evolving. Think of it as an ever-morphing landscape of ideas and interpretations, where nothing is fixed, and everything is open to reinterpretation. Thus, possibilities and conceptualizations are continuously redefined and transformed.
    - **Abstract Machinic Phyla (M)**: This represents the "Actual" as potential futures or pathways that are more defined and have a clearer trajectory. It is like having a map

---

[10] In his research on trust among healthy volunteers in clinical drug trials, Mwale (2020) delves into the process of becoming a repeat volunteer, where potentialities within the clinical trial landscape are actualized through the volunteers' participation and each participation experience introduces variations and influences future decisions. The clinical trial acts as an 'event' that transforms volunteers, influencing their perceptions, decisions, and identities, in a continuous process of becoming, shaped by the interactions and relations within the trial ecosystem. Volunteers manage their identities and navigate between being treated as subjects versus objects in trials in a dynamic interplay between the Real (current experiences and realities of the trial) and the Possible (future pathways and outcomes). Trust is continuously negotiated, where each repetition (or trial participation) generates new differences and possibilities.

[11] The attentive reader will notice that the functorial path described in Fig. 12.1 is topologically identical to Fig. 2.7. The former is an untangled instance of the latter, wherein the 3-dimensional geometry of the path might make it visually challenging to recognize the dynamics of difference and repetition.

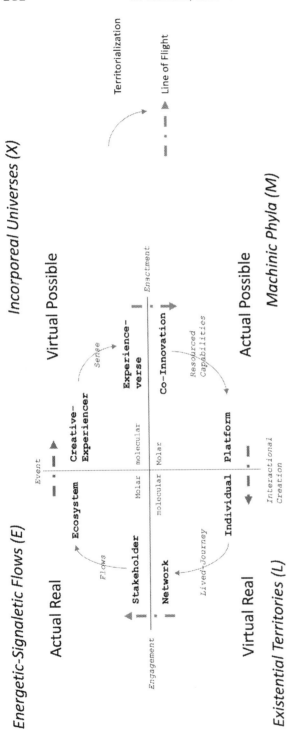

FIGURE 12.1 Differentiation from virtual realities to actual possibilities.

with specific routes marked out, each leading to a particular future or outcome. This concept captures the notion of possibilities that are more tangible and within reach, though they have not yet fully materialized.[12]

- **Dynamics Between Guattari's Domains Using Transformations of Difference and Repetition**:
  - **From L to E:** The transition could involve the inversion operator acting on the virtual, representing the direct emergence of difference as subjective experiences actualize into tangible flows. Alternatively, the latent operator could mobilize the hidden dynamics that subtly push subjective experiences toward tangible manifestations, symbolizing the unseen undercurrents guiding the actualization process. If implicated, the dominance/bifurcation operator embodies the tension between the virtual and the actual, marking the boundary where subjective experiences meet tangible manifestations. This transition from **L** to **E** is characterized by a stable phase of repetition.
  - **From E to X:** The transition might involve the inversion operator acting on the Actual, symbolizing the transformation of tangible flows into abstract landscapes of meaning. Additionally, the latent operator would capture those latent processes representing the unseen transformation of material realities into intangible potentials. The dominance/bifurcation operator represents the tension between tangible realities and abstract meanings, marking the boundary between what is and what could be. A varying phase, capturing the nuances and variations in repetition mark this transition from **E** to **X**, as tangible flows influence our understanding of the world.
  - **From X to M:** The transition possibly involves the inversion operator acting on the Virtual, representing the actualization of abstract potentials into concrete trajectories of change. The latent operator could drive the hidden dynamics guiding abstract meanings toward potential evolutionary paths, representing the invisible hand shaping potential futures from abstract landscapes. If implicated, the dominance/bifurcation operator embodies the tension between abstract potentials and concrete evolutionary paths, marking the boundary where intangible potentials meet tangible futures. The transition from **X** to **M** introduces a distinct phase, capturing the specific trajectory of repetition leading to potential futures.
  - **From M to L:** The transition could involve the inversion operator acting on the Actual, symbolizing the tangible potential futures interacting with and reshaping the realm of unactualized potentialities. Here, the latent operator captures the underlying processes guiding potential futures back into the realm of unactualized potentialities, representing the unseen forces reshaping existential territories. The dominance/bifurcation operator represents the tension between potential futures and the realm of unactualized potentialities, marking the boundary where the future meets the

---

[12] Mwale's (2020) findings about repeat healthy volunteers in clinical drug trials can be mapped onto Guattari's domains as follows. Volunteers' potential motivations and initial decisions to participate (**L**) reflect the unactualized potentialities within the Virtual realm. The actual processes and experiences of participating in the trials (**E**) embody the Actual, where potentialities are transformed into lived experiences. The changing perceptions, values, and trust levels among volunteers (**X**) highlight the fluid potentialities of the Virtual, influenced by various factors. The more defined pathways and future decisions regarding participation in trials (**M**) illustrate the Actual as potential futures, mapping the trajectory of volunteers' ongoing involvement.

present. From **M** to **L** returns to a constant phase, indicating a return to the foundational phase of repetition.[13]

- **Dynamics Governed by the Lagrangian**: The Lagrangian **L** captures the dynamics of differentiation, repetition, and actualization between the dimensions **L**, **E**, **X**, and **M**.
  - **Kinetic Terms**: These describe the metaphorical "energy" associated with the active processes of differentiation and repetition. As entities move between states and dimensions, they undergo differentiation, and this process has its own dynamism and "energy." The nuances introduced by repetition, where each repetition brings about a new difference, also contribute to this kinetic energy.

    From L to E: The kinetic term captures the energy associated with the actualization of subjective experiences from existential territories (**L**) into tangible, material, and semiotic flows (**E**). This is the energy of the virtual becoming actual.

    From E to X: The kinetic term represents the energy of material and semiotic flows (**E**) giving rise to new meanings and potentials, leading to the formation of incorporeal universes (**X**). This is the energy of tangible flows influencing our subjective understanding.

    From X to M: The kinetic term captures the energy of abstract meanings in the incorporeal universes (**X**) giving rise to potential evolutionary paths (**M**). This is the energy of abstract potentials becoming concrete trajectories.

    From M to L: The kinetic term represents the energy of potential futures (**M**) interacting with and reshaping existential territories (**L**). This is the energy of tangible futures influencing the realm of unactualized potentialities.
  - **Potential Terms**: These capture the "energy" or tension associated with the potential of the virtual and its process of actualization. The virtual, being a realm of potentialities, has an inherent tension as it holds numerous possible futures. As these

---

[13] Returning to Mwale's (2020) exploration on repeat healthy volunteers in clinical drug trials.

- From **L** to **E**: Participants' decisions to become involved and repeatedly volunteer in clinical trials are, beyond apparent financial incentives, deeply embedded in their social networks, experiences, and the embodied nature of trust, suggesting a latent operator at work. The stable phase of repetition here could be the consistent pattern of trust and decision-making processes across different clinical trials.
- From **E** to **X**: Positive or negative experiences within trials bifurcate participants' paths towards either continued participation or heightened caution, demonstrating how experiential knowledge becomes a dominant factor in shaping abstract attitudes towards trust in clinical drug trials. The varying phase of repetition marks the nuanced understanding and perception changes of participants regarding the role and impact of their involvement over time.
- From **X** to **M**: Despite the abstract understanding of risks and trust developed through previous experiences, participants often invert these considerations based on their embodied experiences and socio-economic needs, evident in how participants may rationalize participating in higher-risk trials. This transition introduces a distinct phase of repetition, capturing the specific trajectory of changes and adaptations in research methodologies based on trust dynamics.
- From **M** to **L**: The trust developed or eroded in the trial process, influenced by trial experiences, subtly informs their broader life choices and social interactions, reflecting a latent transformation. Returning to a constant phase indicates a foundational phase of repetition, where the foundational trust dynamics and ethical considerations in clinical research are revisited and reinforced.

potentials actualize, they release or transform this energy. The energy associated with the differentiation process itself, the very act of becoming different, is also part of this potential term.

For L: The potential term captures the inherent tension within the existential territories, representing the unactualized potentialities waiting to be realized.

For E: The potential term represents the tension between the actual processes and their potential to influence and shape incorporeal universes.

For X: The potential term captures the tension of abstract meanings and values, representing the potential paths of transformation they can inspire.

For M: The potential term represents the tension between the potential futures and their realization in existential territories.[14]

- **Interactions and Transformations between Difference and Repetition**
  - **Inversion Triggering New Cycles of Repetition**: A groundbreaking medical device is introduced for breast cancer screening, revolutionizing the traditional mammography process by employing AI for early cancer detection. This inversion signifies a direct emergence of difference, transitioning from Existential Territories (**L**) to Energetic-Signaletic Flows (**E**). The kinetic term of the Lagrangian, capturing the dynamic energy of this transition, propels the transformation of screening contexts. As a result, patients undergo procedures under novel conditions, embodying a changing the context or conditions of repetition. The potential term further intensifies this shift, emphasizing the tension between the new and established screening methods.
  - **Latent Processes of Differentiation Redirecting the Path of Repetition**: Behind the scenes, dedicated researchers refine the AI algorithms of the new device. This transformation captures the hidden processes of differentiation, transitioning from Energetic-Signaletic Flows (**E**) to Incorporeal Universes (**X**). The kinetic term of the Lagrangian, reflecting the dynamics of differentiation, steers the subtle alterations in the diagnostic process. Consequently, the AI algorithm's continuous updates represent a qualitative transformation in repetition. The potential term underscores the tension between evolving and traditional diagnostic methods.
  - **Dominance/Bifurcation Modulating the Frequency of Repetition**: As the AI's accuracy improves, a tension emerges between traditional biannual screenings and potential, more infrequent, AI-driven screenings. This tension emphasizes the distinction between the virtual and the actual, transitioning from Incorporeal Universes (**X**) to Abstract Machinic Phyla (**M**). The kinetic term of the Lagrangian,

---

[14] In Mwale's (2020) study, the kinetic terms are embodied in the transformations of subjective experiences of trust (**L**) into tangible actions and decisions within the clinical trial environment (**E**), material interactions and experiences in the trials (**E**) into new meanings and understandings about trust and clinical research (**X**), abstract meanings and values about trust and ethical conduct in research (**X**) into potential evolutionary paths in clinical trial methodologies and policies (**M**), and potential futures in research ethics and trust mechanisms (**M**) into existential territories of new participants (**L**). As for the potential terms, they represent the tension and anticipation of unactualized potentialities within individuals considering trial participation (**L**), the tension between current practices and their capacity to foster or hinder trust in the tangible processes of clinical trials (**E**), potential paths of transformation inspired by collective aspirations and concerns about trust in research (**X**), and the tension in navigating trust within evolving research landscapes (**M**).

capturing the dynamic energy of this distinction, influences the potential change in screening frequency, marking a temporal shift in repetition. The potential term highlights the tension between the two screening frequencies.

- **Phase Shift in Repetition Inducing Inversion in Differentiation**: Patient feedback indicates a desire for a more holistic treatment approach, which embodies a shift in perspective or interpretation represented by a phase shift in repetition. This phase shift then induces a transformation in the medical community. Driven by inversion operator, there is a push toward integrating complementary therapies. The shift toward a more holistic approach, driven by patient feedback, represents a move from potential futures (**M**) back to the realm of lived experiences and subjective territories (**L**). The kinetic term of the Lagrangian captures the dynamic energy propelling this inversion toward a holistic approach. As the medical community grapples with this change, the potential term of the Lagrangian emphasizes the tension between holistic and traditional treatments, reflecting the challenges and opportunities of integrating new methods into established practices.

- **Phase Shift in Repetition Modulating Latent Processes of Differentiation**: As more hospitals adopt the AI-driven device, training protocols for radiologists undergo a change in the character or quality of the repeated training sessions, indicative of a phase shift in repetition. This shift then modulates the underlying processes of differentiation in the medical community about nature of radiologist roles and training, transitioning from Existential Territories (**L**) to Energetic-Signaletic Flows (**E**). The kinetic term of the Lagrangian captures the dynamic energy of this differentiation, while the potential term underscores the tension between the evolving training protocols and the traditional methods, highlighting the challenges of integrating new technologies into established practices.

- **Phase Shift in Repetition Influencing the Dominance/Bifurcation in Differentiation**: As the AI-driven device becomes standard, there is a temporal shift in repetition in the adoption rates of the technology across hospitals. This shift, indicative of a phase change in repetition, marks a change in the timing or rhythm at which hospitals are integrating the new technology. This temporal shift then accentuates the distinction and tension between technologically advanced hospitals and those lagging behind, symbolized by the dominance/bifurcation operator. As they transition from the tangible, material flows of technology (**E**) to the more abstract realms of reputation and potential (**X**), the kinetic term of the Lagrangian captures the dynamic energy of this distinction between the potential benefits the AI promises (virtual) and its real-world implementation (actual). The potential term underscores the tension between hospitals with advanced AI, perceived as cutting-edge, and those without, emphasizing the challenges of technological disparities in healthcare.[15]

Dynamic relationality #10: Immanence of emergence and evolution

For Deleuze and Guattari, Plane of Immanence (**P**) is a central concept representing potentiality. It is not just a static space but a dynamic realm of pure becoming, where entities are not mere fixed identities but are in a constant state of transformation. This plane, teeming with potential, is a primordial category where the objects are the myriad entities of the world, and the morphisms are the myriad transformations they undergo. However, the richness of **P** is such that it does not remain a singular entity. It undergoes a process of differentiation, fracturing into four distinct subcategories: **L**, **E**, **X**, and **M**. This is not a mere division but a creative emergence, where the vast potential of **P** births new forms, each with its unique characteristics and dynamics.

The Existential Life Territories (**L**) emerge as spaces where new entities find their identity within **P**. Represented as topological spaces, **L** is where entities carve out their territories of existence. Each point in this space is an entity, and the topology captures the intricate relationships between them. It is the realm where identity is forged, where entities define themselves and their relationships with others. Driving the internal dynamics of **P** are the Energetic-Signaletic Ecosystem Flows (**E**). Represented as smooth manifolds, **E** is the force behind the continuous transformations of entities. Every point on this manifold is a potential state of an entity, and the paths on it represent the possible transformations. **E** is the heartbeat of **P**, the engine that propels entities along their transformative journeys. Guiding these transformations are the Incorporeal Experience Universes (**X**). These are the abstract realms, represented as vector fields, where each vector signifies a force or direction influencing the transformations. **X** is

---

[15] Mwale's (2020) study illustrates how trust is not static but is continuously negotiated and managed by participants.

- Inversion triggering new cycles of repetition, e.g., decision to participate in new trials or continue participation after experiencing or learning about adverse effects reflects the dynamic energy of transitioning from a state of potential harm (virtual) to actual participation (actual).
- Latent processes of differentiation redirecting the path of repetition, e.g., volunteers' trust evolves through unseen personal and relational dynamics creating tension between newly influenced paths of participation and existing cycles of involvement.
- Dominance/bifurcation modulating the frequency of repetition, e.g., how participants discern between their perceived risk and the actual experience of trial participation influences their decision to continue or alter their frequency of participation.
- Phase shift in repetition inducing inversion in differentiation, e.g., external events or significant experiences (e.g., learning about a trial's adverse effects) lead to a reevaluation of trust.
- Phase shift in repetition modulating latent processes of differentiation, e.g., volunteers' trust is silently shaped by their experiences and interactions with the trial process.
- Phase shift in repetition influencing the dominance/bifurcation in differentiation, e.g., new aspects of repetition (e.g., additional knowledge or changed trial conditions) impact volunteers' discernment between perceived and actual risks, stabilizing or destabilizing their trust and participation patterns.

the abstract space, the realm of ideas and concepts that shape the trajectories of entities within **P**. Lastly, the Abstract Machinic Phyla (**M**) emerge as spaces where entities within **P** interact and merge. Represented as fiber bundles, **M** is where entities intertwine, forming complex interconnections. Each fiber is a potential set of connections between entities, making **M** the realm of interaction, where entities unite to form intricate systems and structures.

This differentiation of **P** into **L**, **E**, **X**, and **M** is not a mere division but a profound process of emergence and co-evolution. It is guided by the double articulation of substance and form, content and expression, which shapes the formation and understanding of these entities. This process, as elucidated by Deleuze, is the articulation of these domains, involving the formation of both physical and social assemblages. Each domain is characterized by a unique blend of these assemblages, reflecting its specific dynamics. In this theoretical framework, **P** is the overarching category encompassing all entities and their potential transformations. **L**, **E**, **X**, and **M** are its subcategories, each focusing on specific facets of these entities and transformations. The objects and morphisms in these subcategories are derived from **P**, reflecting their specific focus. The double articulation process in each category involves the interplay between content and expression, shaping the formation and understanding of these entities and transformations. In the language of category theory and differential topology, this fracturing of **P** into **L**, **E**, **X**, and **M** can be described in intricate detail. From the Existential Territories to the Economy of Flows, from the Incorporeal Universes to the Machinic Phyla, each transition is marked by a functor that maps entities and their transformations from one domain to another, reflecting the dynamic interplay and transformations within each realm.

The Plane of Consistency/Immanence (**P**) embodies a dynamic ecological matrix, continuously evolving through an intricate interplay of virtual potentialities and actualized realities within the subcategories **L**, **E**, **X**, and **M**. The Existential Territories (**L**) symbolize a realm where potentialities exist in a state of latent possibility, while the Energetic-Signaletic Flows (**E**) represent the actualization of these potentialities into tangible processes. The Incorporeal Universes (**X**), in constant flux, depict the evolving virtual potentials that transform into defined paths within the Abstract Machinic Phyla (**M**), symbolizing the possible futures emerging from the abstract landscapes. In this complex ecological landscape, the dynamics of Folding and Unfolding play a pivotal role. Folding symbolizes the internalization of external complexities, enriching the internal landscape and enhancing well-being and empowerment. Conversely, Unfolding represents the externalization of internal complexities in response to existential shifts, generating wealth and fostering welfare by revealing new potentials and evolutionary paths. The gauge-invariant Lagrangian, with its kinetic and potential terms, quantifies the dynamics of these transformations, capturing the metaphorical 'energy' and tension inherent in the process of differentiation and actualization. In essence, this perspective provides a rich, dynamic portrayal of an ecological system where deep structures, subjective experiences, tangible processes, and abstract potentials interact in a ceaseless dynamic of creation and transformation.

The ongoing evolution of the Plane of Consistency/Immanence (**P**) is not a linear progression but a complex process of differentiation and actualization, driven by the interplay between the virtual and the actual. The four dimensions—**L**, **E**, **X**, and **M**—represent distinct stages in this evolutionary journey, each characterized by its unique dynamics and interactions. The Existential Territories (**L**) and Energetic-Signaletic Flows (**E**) capture the realms of the virtual real and the actual real, respectively. They represent the foundational layers of this ecosystem, where potentialities are birthed and then actualized into tangible processes. On the other hand, the Incorporeal Universes (**X**) and Abstract Machinic Phyla (**M**) delve into the realms of the virtual possible and the actual possible. These dimensions explore the myriad pathways through which potentialities can evolve, either remaining in the realm of abstract meanings (**X**) or actualizing into concrete trajectories (**M**). The intricate dynamics between these dimensions, as represented by transformations between difference and repetition, showcase the fluidity and complexity of this ecosystem's evolution. The Lagrangian, with its kinetic and potential terms, further elucidates the forces driving these transformations. It captures the energy and tension inherent in the process of differentiation, repetition, and actualization. In essence, this theoretical framework paints a vivid picture of an ever-evolving ecosystem, where virtual realities continuously morph into actual possibilities, shaping the trajectory of entities and transformations within the Plane of Consistency/Immanence.

In conclusion, the Plane of Consistency/Immanence (**P**) and its fracturing into **L**, **E**, **X**, and **M** offer a profound theoretical framework to understand the dynamic processes of emergence and transformation in ecosystems. It provides a lens to view the world not as static entities but as a vibrant performance of potentialities and transformations, guided by the intricate interplay of substance and form, content and expression. This framework, enriched by transformations between difference and repetition and the Lagrangian dynamics, paints a vivid picture of an ever-evolving ecosystem where virtual realities continuously morph into actual possibilities. It underscores the fluidity and complexity of the evolutionary journey, emphasizing the continuous shift from potential to actual, from abstract meanings to concrete trajectories.

From analysis and modeling to strategy development

Let us incorporate all the insights and concepts we have discussed into a **step-by-step analytical and modeling procedure**. In each category, the monadic structure provides a framework for understanding the transformations that occur as a result of the interaction between the Machinic Assemblages of Bodies (MABs) and the Collective Assemblages of Enunciation (CAEs). This understanding can provide valuable insights for strategic analysis and development. Let us delve deeper into the concept of monadic structure in the context of the four subcategories (**L, E, X, M**) and how it can be applied in the strategic analysis and development process.

- **Mapping the Plane of Consistency/Immanence (P)**: Utilize the Unified Gauge Theory of Folding and Ecosophy to explore the interconnectedness of various sectors (Nature, Economy, Society, Technology) within the healthcare ecosystem. Begin by identifying the broader ecosystem of practices and experiences that intersect with the healthcare organization. This includes individual, communal, and organizational practices and experiences related to healthcare, public health, and societal responses to health crises like COVID-19. Apply these theories to assess how changes in one sector impact the others, creating a dynamic, interconnected system. This plane is not confined to the organization but extends beyond it, encompassing a larger field of potentialities and interactions.

- **Identifying Existential Territories (L):** Identify the specific contexts or 'territories' within which the healthcare organization operates. This could include different patient populations, healthcare markets, regulatory environments, etc. Here, apply Unified Gauge Theory of Differentiation to understand the micro-transformations within these territories, focusing on how abstract potentials evolve into tangible realities. Use the Lagrangian to analyze the energy and tension in the transition from virtual to actual within these territories. For each territory, identify the key practices and experiences (objects) and how they change over time or due to specific events (morphisms). The MABs here could be the individuals themselves, and the CAEs could be the personal narratives and discourses they create. The double articulation process here involves the interplay between the physical health status of individuals (substance and form of content) and their personal experiences and narratives of their health (substance and form of expression). In terms of differential topology, consider each individual's health status as a point in a topological space, and track the creation of new points as individuals navigate their health experiences during the pandemic.
  * The monadic structure here involves the individual health experiences of people navigating the pandemic. The unit transformations of the monad could be seen as the creation of new personal narratives as individuals navigate the pandemic. For example, when a person gets vaccinated, a new point representing their postvaccination health status is created in the topological space. This monadic structure allows us to track the evolution of individual health experiences over time and in response to different events, providing valuable insights for understanding the changing needs and experiences of different patient populations.
- **Analyzing Energetic-Signaletic Flows (E):** Employ Unified Gauge Theory of Differentiation to delve into how tangible processes within the healthcare organization respond to and influence the existential territories. This could include processes like patient care, research and development, supply chain management, etc. Use the Unified Gauge Theory of Folding and Ecosophy to contextualize these processes within the broader ecosystem, assessing their ripple effects across different sectors. For each process, identify the current state (object) and how it transforms in response to changes (morphisms). The MABs here could be the healthcare workers, hospitals, and other entities involved in these processes, and the CAEs could be the policies, guidelines, and protocols that guide them. The double articulation process here involves the interplay between the actual processes and transformations in the healthcare system (substance and form of content) and the interpretation, representation, and discourse around these processes (substance and form of expression). In terms of differential topology, consider each state of the healthcare system as a point on a smooth manifold, and track the transformation of points as the healthcare system adapts to the pandemic.
  * The monadic structure in this category involves the dynamic processes within the healthcare system as it responds to the pandemic. The endofunctor of the monad could be seen as the transformation of these processes as the pandemic evolves. For example, as the healthcare system adapts to the pandemic, the state of the system changes, moving along a path on the smooth manifold. This monadic structure allows us to track the evolution of healthcare processes over time and in response to different events, providing valuable insights for improving the efficiency and effectiveness of these processes.

- **Understanding Incorporeal Universes (X):** Apply Unified Gauge Theory of Differentiation to explore the guiding principles, goals, and values that influence the organization's responses to ecosystem changes. Identify the guiding principles, goals, and values that shape the healthcare organization's response to changes in the existential territories and energetic-signaletic flows. This could include principles like patient-centered care, evidence-based practice, etc. Integrate insights from the Unified Gauge Theory of Folding and Ecosophy to align these principles with broader ecological and societal considerations. For each principle, identify how it influences the organization's responses (morphisms). The MABs here could be the public health officials, scientists, and policymakers who articulate these principles, goals, and values, and the CAEs could be the public discourses, debates, and narratives that shape them. The double articulation process here involves the interplay between these abstract principles, goals, and values (substance and form of content) and their concrete impacts on the healthcare system and societal understanding of health (substance and form of expression). In terms of differential topology, consider each principle, goal, or value as a vector in a vector field, and track the influence of these vectors on the healthcare system's response to the pandemic.
  - The monadic structure in this category involves the guiding principles, goals, and values that shape the response to the pandemic. The vectors in the vector field could be seen as the "force" or "direction" that these principles, goals, and values provide in guiding the response to the pandemic. For example, the principle of public health could be a vector that guides the healthcare system's response to the pandemic, influencing the transformations represented by the endofunctor and the multiplication transformation. This monadic structure allows us to track the influence of different principles, goals, and values on the healthcare system's response to the pandemic, providing valuable insights for aligning the response with these guiding principles.
- **Examining Abstract Machinic Phyla (M):** Use Unified Gauge Theory of Differentiation to examine the evolving structure and organization of the healthcare system, focusing on potential futures and their interactions with existing territories. This could include the organization's network of relationships with other healthcare providers, suppliers, regulators, etc. Examine the structure and organization of the healthcare organization and how it adapts to changes in the existential territories, energetic-signaletic flows, and incorporeal universes. Utilize the Unified Gauge Theory of Folding and Ecosophy to assess how these structural changes align with and impact broader ecological dynamics. For each part of the organization's structure or organization, identify the current state (object) and how it changes in response to transformations (morphisms). The MABs here could be the administrative bodies, organizations, and systems that structure and organize the healthcare system, and the CAEs could be the policies, procedures, and protocols that guide them. The double articulation process here involves the interplay between the actual structure and organization of the healthcare system (substance and form of content) and the understanding, representation, and discourse around this structure and organization (substance and form of expression). In terms of differential topology, consider each connection within the healthcare system as a fiber in a fiber

bundle, and track the creation of new fibers as the healthcare system adapts to the pandemic.

  * The monadic structure in this category involves the structure and organization of the healthcare system as it responds to the pandemic. The multiplication transformation of the monad could be seen as the coordination and integration of these structures and organizations as they respond to the pandemic. For example, as new healthcare providers join the network of those treating COVID-19 patients, new fibers representing these new connections are created in the fiber bundle. This monadic structure allows us to track the evolution of the healthcare system's structure and organization over time and in response to different events, providing valuable insights for enhancing the coordination and integration of the healthcare system.

- **Developing Strategic Responses:** Integrate insights from both perspectives to develop holistic strategies that enhance the organization's adaptability and responsiveness to systemic changes and sectorial transformations. These responses should aim to enhance the organization's ability to adapt and transform in response to new threats and opportunities like COVID-19. The development of these strategic responses should involve a double articulation process, where the strategies (substance and form of content) are shaped by and in turn shape the organization's MABs and CAEs (substance and form of expression).

- **Implementing and Monitoring the Strategy:** Employ both frameworks in implementing and adjusting strategies, ensuring they are responsive to changes in the broader ecosystem as well as within specific sectors. Adjust the strategy as needed based on feedback and changes in the existential territories, energetic-signaletic flows, incorporeal universes, and abstract machinic phyla. The implementation and adjustment of the strategy should also involve a double articulation process, where the changes in the strategy (substance and form of content) are shaped by and in turn shape the organization's MABs and CAEs (substance and form of expression).

In each step, the differential topological concepts provide a mathematical framework for understanding the transformations that occur as a result of the interaction between the Machinic Assemblages of Bodies (MABs) and the Collective Assemblages of Enunciation (CAEs). This understanding can provide valuable insights for strategic analysis and development.

Epilogue

In this concluding chapter, we will first retrace the genealogical roots of DRT, using the Klein Bottle as a metaphor, to identify a shift from theoretical constructs to pragmatic solutions, culminating in an enriched framework that harmonizes diverse theories to address both internal and external dynamics of living systems. We then transition to discuss DRT's multifaceted lens—spanning assemblages, actor-networks, autopoiesis, and more—to illuminate the roles of human and nonhuman actors in the adaptive, self-organizing capabilities of Machinic Life-Experience Ecosystems (MLXEs). Next, we delve into the implications of DRT for understanding the interconnectedness of key sectors like food, health, education, and mobility, especially in the context of the Anthropocene, highlighting the theory's applicability to planetary sustainability. We conclude by reflecting on DRT's impact on well-being, wealth, empowerment, and welfare, positioning these as crucial outcomes of the theory's application in navigating the complexities of human, machine, and nature interactions.

Tracing DRT's genealogical topography

Deleuze and Guattari's Assemblage Theory provide foundational insights into the intricate processes inherent "inside" Living Systems.[1] Their seminal work set the stage for a deeper exploration of the complex interrelations and emergent properties characteristic of these systems. Building upon this, DeLanda offered a refined perspective by integrating Differential Topology with Assemblage Theory, further elucidating the internal dynamics of Living Systems.[2] Transitioning from pure theory to its tangible applications, Prahalad and Ramaswamy introduced the concept of Co-Creation.[3] This innovative approach presented a practical mechanism to understand and navigate Living Systems, effectively bridging theoretical constructs with real-world scenarios. In our subsequent work, we sought to harmonize Assemblage Theory with the Co-Creation framework, aiming to establish a more interconnected perspective.[4] Through our continued exploration in this volume, we have found Category Theory to resonate with the principles of Assemblage Theory and Differential Topology. While our intention was not to reinvent, but rather to integrate, this integration subtly expanded the theoretical boundaries of the framework, making it relevant to the external dynamics of Living Systems as well. Through this journey, we hope to contribute to a framework that is both theoretically robust and pragmatically actionable in the realm of Dynamic Relationality Theory.

---

[1] See Deleuze and Guattari (1987).

[2] See DeLanda (2002).

[3] See Prahalad and Ramaswamy (2004).

[4] See Ramaswamy and Ozcan (2014).

294   Epilogue

The Klein Bottle, with its intricate topology, serves as a fitting metaphor to visualize the theoretical genealogy and topography of Dynamic Relationality Theory (see Fig. 1). Its unique topology allows for a vivid depiction of the progression from foundational theories to their tangible applications. At the very **core of the interior surface** lies **Differential Topology**. This placement underscores its foundational role, providing the mathematical bedrock upon which our understanding of dynamic systems is constructed. As we journey outward from this core, moving **through the handle and towards the bottle's opening**, we encounter **Assemblage Theory**. Positioned here, it signifies the theory's pivotal role in elucidating the emergent properties and dynamics of systems, especially as they begin their interactions with the external world. The **round rim of the bottle**, that intriguing transitional zone between the inside and outside, is adorned with **Category Theory**. This strategic placement emphasizes Category Theory's bridging role. With its abstract and encompassing nature, it captures insights from both the internal dynamics and the external manifestations, serving as a theoretical conduit between the two realms. On the **exterior surface**, the tangible and observable manifestations of our theories come to life. Life-experience of **Living Systems** finds its place on the outermost layer, symbolizing the real-world applications and implications of our theoretical constructs. Meanwhile, the handle's exterior surface is home to the **Co-Creation Paradigm**. Its proximity to Assemblage Theory, albeit on the opposite side of the bottle, suggests a direct application of theoretical insights to real-world practices. It represents our hands-on approach to managing and interacting with Living Systems, emphasizing the practicality embedded in our theoretical framework. Hence, each theory, while distinct, is

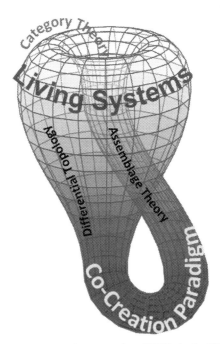

FIGURE 1   Genealogy and topography of DRT via the Klein bottle.

part of a cohesive whole, much like the bottle itself—a single surface with no boundaries. Each section of the bottle represents a distinct theoretical domain, yet they all seamlessly integrate, much like the interconnected nature of the theories themselves.

In conclusion, while Category Theory provides a nuanced lens to understand the theoretical intricacies of Assemblage Theory and Differential Topology, the **Life-Experience Co-Creation Paradigm** stands as a testament to the practical applicability of these theories in understanding the multilayered confluence of relationships of living systems and its ecosystem dynamics.

DRT as a platform of thought and practice for creative transformation

Drawing on the intellectual resources of assemblage theory, actor-network theory, phenomenology, autopoiesis, and philosophical stances on nondualism and dual-aspect monism, Dynamic Relationality Theory (DRT) offers a multifaceted lens through which the dynamic, processual nature of Machinic Life-Experience Ecosystems (MLXEs) can be understood. DRT illuminates the roles of both human and nonhuman actors, the importance of experiential processes, and the adaptive, self-organizing capabilities of digitalized environments while also challenging traditional dichotomies and embracing the complexity of reality. These perspectives offer a unique lens for understanding the fluid nature of relations, the role of technology, the lived experiences of stakeholders, the adaptive nature of digital ecosystems, and the interconnected nature of relations within these MLXEs:

- **Assemblages and individuation**: Deleuze and Guattari's concept of assemblages and Simondon's theory of individuation both emphasize the dynamic, processual, and relational nature of entities. These perspectives in DRT provide a unique lens for understanding the fluid and contingent nature of relations within digital ecosystems.[5]
- **Actor-networks and translation sociology**: Latour's Actor-Network Theory and Callon's Sociology of Translation both emphasize the role of nonhuman actors and the dynamic networks of relations that constitute social reality. These perspectives in DRT provide a unique understanding of the role of technology and other nonhuman actors within digital ecosystems.[6]
- **Experience, process, and change**: Uniting the thoughts of Bergson and Whitehead, this perspective underscores the centrality of experiential processes in understanding reality, enriching our comprehension of stakeholder experiences in digitalized environments.[7]
- **Autopoiesis and active inference**: Varela's work on autopoiesis and Friston's active inference framework both emphasize the role of self-organization and the minimization of free energy in the behavior of living systems. These perspectives in DRT provide a unique understanding of the adaptive and self-organizing nature of digitalized ecosystems.[8]

---

[5]See Deleuze and Guattari (1987) and Simondon (2020).

[6]See Latour (2005) and Callon (1986).

[7]See Deleuze (1988) and Hansen (2015).

[8]See Varela (2016) and Parr (2022).

- **Embodied phenomenology**: Both Heidegger's phenomenology and Merleau-Ponty's existential phenomenology emphasize the embodied and situated nature of human experience. These perspectives in DRT provide a unique understanding of the embodied experiences of stakeholders within digitalized ecosystems.[9]
- **Nondualism and dual-aspect monism:** By positing that the essence of entities emerges from their relational dynamics, transcending traditional dichotomies between mind and matter, subject, and object, DRT reflects an intellectual affinity with nondualism's emphasis on the inseparability of phenomena and dual-aspect monism's view that reality's physical and mental aspects are two complementary descriptions of a unified underlying reality.[10]

These diverse theoretical and philosophical perspectives help advance a comprehensive and nuanced understanding of creative transformation of MLXEs, especially their dynamics, and the role of stakeholders within them:

- Interdisciplinary integration: DRT unifies concepts from social theory, mathematics, digital ecosystem studies, and nondual philosophies. This integration offers a holistic understanding of complex phenomena.
- Dynamic relations: DRT emphasizes the fluid and contingent nature of relations, which provides a fresh lens to understand complex systems and processes.
- Category theory: DRT uses category theory to formalize assemblages and understand digital ecosystems' dynamics. This mathematical framework provides a systematic understanding of complex interactions.
- CARE architecture: The development of the Complex Adaptive Relational Event-Sense (CARE) architecture is a unique contribution, offering a holistic understanding of digital ecosystems and their impact.

DRT's novelty of co-creative transformation of MLXEs thus lies in its interdisciplinary integration, emphasis on dynamic relations, incorporation of non-dual philosophies, application of category theory, development of the CARE architecture, and the integration of diverse social theoretical concepts. This approach provides a comprehensive and nuanced understanding of digital ecosystems, their dynamics, and the role of stakeholders within them.

Epistemological status of DRT as a scientific theory

While DRT presents a sophisticated and comprehensive approach to understanding dynamic relational systems, it faces potential criticisms related to empirical validation, operationalization, complexity, and causal clarity. Addressing these criticisms requires ongoing refinement, empirical research, and dialogue across various academic and practical domains. Here are some issues to consider:

---

[9]See Heidegger (1996) and Merleau-Ponty (1968).
[10]See Priest (2018) and Atmanspacher (2012).

- **Empirical validation:** Since DRT deals with complex, abstract concepts like morphisms and functors in the context of relational dynamics, empirically testing these concepts in real-world scenarios might seem challenging. However, these concepts encourage looking at relational dynamics in new ways, generating hypotheses about complex systems, and fostering innovative empirical studies that can validate and refine the theory's applications.
- **Accessibility:** DRT's complexity is a reflection of the intricate systems it seeks to understand. By embracing this complexity, DRT offers a more nuanced and comprehensive view of relational dynamics than simpler models. Interdisciplinary efforts can translate these complex ideas into more accessible formats, broadening the theory's reach and impact.
- **Operationalization:** The theory's concepts, such as morphisms and functors, might be criticized for being too vague or not well-defined in operational terms. However, this openness is a strength, allowing DRT to adapt and incorporate new insights, making it increasingly relevant and applicable to real-world scenarios.
- **Causal framework:** While DRT may not provide a conventional causal framework, it offers a relational perspective that reveals how entities influence each other in complex systems. This perspective broadens our understanding of causality, incorporating emergent properties and nonlinear interactions that traditional causal models may overlook.
- **Predictive ability:** Although DRT may face challenges in making precise predictions in rapidly changing environments, its value lies in framing our understanding of systemic transformations and potential trajectories. This framing is crucial for strategic planning and policy-making, offering insights that inform adaptive and resilient approaches.

DRT has a plausible claim to be considered as a theory within the philosophy of science and scientific methodology due to its well-defined conceptual framework, predictive capabilities, methodological rigor, empirical relevance, potential for falsifiability, progressive development, and its coherent and comprehensive nature.

- **Conceptual framework and constructs**: A fundamental attribute of a scientific theory is the presence of a well-defined conceptual framework. DRT, as detailed in this book, presents a comprehensive conceptual framework that integrates diverse disciplines and philosophies. For example, it builds on the philosophical works of Gilles Deleuze and Félix Guattari, integrating their ideas on relationality, assemblages, and the virtual/actual dichotomy, into a structured theory. This integration can be seen in our discussions about the interplay between existential life territories, energetic-signaletic ecosystem flows, incorporeal experience universes, and abstract machinic phyla.
- **Theoretical constructs and predictive capability**: A theory is expected to offer predictive capabilities or explanatory power. DRT extends beyond mere description to explain and predict the dynamics of complex systems, particularly in digitally driven ecosystems. It elucidates how entities interact within these systems, influenced by a myriad of internal and external dynamics. This predictive aspect is illustrated in DRT's application to various sectors, including healthcare, where it aids in understanding the complex interplay of factors influencing a patient's health.
- **Methodological rigor**: DRT exhibits methodological rigor, another hallmark of scientific theories, by employing analytical tools from category theory and differential topology.

This approach allows for a structured and systematic exploration of relational dynamics, echoing the scientific method's emphasis on rigor and reproducibility.
- **Empirical relevance**: While DRT is heavily theoretical, it demonstrates empirical relevance by offering applications in real-world contexts. For instance, in the healthcare sector, DRT's principles are applied to understand the dynamics of patient care and the interplay of various health-related factors. This practical applicability is crucial in establishing the relevance of a theory in scientific discourse.
- **Falsifiability**: A key criterion in the philosophy of science, as posited by Karl Popper, is falsifiability—the ability of a theory to be tested and potentially refuted. DRT, with its clear constructs and predictive assertions, lays the groundwork for empirical testing. For example, its assertions about the dynamics of machinic life-experience ecosystems can be empirically investigated in various technological and organizational contexts.
- **Progressive development**: DRT is not static; it allows for progressive development, incorporating new insights and adapting to new contexts. This evolutionary nature is in line with Thomas Kuhn's view of scientific theories as evolving paradigms.
- **Coherence and comprehensiveness**: DRT is coherent and comprehensive, integrating various aspects of human and technological interactions into a unified framework. This holistic approach is evident in its application across different domains, from cognitive science to public policy.

The scientific and epistemological status of DRT in its current iteration is reflective of an evolving, flexible, and heuristic theory. It embodies the provisional and pluralistic nature of scientific theorizing, especially in fields dealing with complex and multifaceted phenomena like MLXEs.

- **Developmental stage of theory**: The current iteration of DRT, where the roles of morphisms and functors are not precisely defined, suggests that it is in a developmental or evolving stage. Scientific theories often undergo periods of refinement and elaboration as they mature. The lack of precise definition of these roles indicates an openness to future research and refinement, which is a natural part of the scientific process.
- **Theoretical pluralism**: The current status of DRT aligns with the concept of theoretical pluralism in the philosophy of science, which advocates for the coexistence and consideration of multiple theoretical approaches. This pluralism can be particularly useful in fields dealing with complex phenomena, where different theoretical lenses may provide complementary insights.
- **Epistemological flexibility**: From an epistemological standpoint, this stage of DRT reflects flexibility and adaptability. The theory is potentially accommodating a wide range of interpretations and applications, allowing it to be applicable in various contexts. This openness can be seen as a strength, as it enables DRT to be responsive to new data and theoretical advancements.
- **Methodological constructivism**: DRT's current iteration could be seen as an example of methodological constructivism, where the theory is part of the ongoing construction of knowledge. The lack of precise definitions allows for a variety of methodological approaches to be employed in understanding and applying the theory.
- **Provisional nature of scientific theories**: In line with Karl Popper's philosophy of science, scientific theories are always provisional and subject to revision. The current iteration of

DRT, with its undefined roles of morphisms and functors, exemplifies this provisional nature, open to being tested, falsified, and revised.
- **Framework theory**: DRT at its current stage can be considered a framework theory, providing a broad conceptual framework within which more specific hypotheses and models can be developed. In this view, the vagueness in the roles of morphisms and functors is not a drawback but rather an invitation for further specification and elaboration within specific contexts or studies.
- **Heuristic value**: The current form of DRT may serve a significant heuristic function. The theory can guide inquiry and research, stimulating new questions and investigations into the nature of complex systems like MLXEs. The exploration of the roles of morphisms and functors could lead to a deeper understanding of the dynamics within these systems.

Extending DRT for planetary sustainability in the anthropocene

Sectors are not static; they evolve and differentiate over time, and in doing so, they exhibit characteristics and dynamics of *all* four domains of MLXE. This reflects the interconnectedness and interdependence of sectors and domains, and it underscores the dynamic, evolving nature of these systems.

Food systems, i.e., food and agriculture, are central to our existential territories, shaping both our physical spaces and immaterial aspects of life, such as cultural practices and societal structures. Historically, the food sector has been pivotal to human survival, influencing societal organization, migration patterns, and exploration of the biosphere and beyond. The development of agriculture marked a significant turning point, leading to settled communities, population growth, and the rise of complex societies. Our food situation continues to determine our livelihoods, with the pressure for better and more resources driving historical and contemporary migrations, conflicts, and explorations. As such, the food sector serves as a foundational domain from which existential territories continue to emerge, influencing our interactions with the natural environment and each other.

- In the context of gauge theory, consider the base space of the food sector. This is where the conditions or states of the system and the entities within the Territorial Spacetime (L-Spacetime) are defined, much like the spacetime in gauge theory. However, these states are not static; they are influenced by the dynamics of other sectors and domains.
  * Consider the health sector, whose gauge fields can be seen as acting on the base space of the food sector. The health conditions of populations, influenced by their diet, can be seen as the values these fields take at each point in the base space. Changes in health conditions can lead to transformations in the food sector, such as shifts in dietary preferences leading to changes in farming practices.
  * Consider the education sector, whose forces also act on these fields. The knowledge and understanding about food and agriculture, disseminated through education, act as forces that influence the state of the health field. This influence can be seen in how education about nutrition and health can lead to changes in dietary habits, which in turn can influence farming practices and food production.

- Consider the mobility sector, wherein ability to move and connect, physically and virtually, can change the internal states of the health field. For instance, the spread of information about health and nutrition can lead to changes in dietary habits without directly causing observable transformations in the health sector. However, these changes in the health field can indirectly influence the food sector by altering demand for certain types of food, which can lead to changes in farming practices and food production.
- While the food sector initially forms the basis of human survival and societal development, as it evolves:
  - The health and vitality of populations become increasingly tied to the quality and safety of food products, and the nutritional information becomes a key part of food packaging and marketing.
  - Societal values and educational initiatives around sustainable farming and ethical food production influence agricultural practices.
  - Technological advancements, such as precision farming and genetically modified crops, transform the food sector.

Health, in its broadest sense, represents the dynamic processes of life.[11] It involves the complex interplay of biological, psychological, and social factors. Health is not a static state but a dynamic process of maintaining balance and adapting to changes in the internal and external environment. It is also a signaletic entity, communicated and interpreted through various signs and symbols, such as symptoms, diagnoses, and health behaviors. Moreover, health should be conceived in the larger sense of the health of life of our planet, including humans. This perspective emphasizes that our health is interconnected with the health of our ecologies. This holistic and integrated view of health extends beyond the achievement of bodily or mental health (homeostatic view) to thriving in changing environments (allostatic view). Health shapes and is shaped by our interactions with others and with the environment. Thus, health shapes and is shaped by the energetic-signaletic flows of our existence.

- In the context of gauge theory, the states of health and disease can be seen as energetic-signaletic flows that exist at every point in the existential territories, and its values determine the transformations experienced by entities.
  - Food can be seen as the base space on which the health fields act. The states of the food sector influence the values that the health fields can take at each point in the base space.
  - Education can be seen as the forces that act on the health fields. The knowledge and understanding about health and disease, disseminated through education, act as forces that influence the state of the health fields.

---

[11]Health can be seen as a form of economy in the Aristotelian sense. Aristotle's notion of economy was based on the idea that economic actions are singular and individual, aimed at using things necessary for life and for the good life (Aristotle, 2011). In this sense, health can be seen as a form of "wise usage" that results in the "produce" of well-being, thus increasing the wealth of life. Just as Aristotle emphasized that expenditures cannot exceed income in the economy, in health too, the "expenditures" of energy and resources in maintaining and enhancing health cannot exceed the "income" of resources available to the individual and the ecosystem.

- Mobility can be seen as the internal degrees of freedom of the health fields. The ability to move and connect, physically and virtually, can change the internal states of the health fields without changing the observable transformations in the health sector, reflecting a fundamental symmetry of the system.
- While the health sector fundamentally embodies Energetic-Signaletic Flows (**E**), representing the dynamic processes of life, as it evolves:
  - It can reflect geographical disparities in health outcomes become apparent, and the physical environment's impact on health becomes a key focus of public health interventions.
  - Health education and promotion campaigns can shape societal attitudes towards health and wellness.
  - Digital health technologies, such as telemedicine and health tracking apps, become integral to healthcare delivery.

Education plays a crucial role in shaping our understanding of the world and our place in it. It provides us with the knowledge, skills, and values we need to navigate our lives. Education influences our beliefs, attitudes, and behaviors, shaping our perceptions and interpretations of reality. It also shapes our social structures and cultural practices, influencing our interactions with others and with the environment.

- In the context of gauge theory, education can be seen as the forces that act on the fields, engendering transformations within the entities from **P.** The knowledge and understanding disseminated through education act as forces that influence the state of the other domains.
  - Food is influenced by the forces of education. The knowledge and understanding about food and agriculture, disseminated through education, act as forces that influence the state of the food sector. This influence works by shaping the understanding and practices around food production, distribution, and consumption.
  - Health is influenced by the forces of education. The knowledge and understanding about health and disease, disseminated through education, act as forces that influence the state of the health fields.
  - Mobility can change the internal states of the health fields without changing the observable transformations in the health sector, reflecting a fundamental symmetry of the system.
- Education shapes our perceptions and interpretations of reality. However, as it evolves:
  - It can reflect the physical environment of learning spaces (from classrooms to virtual learning environments) impacts educational outcomes.
  - The health and wellbeing of students become recognized as crucial factors influencing learning outcomes.
  - Digital technologies transform teaching and learning methods, and as the education sector becomes more interconnected with other sectors through online platforms and resources.

Mobility, encompassing both physical transportation and virtual connections, is central to our interactions with the world.[12] It enables us to move across physical space, connect with others, access resources, and explore new ideas and possibilities. Mobility is also a machinic entity, mediated and facilitated by various technologies, from vehicles and infrastructure to digital networks and devices. Mobility shapes and is shaped by our social structures, cultural practices, and economic systems.

- In the context of gauge theory, each point in the base space of mobility can be seen as representing the internal states of the entity at that point. The ability to move and connect, physically and virtually, can change the internal states of the fields without changing the observable transformations in the other domains.
    * Food can be seen as the existential conditions that can influence the internal degrees of freedom represented by mobility. The state of the food sector can change the internal states of the abstract machinic phyla associated with mobility without changing the observable transformations in the health sector.
    * The internal states of health are influenced by mobility. The ability to move and connect, physically and virtually, can change the internal states of the health fields without changing the observable transformations in the health sector.

---

[12]Mobility, both in the physical and virtual sense, is a key archetype for Guattari's concept of Abstract Machinic Phyla (**M**), which represents the machinic and connective aspects of our existence. This is because mobility, like the Abstract Machinic Phyla, is fundamentally about processes of deterritorialization and reterritorialization. The theories of Deleuze and Guattari, Manuel Castells, and David Harvey and provide compelling arguments for this association.

- Deleuze and Guattari (1987) explain the deterritorializing effect of the thumb in the emergence of embodied technologies. They argue that the thumb is a "deterritorialized" organ because it is not tied to a specific function or task. Instead, the thumb is capable of performing a wide range of tasks, making it a versatile tool for humans. This deterritorialization of the thumb allowed humans to develop new embodied technologies, such as the use of tools and language, which in turn led to the emergence of new forms of social organization. This mirrors the concept of Abstract Machinic Phyla, which is about the emergence and transformation of new forms and systems.
- Castells (2010) discusses how the emergence of the network society is characterized by the widespread use of information and communication technologies. These technologies have facilitated the flow of information, capital, and people across geographical boundaries, leading to a deterritorialization of social and economic processes. This mirrors the concept of Abstract Machinic Phyla, which is about the flow and transformation of entities across different territories.
- Similarly, Harvey (2006) explores the relationship between capitalism, space, and mobility, emphasizing how the increased mobility of capital has necessitated the management and control of finance across different territories. This process of deterritorialization and reterritorialization is driven by the global circulation of capital and the creation of transnational networks. This aligns with the concept of Abstract Machinic Phyla, which is about the creation and transformation of networks and systems.

In summary, mobility, with its inherent processes of deterritorialization and reterritorialization, is a key archetype for Abstract Machinic Phyla. It embodies the dynamic, transformative, and connective aspects of this concept, making it a fitting representation of the machinic and connective aspects of our existence.

- Education shapes the internal degrees of freedom associated with mobility. The knowledge and understanding about mobility, disseminated through education, are integral aspects of the internal states of mobility, shaping and being shaped by them in a continuous process of mutual becoming.
- Mobility enables movement and connection. However, as it evolves:
  - The physical infrastructure of transportation networks shapes the mobility options available to individuals and communities.
  - The health impacts of different modes of transportation (e.g., active transport like walking and cycling vs. motorized transport) become recognized.
  - Societal values and education around sustainable transportation influence mobility choices and policies.

Food, health, education, and mobility can be viewed as archetypical MLXEs in planetary society:

- **Universality and fundamental nature**: Food, health, education, and mobility are universal and fundamental aspects of human existence. They are applicable across all cultures and societies and have been central to human life throughout history. These sectors are relevant to all societies, and changes in these sectors can have significant impacts on people's lives.
- **Interconnectedness**: Food, health, education, and mobility are deeply interconnected, and changes in one can have profound effects on the others. For example, improvements in health can lead to improvements in education, which can lead to improvements in mobility, which can lead to improvements in food, and so on. This interconnectedness mirrors the interconnectedness of **L**, **E**, **X**, and **M**, in Guattari's model.
- **Demonstration of Gauge Theory Logic/Geometry**: Food, health, education, and mobility provide a clear and intuitive demonstration of gauge theory logic/geometry. They allow us to visualize the concepts of base space, fields, forces, and internal degrees of freedom in a concrete and tangible way, with each sector representing a different aspect of the gauge theory framework.[13]

---

[13] Ong et al. (2021) offers a concrete example of how changes in the food environment can have far-reaching effects on the health and well-being of communities, thereby emphasizing the need for holistic approaches to address these challenges. This study:

- posits food systems as crucial to our existential territories, significantly affecting human survival, societal organization, and our interaction with the natural world.
- illustrates how gentrification intensifies food insecurity and health disparities, latter being directly influenced by populations' dietary habits and access to culturally appropriate foods.
- underscores the role of education in shaping transformations within these contexts, advocating for education around food justice and equitable food systems planning.
- emphasizes mobility as a critical factor in accessing resources, highlighting the dynamic interplay between physical and virtual connections and their influence on sectors like food and health.

Interconnectedness of food, health, education, and mobility in the anthropocene

In the Anthropocene, an epoch marked by significant human impact on Earth's geology and ecosystems, when the boundaries between human, machine, and nature are increasingly blurred, and their interdependencies become more evident and critical to our survival and prosperity, the interconnectedness of key human-industrial sectors becomes ever more crucial to our understanding of life's complex web. These sectors—food, health, education, and mobility—not only stand as vital components of human existence but also embody the intricate interplay of Guattari's fourfold domains: Existential Territories, Energetic-Signaletic Flows, Incorporeal Universes, and Abstract Machinic Phyla. Each of these domains plays a pivotal role in shaping the primordial matrix of life on Earth, contributing to the evolution of human-machine-nature interactions.

At the heart of this evolution is the co-creative process between humans, machines, and nature. These sectors exemplify the various stages of this co-evolution, from the early divergence of our species from the primordial matrix to the ongoing dynamics of this creative relationship. The Anthropocene, with its distinctive blending of natural and technological elements, offers a macro-historical perspective on this journey. It reveals how our species has evolved alongside machines and nature, with these four sectors grounding the myriad ways in which this evolution continues to unfold.

In this era, the sectors of food, health, education, and mobility are not only ripe for disruption and radical transformation but also serve as archetypal examples of the functioning of Guattari's fourfold domains. They demonstrate how each domain, when seen through the lens of a gauge theory logic and geometry, contributes to the global interactional creation of life experiences. By examining these sectors, we gain insights into the complex dynamics that govern our co-existence with machines and the natural world, and how these interactions shape our shared history and future possibilities. Through their interconnectedness, these sectors collectively embody the existential territories that nurture us, the energetic-signaletic flows that sustain us, the incorporeal universes that enlighten us, and the abstract machinic phyla that mobilize us (see Fig. 2).

The **agriculture and food** sector forms the existential territories of our Anthropocene narrative. It is the foundation of our survival and well-being, providing the nourishment we need to live and thrive. But it is not just about growing and consuming food. It is about the complex interplay of natural resources, human labor, and technological innovation that brings food from the field to our tables. It is about the policies and practices that determine who gets to eat and who goes hungry. And it is about the impact of our food choices on the health of our planet. The MAB here includes the physical land, the crops and livestock, and the technologies used for farming and food production. The CAE includes the knowledge and practices of farming, the cultural and symbolic meanings of food, and the policies and systems that govern food production and distribution. The existential territories of food are the foundation from which other territories emerge and evolve, shaping our livelihoods and driving our migrations, conflicts, and explorations throughout history.

The **health** sector forms the energetic-signaletic flows of our Anthropocene narrative. It is the pulse of our collective well-being, providing the care and treatment we need to maintain our health. But it is not just about treating illnesses. It is about the complex interplay of biological processes, medical knowledge, and social interactions that determine our health outcomes. It is about the policies and practices that determine who gets access to care and who does not. And it is about the impact of our health choices on the health of our

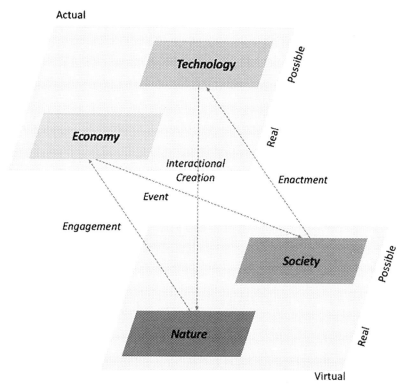

FIGURE 2 Interconnectedness of food, health, education, and mobility.

communities and our planet. The MAB here includes the physical bodies of patients and healthcare providers, the biological processes of health and disease, and the technologies used for diagnosis and treatment. The CAE includes the medical knowledge and practices, the cultural and symbolic meanings of health and illness, and the policies and systems that govern healthcare provision. The energetic-signaletic flows of healthcare are deeply interconnected with the existential territories of food, as our health is directly influenced by our agriculture and food practices. In this more holistic and integrated view of health, we also consider the ability to adapt and thrive in changing environments, reflecting an allostatic view of health.

The **education** sector forms the incorporeal universes of our Anthropocene narrative. It is the compass that guides our collective learning and growth, providing the knowledge and skills we need to navigate our world. But it is not just about acquiring knowledge. It is about the complex interplay of cognitive processes, social interactions, and cultural contexts that shape our learning experiences. It is about the policies and practices that determine who gets access to education and who does not. And it is about the impact of our learning choices on the development of our societies and our planet. The MAB here includes the physical schools and universities, the cognitive processes of learning, and the technologies used for education. The CAE includes the pedagogical methods and practices, the cultural and symbolic meanings of education, and the policies and systems that govern educational provision. The incorporeal universes of education are interconnected with the existential territories of food, the energetic-signaletic flows of healthcare, and the abstract machinic phyla of mobility, as our learning influences and is influenced by our food, health, and mobility practices.

Mobility, encompassing both physical transportation and virtual connections, forms the abstract machinic phyla of our Anthropocene narrative. The MAB here includes the physical infrastructure of transportation, the technologies used for travel, communication, and data transfer, and the biological processes of movement and perception. The CAE includes the rules and norms of travel and communication, the narratives, and symbols associated with different modes of mobility and connectivity, and the policies and systems that govern mobility. The abstract machinic phyla of mobility are intertwined with the existential territories of food and the energetic-signaletic flows of healthcare, as our ability to move and connect influences both our access to food and our health outcomes. In this context, mobility is not just about moving from one place to another but also about adapting to and navigating through changing environments, reflecting the allostatic view of health.

This process allows the sectors of food, health, mobility, and education to adapt and evolve in response to changes in both the physical, tangible aspects of the system and the discursive, symbolic aspects of the system. Through this complex interchange, these sectors continue to shape our co-evolution with machines and nature, grounding us in our shared history while propelling us into a future of endless possibilities.

DRT for impact: Well-being, wealth, empowerment, and welfare in MLXEs

In this concluding section of the Dynamic Relationality Theory (DRT) book, we reflect upon the profound impacts of wellbeing, wealth, empowerment, and welfare, each serving as a meta-level outcome in the interconnected domains of sectors such as food, health, education, and mobility. These impacts not only encapsulate the essence of the sectors we have explored but also embody the core themes of DRT, tying together the intricate juxtaposition of human-machine-nature interactions within the Anthropocene.

Wellbeing and food: The access to nutritious food, sustainable agricultural practices, and food security form the crux of wellbeing. This is emblematic of the DRT's emphasis on existential territories (**L**) where the transformations in the food sector directly enhance individual and community well-being. For instance, advancements in agricultural practices leading to better food security epitomize the DRT's narrative of dynamic relationality, wherein changes in one domain ripple through the entire ecosystem, influencing human well-being in profound ways.

Wealth and health: Health transcends its traditional confines to emerge as a form of wealth within DRT's framework. This perspective aligns with the energetic-signaletic flows (**E**), where improvements in health bolster economic prosperity. By enhancing the capacity for work and societal participation, health advancements underscore DRT's theme of interconnectivity, illustrating how changes in the health sector can elevate the economic and social wealth of communities.

Empowerment and education: Education is the cornerstone of empowerment, a theme deeply resonant with DRT's focus on incorporeal universes (**X**). The transformative power of education, enabling individuals to make informed choices and drive changes in their lives, mirrors DRT's narrative of co-creative transformation. It exemplifies how the dissemination of knowledge and skills can catalyze profound shifts in societal structures and individual capacities.

Welfare and mobility: Mobility, encompassing both physical and social dimensions, is central to welfare. This reflects the abstract machinic phyla (**M**) within DRT, where the capacity to access resources and navigate social structures is pivotal for welfare. The DRT book's exploration of mobility emphasizes the fluidity of interactions and the necessity of adaptability, highlighting how advancements in this sector can significantly enhance societal welfare.

Through this brief discussion, it becomes evident that further exploration of wellbeing, wealth, empowerment, and welfare, in NEST-networked MLXEs, is needed for a comprehensive and nuanced understanding of the interplay between human, machine, and nature in the Anthropocene. While this exploration would require a separate volume, these themes tie to the broader narrative of interconnectedness and interactional creation integral to DRT, underscoring the importance of recognizing and harnessing the dynamic relationships within and across sectors to navigate the complexities of our time. Therefore, this conclusion does not mark an end, but rather, it serves as an invitation to continue exploring, understanding, and shaping the intricate web of NEST-ed relations that define our world, ensuring a sustainable and thriving future for all.

Appendix—Category theory notational reference table

Categories

C_1 ("Individual")
Objects: Affordances, UserInteractions, Generativity
Morphisms:
f_{11}: Affordances → UserInteractions
f_{12}: UserInteractions → Generativity

C_2 ("Network")
Objects: Technicity, TransindividualRelations, CollectiveIndividuation
Morphisms:
f_{21}: Technicity → TransindividualRelations
f_{22}: TransindividualRelations → CollectiveIndividuation

C_3 ("Stakeholder")
Objects: BlackBoxes, IndividualTechnologyEnvironmentRelations, SocialActorNetworks
Morphisms:
f_{31}: BlackBoxes → IndividualTechnologyEnvironmentRelations
f_{32}: IndividualTechnologyEnvironmentRelations → SocialActorNetworks

C_4 ("Ecosystem")
Objects: AdaptiveStructuration, DigitalNicheConstruction, CoShapingTechnologyOrganizationalPractices
Morphisms:
f_{41}: AdaptiveStructuration → DigitalNicheConstruction
f_{42}: DigitalNicheConstruction → CoShapingTechnologyOrganizationalPractices

C_5 ("Creative-Experiencer")
Objects: FreeEnergyPrinciple, ActiveInference, Phenomenotechnics
Morphisms:
f_{51}: FreeEnergyPrinciple → ActiveInference
f_{52}: ActiveInference → Phenomenotechnics

C_6 ("Experience-verse")
Objects: SenseMaking, SeriesStructure, PoiesisEnchantment
Morphisms:

f_{61}: SenseMaking → SeriesStructure
f_{62}: SeriesStructure → PoiesisEnchantment

C_7 ("Co-Innovation")
Objects: Collaboration, CoDesign, OpenInnovation
Morphisms:
f_{71}: Collaboration → CoDesign
f_{72}: CoDesign → OpenInnovation

C_8 ("Platform")
Objects: DigitalInfrastructure, Platforms, Services
Morphisms:
f_{81}: DigitalInfrastructure → Platforms
f_{82}: Platforms → Services

Functors

F_1: C_1 → C_2 ("Lived Journey")
F_1(Affordances) = Technicity
F_1(UserInteractions) = TransindividualRelations
F_1(Generativity) = CollectiveIndividuation
$F_1(f_{11}) = f_{21}$
$F_1(f_{12}) = f_{22}$

F_2: C_2 → C_3 ("Engagement")
F_2(Technicity) = BlackBoxes
F_2(TransindividualRelations) = IndividualTechnologyEnvironmentRelations
F_2(CollectiveIndividuation) = SocialActorNetworks
$F_2(f_{21}) = f_{31}$
$F_2(f_{22}) = f_{32}$

F_3: C_3 → C_4 ("Flows")
F_3(BlackBoxes) = AdaptiveStructuration
F_3(IndividualTechnologyEnvironmentRelations) = DigitalNicheConstruction
F_3(SocialActorNetworks) = CoShapingTechnologyOrganizationalPractices
$F_3(f_{31}) = f_{41}$
$F_3(f_{32}) = f_{42}$

F_4: C_4 → C_5 ("Event")
F_4(AdaptiveStructuration) = FreeEnergyPrinciple
F_4(DigitalNicheConstruction) = ActiveInference
F_4(CoShapingTechnologyOrganizationalPractices) = Phenomenotechnics
$F_4(f_{41}) = f_{51}$
$F_4(f_{42}) = f_{52}$

$F_5: C_5 \to C_6$ ("Sense")
$F_5(FreeEnergyPrinciple) = SenseMaking$
$F_5(ActiveInference) = SeriesStructure$
$F_5(Phenomenotechnics) = PoiesisEnchantment$
$F_5(f_{51}) = f_{61}$
$F_5(f_{52}) = f_{62}$

$F_6: C_6 \to C_7$ ("Enactment")
$F_6(SenseMaking) = Collaboration$
$F_6(SeriesStructure) = CoDesign$
$F_6(PoiesisEnchantment) = OpenInnovation$
$F_6(f_{61}) = f_{71}$
$F_6(f_{62}) = f_{72}$

$F_7: C_7 \to C_8$ ("Resourced Capabilities")
$F_7(Collaboration) = DigitalInfrastructure$
$F_7(CoDesign) = Platforms$
$F_7(OpenInnovation) = Services$
$F_7(f_{71}) = f_{81}$
$F_7(f_{72}) = f_{82}$

$F_8: C_8 \to C_1$ ("Interactional Creation")
$F_8(DigitalInfrastructure) = Affordances$
$F_8(Platforms) = UserInteractions$
$F_8(Services) = Generativity$
$F_8(f_{81}) = f_{11}$
$F_8(f_{82}) = f_{12}$

Appendix—Healthcare case: Details of category theory analysis

Category C_3 ("Stakeholder")

In the exploration of Category C_3 ("Stakeholder") within the MLXE, ideal types across BlackBoxes ("BB"), IndividualTechnologyEnvironmentRelations ("ITER"), and Social-ActorNetworks ("SAN") serve as essential constructs for simplifying the complexity of global processes, facilitating a clearer understanding of potential configurations and interactions. These ideal types offer a foundational framework to dissect the nuances of stakeholder engagements and technological integrations in healthcare, specifically in the realm of breast cancer care within US healthcare systems:

- **Transparent BB**, like mammography or chemotherapy, are well-understood by users, ensuring that healthcare professionals and patients are well-informed about procedures and outcomes. Conversely, **opaque BB** represent the realm of newer or experimental technologies such as AI-driven diagnosis or treatments, where the complexity and uncertain outcomes might not be fully comprehensible to all users.
- The dichotomy between **technology-dependent ITER** and **technology-independent ITER** further enriches our understanding of individual–environment interactions. While the former highlights a reliance on advanced technologies for diagnosis, treatment, and monitoring, the latter emphasizes traditional, less technology-centric approaches, underscoring the variability in patient care pathways.
- The contrast between **hierarchical SAN** and **collaborative SAN** illustrates the evolving dynamics of decision-making in healthcare. Hierarchical SAN depicts a traditional top-down approach with physicians at the helm, whereas collaborative SAN advocates for a more egalitarian, multistakeholder engagement model, facilitated by digital platforms, enhancing patient involvement and interdisciplinary collaboration.

Thus, we can define the open sets in the topological space based on different combinations of these ideal types. Each set represents unique interactions and reliance on technology within healthcare ecosystems, delineating the diverse ways in which technology and social structures intertwine in the context of US healthcare systems. In the exploration of the topological space within the MLXE for breast cancer prevention, diagnosis, and treatment, we can identify 12 distinct open sets based on combinations of ideal types across BB, ITER, and SAN:

- **Open Set 1**—Hierarchical SAN and Technology-Dependent ITER, like AI in mammography, demonstrate top-down technology integration.
- **Open Set 2**—Hierarchical SAN and Technology-Independent ITER, showcasing traditional face-to-face consultations, where technology plays a minimal role.

- **Open Set 3**—Collaborative SAN and Technology-Dependent ITER, exemplified by patient portals in care settings, enhancing shared decision-making.
- **Open Set 4**—Collaborative SAN and Technology-Independent ITER, focusing on traditional communication methods for decision-making.
- **Open Set 5**—Hierarchical SAN and Transparent BB, such as mammography, indicating a preference for transparent technological interventions.
- **Open Set 6**—Hierarchical SAN and Opaque BB, like AI for diagnosis, highlight the challenges of integrating complex technologies.
- **Open Set 7**—Collaborative SAN and Transparent BB, such as patient portals, facilitating open and informed collaboration.
- **Open Set 8**—Collaborative SAN and Opaque BB, like experimental AI tools, reflecting the challenges in understanding and trust.
- **Open Set 9**—Technology-Dependent ITER and Transparent BB, showcasing reliance on well-understood tools like mammography for diagnosis.
- **Open Set 10**—Technology-Dependent ITER and Opaque BB, indicating reliance on complex, less understood tools for care
- **Open Set 11**—Technology-Independent ITER and Transparent BB, representing scenarios where well-understood methods like face-to-face consultations are preferred.
- **Open Set 12**—Technology-Independent ITER and Opaque BB, such as experimental treatments, where technology plays a minimal role.

In the US healthcare system's context of breast cancer prevention, diagnosis, and treatment, seven distinct base sets represent various global processes within Category C_3. These sets encompass the dynamic interplay between technological advancements, social actor networks, and individual-environment relations, highlighting the multifaceted nature of healthcare delivery and policy.

- **Physician—patient—nurse triad field** emphasizes traditional hierarchical interactions, where decision-making primarily resides with physicians. This field exemplifies how established healthcare practices navigate between technology-dependent and independent scenarios, impacting patient care delivery in hospital settings.
- **Healthcare technology innovators field** showcases the collaborative effort in developing and implementing cutting-edge technologies like AI and telehealth. This field bridges hierarchical and collaborative networks, underlining the pivotal role of innovation in transforming healthcare practices.
- **Patient advocacy and empowerment field** focuses on enhancing patient care through collaborative efforts between patients, advocacy groups, and providers. It underscores the shift towards patient-centric care models, advocating for both technology-dependent and independent approaches to improve healthcare outcomes.
- **Healthcare policy and regulation field** delves into the collaborative and hierarchical dynamics of policy formulation and regulation, crucial for governing healthcare practices. This field generates discussions around transparent and opaque technologies, reflecting on the regulatory challenges and opportunities in healthcare.
- **Healthcare research and education field** highlights the synergy between research, education, and clinical practice. It fosters a collaborative environment conducive to transparency and innovation, crucial for advancing healthcare knowledge and practices.

- **Healthcare data and analytics field** illustrates the critical role of data in healthcare, where technology-dependent interactions enhance patient care through analytics. This field bridges the gap between transparent and opaque technological understandings, emphasizing the importance of data-driven decision-making.
- **Healthcare access and equity field** addresses the imperative of equitable healthcare, focusing on improving access and reducing disparities. It reflects on the interaction between technology-independent practices and transparent processes, highlighting the ongoing efforts to achieve healthcare equity.

Category C_4 ("Ecosystem")

In the US healthcare system's context of breast cancer prevention, diagnosis, and treatment, Category C_4 ("Ecosystem") encompasses the dynamics of AdaptiveStructuration ("AS"), DigitalNicheConstruction ("DNC"), and CoShapingTechnologyOrganizationalPractices ("CSTOP"). These frameworks illustrate the interplay between technological advancements and organizational practices involving key stakeholders: physicians, patients, and nurses.

- **AS** underscores the evolution of decision-making paradigms from **centralized**, where physicians predominate, to **decentralized** models. The latter democratizes the decision-making process, leveraging technologies like AI and telehealth to distribute responsibilities and empower patients and nurses alongside physicians. This shift enhances personalized care through digital twins and broader access to specialized treatments and clinical trials.
- **DNC** examines the ecosystem's receptiveness to technological integration. The **technology-receptive** ideal type embraces innovations such as CRISPR, AI, and robotic surgery, integrating them into breast cancer care. Conversely, the **technology-resistant** type highlights the challenges of adopting new technologies, attributed to resource limitations, ethical concerns, or lack of training.
- **CSTOP** delves into the catalysts for change within healthcare practices. The **technology-driven** approach showcases how innovations like robotic surgery can revolutionize traditional practices, leading to improved patient outcomes. Meanwhile, the **practice-driven** type illustrates how practical needs, such as the exigencies of pandemic-era care, can spur technological adaptations, notably in telehealth.

Let us define open sets based on the interplay between AS, DNC, and CSTOP:

- **Open Set 1**—Centralized AS and Technology-Receptive DNC, like the National Cancer Institute's adoption of CRISPR and AI, maintaining a hierarchy in decision-making but welcoming technology.
- **Open Set 2**—Decentralized AS and Technology-Receptive DNC, like telehealth consultations, where technology integration is distributed and receptive across healthcare networks.

- **Open Set 3**—Centralized AS and Technology-Driven CSTOP, where new technologies, such as cryo-electron microscopy, drive the evolution of organizational practices, emphasizing top-down technology changes.
- **Open Set 4**—Decentralized AS and Technology-Driven CSTOP, like the use of the Infinium Assay, showing how technology reshapes practices across various organizations.
- **Open Set 5**—Technology-Receptive DNC and Technology-Driven CSTOP, where the receptiveness to new technologies, such as robotic surgery, drives changes in surgical practices, highlighting the technology's impact on niche surgical procedures.
- **Open Set 6**—Technology-Receptive DNC, Decentralized AS, and Technology-Driven CSTOP, exemplified by networks leveraging telehealth technologies for remote care.
- **Open Set 7**—Centralized AS and Technology-Resistant DNC, like the DEA's initial resistance to telemedicine, demonstrating a shift toward technology acceptance amid necessity.
- **Open Set 8**—Decentralized AS and Technology-Resistant DNC, like telehealth's equitable use challenges, indicating resistance to full technology adoption.
- **Open Set 9**—Centralized AS and Practice-Driven CSTOP, like AI's development for "digital twins" in cancer care, evolving with healthcare provider practices.
- **Open Set 10**—Decentralized AS and Practice-Driven CSTOP, with organizations like HIMSS supporting health transformation through digital tools and internal innovation.
- **Open Set 11**—Technology-Resistant DNC and Practice-Driven CSTOP, like the NHS's initial telehealth resistance, eventually integrating technology into service delivery.
- **Open Set 12**—Technology-Resistant DNC, Decentralized AS, and Practice-Driven CSTOP, showing the complex dynamics of adopting new technologies like robotic surgery in various healthcare settings.

The tentative base for the topology within the healthcare ecosystem, as outlined, can be summarized as follows:

- **AI and data analytics field**: This field could include practices that involve AI, machine learning, and data analytics. Practices like AI for improved diagnosis and personalized medicine, analyzing population-based cancer data, and predictive analytics could be part of this group.
- **Gene editing and molecular understanding field**: This field could encompass practices that involve genetic manipulation and molecular understanding. Practices like the use of CRISPR for gene editing in cancer treatment and the Infinium Assay for genotyping and understanding genetic variations could be included.
- **Remote healthcare delivery field**: This field could include practices that involve remote delivery of healthcare services. Practices like telehealth for remote care and treatment, and digital health platforms like Accelerate from HIMSS could be part of this field.
- **Advanced imaging and surgical procedures field**: This field could encompass practices that involve advanced imaging techniques and surgical procedures. Practices like cryo-EM for generating high-resolution images of molecular behavior and robotic surgery for precise, minimally invasive surgeries could be included.
- **Healthcare innovation and strategy field**: This field could include practices that involve the development of internal innovation programs, the assessment of the potential of

employee-led projects, and looking to the market to find solutions to implement or invest in, such as those driven by Healthbox, a HIMSS Solution and healthcare advisory firm.
- **Patient-centered health and wellness field**: This field could encompass practices that focus on patient/consumer-centered health, wellness, and disease prevention, such as those advanced by the Personal Connected Health Alliance (PCHAlliance), a membership-based HIMSS Innovation Company.

Category C_5 ("Creative-Experiencer")

In the healthcare context of breast cancer prevention, diagnosis, and treatment, Category C_5 ("Creative-Experiencer") focuses on the conceptual frameworks of the Free-EnergyPrinciple ("FEP"), ActiveInference ("AIF"), and Phenomenotechnics ("PT"). These frameworks represent the interplay between predictability, data reliance, and the subjective experience of healthcare, offering insights into personalized and technologically integrated care strategies.

- The **FEP** is explored through entropy-reducing and entropy-increasing practices. **Entropy-reducing** practices aim to diminish uncertainty in healthcare outcomes, utilizing predictive models and AI for precise disease progression forecasts and treatment efficacy. Conversely, **entropy-increasing** practices recognize the inherent unpredictability within healthcare, embracing personalized medicine to accommodate individual variability in patient health, acknowledging the unique interplay of genetic and environmental factors in each patient's condition.
- **AIF** is delineated into data-driven and experience-driven approaches. **Data-driven** active inference leverages extensive data analysis and evidence-based decision-making, employing AI to enhance diagnosis accuracy and treatment planning. **Experience-driven** active inference prioritizes clinical intuition and patient-reported outcomes, emphasizing the significance of telehealth in fostering patient–physician communication and collaborative decision-making.
- **PT** is conceptualized through technology-mediated and direct-experience perspectives. **Technology-mediated** phenomenotechnics underscore the role of digital platforms in enriching the patient care experience, enhancing engagement, and streamlining care coordination. **Direct-experience** phenomenotechnics focus on traditional care delivery models, valuing in-person interactions and physical examinations as essential components of diagnosis and treatment processes.

The open sets in the topological space of Category C_5 illustrate the dynamic interplay between entropy, data, experience, and the role of technology in shaping healthcare practices, specifically in the context of breast cancer prevention, diagnosis, and treatment:

- **Open Set 1**—Entropy-Reducing FEP and Data-Driven AIF, focusing on using AI for predictive models in breast cancer recurrence, showcasing a synergy between reducing uncertainty and leveraging data.

- **Open Set 2**—Entropy-Reducing FEP and Experience-Driven AIF, valuing clinical experience and patient reports, using telehealth for follow-ups in breast cancer care.
- **Open Set 3**—Entropy-Increasing FEP and Data-Driven AIF, embracing genetic variability and data analysis, like genotyping tools for individual variations.
- **Open Set 4**—Entropy-Increasing FEP and Experience-Driven AIF, prioritizing patient–physician communication through telehealth platforms for shared decision-making.
- **Open Set 5**—Entropy-Reducing FEP and Technology-Mediated PT, using digital platforms like HIMSS's Accelerate to enhance patient engagement and care coordination.
- **Open Set 6**—Entropy-Reducing FEP and Direct-Experience PT, focusing on traditional in-person consultations and examinations for breast cancer treatment.
- **Open Set 7**—Entropy-Increasing FEP and Technology-Mediated PT, like AI-driven "digital twins" for exploring treatments and predicting outcomes.
- **Open Set 8**—Entropy-Increasing FEP and Direct-Experience PT, maintaining traditional care models while embracing patient variability.
- **Open Set 9**—Data-Driven AIF and Technology-Mediated PT, utilizing AI algorithms for imaging analysis to detect breast cancer early.
- **Open Set 10**—Data-Driven AIF and Direct-Experience PT, involving AI analysis for personalized care combined with in-person consultations.
- **Open Set 11**—Experience-Driven AIF and Technology-Mediated PT, leveraging telehealth and digital platforms for symptom monitoring and patient engagement.
- **Open Set 12**—Experience-Driven AIF and Direct-Experience PT, emphasizing traditional care models alongside patient-reported outcomes for decision-making.

We can suggest the following empirically plausible base sets that could generate the open sets listed above:

- **Predictive analytics community**: This community could include practices that involve the use of predictive analytics and machine learning to improve healthcare outcomes. This could involve the use of AI and data analytics to predict the likelihood of breast cancer recurrence, the effectiveness of different treatment options, and the potential side effects of treatment.
- **Personalized medicine community**: This community could encompass practices that involve the use of genomics and personalized medicine to tailor treatment to individual patients. This could involve the use of genotyping tools like the Infinium Assay to understand individual genetic variations in breast cancer patients, and the use of AI to create a virtual model of the patient (a "digital twin") to explore treatments and predict possible outcomes.
- **Patient-centered care community**: This community could include practices that prioritize patient–provider interactions and shared decision-making. This could involve the use of telehealth platforms to facilitate patient–physician communication and shared decision-making in breast cancer care, and the use of digital health platforms to enhance patient engagement.
- **Technological innovation community**: This community could encompass practices that involve the adoption and integration of various digital technologies in healthcare. This could involve the use of AI in creating a virtual model of the patient (a "digital twin"), the

use of telehealth for remote care and treatment, and the use of digital health platforms like accelerate from HIMSS to improve patient engagement and care coordination.
- **Traditional care delivery community**: This community could include practices that prioritize traditional care delivery models and direct patient–provider interactions. This could involve the traditional model of in-person consultations and examinations in breast cancer diagnosis and treatment, and shared decision-making based on patient-reported outcomes.

Category C_6 ("Experience-verse")

In the healthcare ecosystem, particularly within the realm of breast cancer prevention, diagnosis, and treatment, Category C_6 ("Experience-verse") introduces a nuanced understanding through the ideal types of SenseMaking ("SM"), SeriesStructure ("SS"), and PoiesisEnchantment ("PE"). These conceptual frameworks offer insights into the interpretative, organizational, and transformative aspects of healthcare experiences.

- SM, divided into data-driven and experience-based types, reflects the dual pathways through which patients and healthcare providers interpret complex health information. **Data-driven SM** leverages AI and technology to distill insights from medical imaging or genomic data, facilitating tailored treatment strategies. **Experience-based SM**, conversely, relies on personal narratives, intuition, and direct communication, crucial for navigating ambiguities or personalizing care decisions where data alone may not suffice.
- SS encompasses Routine-based and adaptive types, highlighting the organizational dynamics of healthcare practices. **Routine-based SS** adheres to established care pathways, offering predictability through evidence-based protocols. **Adaptive SS**, on the other hand, emphasizes flexibility, allowing care plans to evolve in response to individual patient responses, preferences, and holistic needs, underscoring the importance of patient-centered care.
- PE, through technology-enabled and human-centered types, explores the capacity of healthcare to inspire and transform. **Technology-enabled PE** harnesses cutting-edge innovations like precision medicine for breakthrough treatments, while **human-centered PE** focuses on the profound impact of compassionate care and human connection, enhancing patient well-being and resilience through meaningful healthcare experiences.

The "experience-verse" offers a multifaceted view through twelve distinct open sets, each illustrating how data, technology, and human elements interact dynamically to shape breast cancer care:

- **Open Set 1**—Data-driven SM and Routine-Based SS, using AI for medical interpretations within a standard care pathway.
- **Open Set 2**—Data-Driven SM and Adaptive SS, where AI-informed decisions allow for tailored treatment plans.
- **Open Set 3**—Experience-Based SM and Routine-Based SS, emphasizing personal and communal experiences within fixed treatment protocols.

- **Open Set 4**—Experience-Based SM and Adaptive SS, adapting treatments based on individual experiences and needs.
- **Open Set 5**—Data-Driven SM and Technology-Enabled PE, leveraging AI for diagnosis and employing precision medicine for transformative care.
- **Open Set 6**—Data-Driven SM and Human-Centered PE, where technology-driven decisions are complemented by compassionate care.
- **Open Set 7**—Experience-Based SM and Technology-Enabled PE, personalizing treatment through advanced technologies based on patient and network experiences.
- **Open Set 8**—Experience-Based SM and Human-Centered PE, focusing on empathetic care informed by personal experiences.
- **Open Set 9**—Routine-Based SS and Technology-Enabled PE, following standard protocols while incorporating cutting-edge treatments.
- **Open Set 10**—Routine-Based SS and Human-Centered PE, delivering compassionate care within established treatment pathways.
- **Open Set 11**—Adaptive SS and Technology-Enabled PE, tailoring flexible treatment plans with innovative technologies.
- **Open Set 12**—Adaptive SS and Human-Centered PE, personalizing care plans with a focus on emotional support and holistic care.

Here are some empirically plausible base sets that could generate the open sets in the topology:

- **Clinical decision-making community**: This community would include practices that involve the use of both data and experience in making clinical decisions. This could involve the use of AI and machine learning for interpreting medical imaging or genomic data, but also the use of personal experience and intuition in interpreting these results and making treatment decisions.
- **Patient engagement community**: This community would encompass practices that involve engaging patients in their own care. This could involve the use of digital tools for patient education and engagement, but also the use of shared decision-making strategies and peer support groups.
- **Quality assurance community**: This community would include practices that involve ensuring the safety and effectiveness of care. This could involve the use of standard care pathways and quality assurance processes, but also the use of continuous improvement processes for enhancing the quality and efficiency of care.
- **Personalized care community**: This community would encompass practices that involve tailoring care to the individual patient. This could involve the use of personalized treatment plans that are adjusted based on the patient's response to therapy, their personal preferences, or their social and emotional needs.
- **Technological innovation community**: This community would include practices that involve the use of advanced technologies in care. This could involve the use of precision medicine technologies for personalized treatment, telehealth technologies for remote care and monitoring, and AI and machine learning for interpreting medical imaging or genomic data.
- **Holistic care community**: This community would encompass practices that involve providing holistic care to patients. This could involve the provision of emotional support

and holistic care to patients, the use of patient-centered design principles in developing healthcare services and technologies, and the promotion of patient empowerment and self-efficacy in managing their health.

Please see Fig. 1 for summary of the various topological spaces and base sets in the healthcare case examples used across multiple chapters of this book.

	Collective Assemblages of Enunciation	Machinic Assemblages of Bodies
Molar	C_4 ("Ecosystem") Topological Space Dimensions: • AdaptiveStructuration: centralized vs decentralized decision-making • DigitalNicheConstruction: technology-receptive vs resistant • CoShapingTechnologyOrganizationalPractices: technology- vs practice-driven Base Sets as Fields of Practice: • *AI and Data Analytics* • Gene Editing and Molecular Understanding • *Remote Healthcare Delivery* • Advanced Imaging and Surgical Procedures • Healthcare Innovation and Strategy • *Patient-Centered Health and Wellness*	C_3 ("Stakeholder") Topological Space Dimensions: • BlackBoxes: transparent vs. opaque • IndividualTechnologyEnvironmentRelations: technology-dependent vs -independent relations • SocialActorsNetworks: hierarchical vs. collaborative networks Base Sets as Fields of Practice: • *Physician-Patient-Nurse Triad* • *Healthcare Technology Innovators* • Patient Advocacy and Empowerment • Healthcare Research and Education • Healthcare Data and Analytics • Healthcare Policy and Regulation • *Healthcare Access and Equity*
Molecular	C_6 ("Experience-verse") Topological Space Dimensions: • SenseMaking: data-driven vs experience-based • SeriesStructure: routine-based vs adaptive • PoiesisEnchantment: technology-enabled vs human-centered Base Sets as Communities of Practice: • *Clinical Decision-Making* • Personalized Care • *Patient Engagement* • *Technological Innovation* • Quality Assurance • Holistic Care	C_5 ("Creative-Experiencer") Topological Space Dimensions: • FreeEnergyPrinciple: entropy-reducing vs increasing • ActiveInference: data- vs experience-driven • Phenomenotechnics: technology-mediated vs direct-experience Base Sets as Communities of Practice: • *Traditional Care Delivery* • *Technological Innovation* • *Patient-Centered Care* • Personalized Medicine • *Predictive Analytics*

Italicized base sets used in illustrating sheaf construction in Chapter 7

FIGURE 1   Topological spaces and base sets in the healthcare case

Appendix—Applying gauge theory to social–behavioral phenomena

Gauge theory provides a mathematical framework for understanding the complex interplay of physical fields that permeate the universe, influencing the behavior of particles and governing the fundamental forces of nature. The stage for this intricate performance is Spacetime, a four-dimensional continuum that combines the three dimensions of space with the one dimension of time. Spacetime is modeled as a smooth manifold, a type of topological space that locally resembles Euclidean space. Each point in this manifold represents an event in spacetime, and the manifold's structure allows us to use the tools of calculus to describe changes and motions within this spacetime. The physical fields exist at every point in spacetime. Their values, represented as vector fields, determine the forces experienced by particles. These vector fields are mathematical entities that assign a vector to each point in spacetime, allowing us to represent physical quantities that vary over spacetime.

However, these fields are not just static entities; they possess internal degrees of freedom that can change without affecting the physical phenomena they govern. This is encapsulated in the concept of gauge transformations, which can alter the internal states of the fields without changing the physical observables, reflecting a fundamental symmetry of the physical laws. To capture this additional layer of structure, gauge theory introduces the concept of a fiber bundle. The base space of this bundle is the spacetime manifold, and attached to each point in this base space is a fiber representing the internal states of the field at that point. The fields themselves are seen as sections of this fiber bundle, mathematical entities that assign a value in the fiber to each point in the base space.

Finally, to describe how these fields interact and evolve, gauge theory introduces the concept of a connection on the fiber bundle. This connection allows us to compare the values of the field at different points in spacetime, and its curvature represents the forces experienced by particles. The curvature of a connection represents the failure of a parallel transport around a closed loop to return to the original state, corresponding to the field strength or the force.[1]

---

[1] In the context of gauge theory, the field strength is a measure of the intensity or "strength" of a given field at a point in spacetime. It is a derived quantity that depends on the values of the field and its derivatives at that point. A vector field, on the other hand, is a mathematical object that assigns a vector to each point in a space. In the context of gauge theory, the vector field represents the physical field itself. The vectors assigned by the vector field represent the value of the physical field at each point in spacetime. The relationship between the field strength and the vector field is that the field strength is computed from the values of the vector field and its derivatives.

Gauge Theory*	Social-Behavioral Fields	Symmetry Group(s) and Transformations	Basis States	Kinetic (T) and Potential (V) Terms of Lagrangian
Desiring Machines	Rhizome	SU(2): Conjunction, Blocking/Repulsion, Schizophrenic	Conformity, Resistance	T: Flow or change of desire over time V: Potential for desire to be produced or blocked
Deterritorialization and Reterritorialization	Territory	SU(3): Full or Partial Deterr., Full or Partial Reterr.	Original State, Deterr. State, Reterr. State	T: Rate of change or flow between states of territorialization V: Forces that push a practice towards a particular state
Desire and Territory	Rhizome and Territory	SU(2): Inversion, Latent, Dominance-Bifurcation	Deterr. Desire, Reterr. Desire	T: Active production and flow of desires V: Potential for desires to be captured, channeled, or blocked by territories
Identity and Structure	Becoming and Lines of Flight	U(1) × SU(2): Phase Change, Inversion, Latent, Dominance-Bifurcation	Phase, Conformity, Resistance	T: Active processes of becoming and lines of flight V: Tension and potential for interactions between identity and structure
Differentiation	Difference and Repetition	SU(2) × U(1): Inversion, Latent, Dominance-Bifurcation, Phase Change	Virtual, Actual, Phase	T: Processes of becoming and cyclical changes V: Tension and potential in virtual/actual dynamics
Signification and Transversality	Meaning and Boundary-Crossing	GL(2,C) × B3: Signifying, Countersignifying, Presignifying, Postsignifying, Braiding	Coded, Decoded, Boundary, Crossing	T: Active processes of signification and boundary crossing V: Tension and potential for subjectification and interdisciplinary innovation
Organization	Structure and Scale	SU(2)×SU(2): Inversion, Latent, Dominance-Bifurcation	Chaos, Order, Molecular, Molar	T: Active processes of organization and scaling V: Tension and potential of chaos and scaling
Folding and Ecosophy	Complication and Ecological Interrelation	SU(2)×SU(4): Inversion, Latent, Dominance-Bifurcation, Engagement, Event, Enactment, Interactional Creation	Interiority, Exteriority, L, E, X, M	T: Active processes of folding/unfolding V: Tension and potential of ecological balance and enhancement

* Detailed exposition of the first three theories and synopses of the remaining five can be found in the following pages of this Appendix. For details of the latter five theories, as well as updates to the first three, please visit www.drtbook.com.

FIGURE 1   Gauge theories applied to social–behavioral phenomena.

In this way, the detailed coordination of the physical fields, their symmetries, and their interactions is fully described within the mathematical framework of gauge theory. The fields are governed by a gauge-invariant Lagrangian, a function on the space of all possible field configurations that determines the equations of motion for the fields. The gauge symmetry of the theory reflects the fundamental physical principle that the laws of physics should be the same at every point in spacetime.

Drawing a parallel to Deleuze and Guattari's philosophy, one could metaphorically think of the various conceptual "fields" or "gauges" they discuss as coexisting within the same "fiber bundle" of thought or the plane of immanence (see Fig. 1). Each gauge or field represents a different aspect or dimension of this plane. The "connections" in this philosophical context could be the various relationships, interactions, and transformations between these different dimensions or aspects. In the context of Deleuze and Guattari's philosophy, if we continue with the analogy to gauge theory, the "Lagrangians" would be the underlying principles or dynamics that regulate or drive the behavior of the "gauges" (concepts). These "Lagrangians" would describe the energy or action of the system in terms of the gauges and their interactions.

Gauge theory of desiring machines

We can construct a plausible framework for **desiring-machines** where the gauge is a rhizome and the Lagrangian is the flow of desire and production. In the social and behavioral context, this refers to the way desires are produced and regulated in society, often in a nonlinear and interconnected manner. The "energy" here is the intensity of desire and its potential to produce or be blocked. Imagine a network (rhizome) of interconnected nodes (desiring-machines). The symmetry group would represent the different possible configurations or states of a desiring-machine. Given the nonhierarchical, noncentralized nature of the rhizome, this group might be nonabelian (where the order of operations matters). The group elements could represent different "arrangements" or "assemblages" of desiring-machines. Since the rhizome allows for multiple, nonlinear, and nonhierarchical connections, the group would have a complex structure with many possible elements and operations. The Lie algebra would be associated with the infinitesimal transformations between different states of the desiring-machine. These could be thought of as the "differential" changes in the flow of desire. The generators of the Lie algebra would represent the fundamental operations or shifts that can be applied to a desiring-machine.

Gauge transformations for desiring-machines: In the context of the "desiring-machines" and the "rhizome," a gauge transformation might represent a shift in the way desire is channeled or expressed. Given the rhizomatic nature of desire, these transformations could be local or global changes in the configuration of a desiring-machine, representing different ways in which desire can flow or be blocked. In a mathematical representation, these transformations could be associated with specific matrices or operators within the $SU(N)$ group, where N represents the number of fundamental arrangements of desiring-machines.[2] The noncommutative nature of this group captures the idea that the order of transformations matters and can lead to different outcomes.

For simplicity, consider $SU(2)$ as a starting point, which is the simplest non-trivial special unitary group. This choice is made for illustrative purposes, and the actual dimensionality N of the group in the context of Deleuze and Guattari's work would likely be much larger, given the complexity of their concepts. Some highly plausible candidate transformations are:

- **Conjunction of flows**: This represents the merging or joining of different flows of desire. In group terms, this could be a transformation that combines or merges certain elements or nodes within the desiring-machine network.

---

[2] In the context of quantum mechanics and gauge theories, $SU(N)$ is a specific mathematical group known as the "special unitary group of degree N", where:

- **S**: Stands for "Special." This means that the matrices in this group have a determinant of 1.
- **U**: Stands for "Unitary." A matrix is unitary if its conjugate transpose is also its inverse. In simpler terms, when you take a matrix from this group, find its conjugate transpose, and multiply it with the original matrix, you'll get the identity matrix.
- **N**: This indicates that the matrices in this group are $N \times N$ in size.

* Matrix Representation: $I = \begin{pmatrix} 1 & 0 \\ 0 & 1 \end{pmatrix}$.
* Interpretation: A unity operation, representing the merging or joining of different flows of desire.
- **Blocking or repulsion of flows**: This transformation signifies a process where certain flows of desire are blocked or repelled. Within the group, this might be represented by a matrix operation that "blocks" or "isolates" certain nodes or elements.
  * Matrix Representation: $I = \begin{pmatrix} -1 & 0 \\ 0 & -1 \end{pmatrix}$.
  * Interpretation: An inversion, signifying a blockage or repulsion of flows of desire.
- **Schizophrenic flow**: This represents a nonlinear, erratic flow of desire that doesn't conform to any established pattern. In the group framework, this could be a complex transformation that randomizes or jumbles the connections within the desiring-machine network.
  * Matrix Representation: $I = \begin{pmatrix} 1 & 0 \\ 0 & -1 \end{pmatrix}$.
  * Interpretation: A differential operation, representing the nonlinear, erratic flow of desire.

States in this context represent specific configurations or arrangements of desiring-machines within the rhizome. When we apply one of the transformation matrices (see above) to this state (as column vectors in a 2-dimensional Hilbert space—a complex vector space), we get a new state vector, representing a transformed configuration of the desiring-machine.

- **Conformity state**($|C\rangle$) $= \begin{pmatrix} 1 \\ 0 \end{pmatrix}$ represents a desiring-machine configuration where desires are flowing in alignment with established structures.
- **Resistance state**($|R\rangle$) $= \begin{pmatrix} 0 \\ 1 \end{pmatrix}$ represents a configuration where desires are flowing against or in resistance to established structures.
- **Hybrid state** $(\alpha|C\rangle + \beta|R\rangle) = \begin{pmatrix} \alpha \\ \beta \end{pmatrix}$ as a superposition state representing a desiring-machine configuration that is a mix of conformity and resistance. α and β are complex coefficients that determine the "weight" or "contribution" of each basis state to the overall state. The squared magnitudes $|\alpha|^2$ and $|\beta|^2$ give the probabilities of finding the system in the conformity or resistance state, respectively.

Lagrangian for desiring-machines describes the dynamics of the system, capturing the "energy" or intensity of desire and its potential to produce or be blocked.

- **Plausible Lagrangian**: $L = \frac{1}{2}\left(\frac{d\psi}{dt}\right)^2 - V(\psi)$ where:
    * ψ represents the state of the desiring-machine.
    * The kinetic term $\frac{1}{2}\left(\frac{d\psi}{dt}\right)^2$ captures the flow or change of desire over time.
    * $V(\psi)$ is the potential energy term, representing the potential for desire to be produced or blocked. It could be influenced by external factors like societal norms or internal factors like individual motivations.
    * When the kinetic term is dominant, it indicates that desire is flowing freely and dynamically. When the potential term $V(\psi)$ is dominant, it indicates that there are significant barriers or blockages to the flow of desire.

Gauge theory of deterritorialization and reterritorialization

Let us now delve into the concepts of "**Deterritorialization**" and "**Reterritorialization**," where the gauge is a Territory, and the Lagrangian is the dynamics of escape and capture. In both social and biological contexts, this refers to the way entities break free from established structures (escape) and the way new structures form to capture or channel these flows (capture). The "energy" is the tension between escape and capture. Imagine a cultural practice or idea that can exist in various states of territorialization. The symmetry group would encompass all possible states of deterritorialization and reterritorialization of this cultural practice. Given the dynamic and multifaceted nature of deterritorialization and reterritorialization, a nonabelian group structure would be most appropriate. The group elements could represent different "degrees" or "states" of territorialization. For instance, a cultural practice might be fully integrated within its original culture, partially deterritorialized in a global context, or fully reterritorialized within a new culture. The Lie algebra would be associated with the infinitesimal transformations between different states of territorialization. These could be thought of as the "differential" shifts between deterritorialization and reterritorialization. The generators of the Lie algebra would represent the fundamental processes that drive these shifts. In the context of Deleuze and Guattari's work, these might include processes like "lines of flight" (which drive deterritorialization) and "machinic assemblages" (which can lead to reterritorialization).

Gauge transformations for deterritorialization and reterritorialization: A gauge transformation in this context might represent a shift between states of deterritorialization and reterritorialization of this practice, i.e., moving it from being deeply rooted in one culture to being a global phenomenon, or to being adopted by another culture. Deterritorialization involves breaking away from established structures, norms, or territories, while reterritorialization involves the re-establishment or re-creation of such structures or territories. For instance, in a societal context, a cultural practice might be deterritorialized when it is taken out of its original cultural context and becomes a global phenomenon. Its reterritorialization might occur when it's adopted and adapted by another culture, gaining new meanings and structures in the process. Once again, the $SU(N)$ group, where N represents the number of fundamental states or degrees of territorialization, could be a suitable choice. Consider $SU(3)$ as a starting point. Here are some highly plausible candidate transformations:

- **Full deterritorialization**: Represents a complete break from the original cultural or territorial context. This transformation would shift a cultural practice from being deeply rooted in its original context to a state of complete "unrootedness" or global ubiquity.
    - Matrix Representation: $T_{FD} = \begin{bmatrix} 0 & 1 & 0 \\ 1 & 0 & 0 \\ 0 & 0 & 1 \end{bmatrix}$.
    - Interpretation: This transformation shifts a cultural practice from the Original state to the deterritorialized state.
- **Partial deterritorialization**: Represents a partial break from the original context. The cultural practice might still retain some of its original meanings but is also influenced by global or external factors.
    - Matrix Representation: $T_{PD} = \begin{bmatrix} 0.5 & 0.5 & 0 \\ 0.5 & 0.5 & 0 \\ 0 & 0 & 1 \end{bmatrix}$.
    - Interpretation: This transformation gives a mix of the original and deterritorialized states.
- **Full reterritorialization**: Represents a complete integration into a new cultural or territorial context. The cultural practice is fully adopted, adapted, and given new meanings within a new culture.
    - Matrix Representation: $T_{FR} = \begin{bmatrix} 1 & 0 & 0 \\ 0 & 0 & 1 \\ 0 & 1 & 0 \end{bmatrix}$.
    - Interpretation: This transformation shifts a cultural practice from the deterritorialized state to the reterritorialized state.
- **Partial reterritorialization**: The cultural practice is adopted by a new culture but still retains some of its original meanings or structures.
    - Matrix Representation: $T_{PR} = \begin{bmatrix} 1 & 0 & 0 \\ 0 & 0.5 & 0.5 \\ 0 & 0.5 & 0.5 \end{bmatrix}$.
    - Interpretation: This transformation gives a mix of the deterritorialized and reterritorialized states.

The **state vector** would represent the current "degree" or "state" of territorialization of a cultural practice. For simplicity, let's consider three primary states:

- **Original state** $(|O\rangle) = \begin{bmatrix} 1 \\ 0 \\ 0 \end{bmatrix}$, represents the cultural practice in its original, untouched state, deeply rooted in its native culture.

- **Deterritorialized state** $(|D\rangle) = \begin{bmatrix} 0 \\ 1 \\ 0 \end{bmatrix}$, represents the cultural practice in a state of global ubiquity, free from its original cultural context.

- **Reterritorialized state** $(|R\rangle) = \begin{bmatrix} 0 \\ 0 \\ 1 \end{bmatrix}$, represents the cultural practice fully integrated into a new cultural context.

- **Hybrid state** $\psi = \alpha|O\rangle + \beta|D\rangle + \gamma|R\rangle = \begin{bmatrix} \alpha \\ \beta \\ \gamma \end{bmatrix}$ represents a superposition of these states, where α, β, γ are complex coefficients that give the probability amplitudes of the cultural practice being in each state.

Lagrangian for deterritorialization and reterritorialization should capture the dynamics between the states of escape (deterritorialization) and capture (reterritorialization). The "energy" in this context would represent the tension between these two processes.

- **A plausible Lagrangian is** $L = \frac{1}{2}\left(\frac{d\psi}{dt}\right)^2 - V(\psi)$ where:
  * ψ represents the state of the cultural practice or idea in terms of its territorialization. It is a vector that captures the degree to which a practice or idea is territorialized, deterritorialized, or reterritorialized.
  * The kinetic term $\frac{1}{2}\left(\frac{d\psi}{dt}\right)^2$ captures the rate of change or flow between states of territorialization. A high value indicates rapid shifts between states, while a low value indicates stability or stagnation.
  * $V(\psi)$ is the potential energy term, representing the "forces" or influences that push a cultural practice or idea towards a particular state. This could be influenced by external societal pressures, historical contexts, or internal dynamics within a culture. The potential $V(\psi)$ could be further specified based on the specific dynamics of deterritorialization and reterritorialization. For instance:
    - A deep potential well around the original state might indicate strong forces that keep a cultural practice rooted in its original context.
    - A shallow potential around the deterritorialized state might indicate that while there are forces pushing towards global ubiquity, they aren't overwhelmingly strong.

- A potential barrier between the deterritorialized and reterritorialized states might represent the challenges or resistances faced when a cultural practice tries to integrate into a new cultural context.
* When the kinetic term dominates, it suggests that the cultural practice or idea is dynamically shifting between states. When the potential term $V(\psi)$ dominates, it indicates that there are significant forces or barriers influencing the state of the practice or idea.

Unified Gauge theory of desire and territory

In physics, gauge theories are combined based on their mathematical structures and the physical phenomena they describe. Here, we will attempt a similar approach, but grounded in the philosophical and conceptual terrain of Deleuze and Guattari. A potential approach for a unified gauge theory, let's say, of desire and territory could be to consider a direct product of the two spaces, leading to a 6-dimensional representation. This would be represented by the group $SU(2) \times SU(3)$. Transformations in the unified theory would be represented by 6×6 matrices, which are direct products of the 2×2 and 3×3 matrices for desire and territory, respectively. The state vector for the unified theory would be a tensor product of the state vectors for desire and territory. The Lagrangian for the unified theory would be a combination of the Lagrangians for desire and territory. The kinetic term would capture the combined dynamics of desire and territorial flows, while the potential term would represent the combined influences and forces acting on both desire and territory. This model specification would offer a clear separation of concepts, flexibility in modeling states and transformations, and potential for future extensions. However, challenges include interpretability, over-complexity, understanding interactions between dimensions, and computational load. As our goal is developing intuitions and demonstrating practical applications, we will develop a theory based on the simpler $SU(2)$ group, which is a 2×2 matrix representation.

In the context of our "Unified Gauge Theory of Desire and Territory" inspired by Deleuze and Guattari, the use of $SU(2)$ is meant to represent the intertwined dynamics of desire (from the Rhizome gauge) and territorial flows (from the Territory gauge). In this representation:

- The transformations within the group could represent the various ways in which desire and territory interact, influence each other, and transform within the philosophical framework of Deleuze and Guattari.
- The two basis elements of the $SU(2)$ group could represent the fundamental states or aspects of desire and territory.[3]

---

[3] It is crucial to understand that this use of SU(2) is a conceptual tool, a way to bridge the language of physics with the philosophical ideas of Deleuze and Guattari, and not a literal application of the mathematical group to their thought. For instance, a specific matrix might symbolize a transformation from a state of "deterritorialized desire" to a state of "reterritorialized desire." The exact nature and interpretation of these metaphorical matrices would depend on the philosophical framework and the specific aspects of desire and territory being considered.

The $SU(2)$ group can be represented by 2×2 matrices of the form:

$$\begin{pmatrix} a & b \\ -c^* & a^* \end{pmatrix}$$

where a and b are complex numbers, and a^* and c^* are their complex conjugates, such that $|a|^2 + |b|^2 = 1$.

The Pauli matrices, which serve as a basis for $SU(2)$, are given by:

$$\sigma_1 = \begin{pmatrix} 0 & 1 \\ 1 & 0 \end{pmatrix}, \sigma_2 = \begin{pmatrix} 0 & -i \\ i & 0 \end{pmatrix}, \sigma_3 = \begin{pmatrix} 1 & 0 \\ 0 & -1 \end{pmatrix}$$

Each of these matrices can be thought of as representing a different kind of transformation or operation on the states of desire and territory:

- σ_1: Represents a transformation that swaps the states of deterritorialized and reterritorialized desire.
- σ_2: Represents a more complex transformation where desire becomes intertwined with territory, possibly indicating a state where desire is in the process of being captured by or escaping from a territory.
- σ_3: Represents a transformation that differentiates between the two states, possibly indicating a scenario where one state becomes dominant over the other.

In the context of Deleuze and Guattari's philosophy, these transformations can be thought of as the various ways in which desire and territory interact, influence each other, and transform. The Pauli matrices, as representations of these transformations, allow us to explore the dynamic and multifaceted nature of these interactions in a mathematical framework.

In the context of $SU(2)$, the states can be represented by 2×1 column vectors. These vectors can be thought of as "basis elements" or "basis states" for the space in which our system (in this case, the interplay of desire and territory) exists.

- **Deterritorialized desire ($|C\rangle$)** $= \begin{pmatrix} 1 \\ 0 \end{pmatrix}$: Here, desire is in a state of free flow and is not bound or captured by any specific territory.

- **Reterritorialized desire ($|R\rangle$)** $= \begin{pmatrix} 0 \\ 1 \end{pmatrix}$: In this state, desire is captured or bound by a specific territory, and its free flow is halted or minimized.

These two basis states can be combined with different coefficients to represent more complex states of desire and territory. For instance, a state like $\begin{pmatrix} a \\ b \end{pmatrix}$, would represent a superposition of deterritorialized and reterritorialized desires, with "a" and "b" indicating the respective "amounts" or "intensities" of each state.

A general state in $SU(2)$ is a 2×1 column vector that can be expressed as a linear combination of the basis states. This general state represents a superposition of the basis states, indicating a more complex, mixed state of the system. For example: $|\Psi\rangle = a|D\rangle + b|R\rangle = \begin{pmatrix} a \\ b \end{pmatrix}$, where a and b are complex coefficients.

Let us use these matrices to represent transformations in the "Unified Gauge Theory of Desire and Territory." In this metaphorical framework, the Pauli matrices serve as operations that transform states of desire and territory in various ways, capturing the intricate dynamics proposed by Deleuze and Guattari.

- **Transformation of deterritorialized desire to reterritorialized desire**: Using σ_1 as the transformation matrix, a state of "deterritorialized desire" can be represented as a column vector $\begin{pmatrix} 1 \\ 0 \end{pmatrix}$, where desire is in a state of free flow (represented by the 1) and is not bound or captured by any specific territory (represented by the 0). Applying the transformation, we get:
  * $\sigma_1 \times \begin{pmatrix} 1 \\ 0 \end{pmatrix} = \begin{pmatrix} 0 \\ 1 \end{pmatrix}$.
  * This new state $\begin{pmatrix} 0 \\ 1 \end{pmatrix}$ represents "reterritorialized desire," where desire is captured or bound by a specific territory (represented by the 1), and the free flow of desire is halted or minimized (represented by the 0).
  * **Illustration**: Imagine a plane where the x-axis represents the intensity of "deterritorialized desire" and the y-axis represents the intensity of "reterritorialized desire." Initially, the state is at the point (1,0), indicating pure "deterritorialized desire." After the transformation using σ_1, the state moves to the point (0,1), indicating a shift to pure "reterritorialized desire." This binary representation is a simplification for the sake of the metaphor and to align with the mathematical framework. In Deleuze and Guattari's work, the concepts of deterritorialization and reterritorialization are more nuanced and fluid, not strictly binary. But for the purpose of this mathematical analogy, we are using this binary representation to capture the essence of these concepts.
- **Interplay between desire and territory**: Using σ_2 and σ_3, we can represent more complex transformations that capture the interplay between desire and territory. For instance, σ_2 might represent a transformation where desire becomes intertwined with territory, while σ_3 could represent a balancing act where desire and territory counteract each other.
  * **Transformation using σ_2**:
    - $\sigma_2 \times \begin{pmatrix} 1 \\ 0 \end{pmatrix} = \begin{pmatrix} 0 \\ i \end{pmatrix}$.
    - The resulting state $\begin{pmatrix} 0 \\ i \end{pmatrix}$ is a complex vector, suggesting a more intricate relationship between desire and territory. The imaginary component can be metaphorically interpreted as the intertwining of desire with territory, where the desire is

neither purely deterritorialized nor reterritorialized but exists in a complex, intertwined state. Further exotic variations on complex vectors are also possible:

* $\begin{pmatrix} -i \\ 0 \end{pmatrix}$: This can be thought of as a phase-shifted version of the "deterritorialized desire." This phase-shifted state might represent a nuanced or altered form of deterritorialized desire. The negative imaginary component could symbolize a state of desire that's not just free-flowing but has an additional "twist" or "complexity" to it, perhaps indicating a form of desire that's introspective, reflexive, or undergoing some internal transformation.

* $\begin{pmatrix} i \\ 0 \end{pmatrix}$: This is another phase-shifted version of the "deterritorialized desire." The positive imaginary component, in contrast to the negative imaginary component we discussed earlier, could represent a form of desire that's outwardly expansive, extroverted, or projecting itself into new territories. It's still a form of deterritorialized desire but with a different "direction" or "orientation" in its complex phase space. Interpretation: A state of desire that's actively seeking new territories or realms of expression, perhaps indicative of exploratory or pioneering tendencies in the subject.

* $\begin{pmatrix} 0 \\ -i \end{pmatrix}$: This is a phase-shifted version of the "reterritorialized desire." The negative imaginary component in the context of reterritorialization could symbolize a state of territory that's introspectively consolidating or solidifying its boundaries, perhaps undergoing an internal reevaluation or transformation. Interpretation: A state of territory that's not just established but is introspectively reinforcing or redefining its boundaries, perhaps indicative of a society or individual that's reflecting on its norms, values, and structures.

* **Transformation using σ_3:**

    ■ $\sigma_3 \times \begin{pmatrix} 1 \\ 0 \end{pmatrix} = \begin{pmatrix} 1 \\ 0 \end{pmatrix}$.

    ■ The resulting state remains unchanged, $\begin{pmatrix} 1 \\ 0 \end{pmatrix}$. This can be interpreted as a balancing act where the forces of desire and territory counteract each other, leading to no net change in the state of desire. It's as if the territory's influence on desire is perfectly balanced, keeping the desire in its deterritorialized state.

Our metaphorical **Lagrangian** could be written as: $L = T(\psi,\dot{\psi}) - V(\psi)$ where:

- ψ is a state vector representing the current configuration of desire and territory in the system.
- $\dot{\psi}$ represents the rate of change of this state, capturing the dynamics of desire production.

- T and V are functions that capture the kinetic and potential "energies" respectively.
  - **Kinetic energy T (desire production energy)**: This represents the active production and flow of desires. It's the "motion" of desires as they manifest, propagate, and interact within the system. Mathematically, we might represent this with terms involving rates of change of desire states, perhaps using derivatives. It might involve terms like $\langle \dot\psi | \dot\psi \rangle$, where $\langle \cdot | \cdot \rangle$ denotes the inner product, capturing the "motion" or "activity level" of desires. When this term is large, it indicates a state of high desire production and flow, a situation where desires are actively being produced and are in a state of flux. This could be likened to Deleuze and Guattari's concept of "deterritorialization," where desire escapes established territories and flows freely.
  - **Potential energy V (territorial energy)**: This represents the potential for desires to be captured, channeled, or blocked by territories. Territories can be seen as "potential wells" or "barriers" that can either attract or repel desires. Mathematically, this could involve terms that describe the interaction potential between different desire states and territories. It might involve terms that describe the interaction potential between different desire states and territories, perhaps something like $V(\psi) = V_0 \langle \psi | U | \psi \rangle$, where U is a matrix representing the "territorial landscape" and V_0 is a constant. When this term is large, it indicates a state where desires are strongly bound or captured by territories. This could be likened to Deleuze and Guattari's concept of "reterritorialization," where desire is captured and organized within stable territories.

The equations of motion for the fields (in this case, the states of desire and territory) are derived from the Lagrangian through the Euler-Lagrange equation:

$$\frac{\partial L}{\partial \psi} - \frac{d}{dt}\left(\frac{\partial L}{\partial \dot\psi}\right) = 0$$

These equations describe how the states of desire and territory evolve over time under the influence of the kinetic and potential "energies." These equations govern the dynamics of the system, dictating how desire and territory interact and transform. For example, in a situation where the kinetic term dominates, we might expect rapid changes in the state of desire, corresponding to intense flows and deterritorializations. When the potential term dominates, we might expect the system to evolve towards stable configurations where desire is captured within territories, corresponding to reterritorializations.

Unified Gauge theory of identity and structure

The Unified Gauge Theory of Identity and Structure leverages the mathematical groups $U(1) \times SU(2)$ to model the transformation of identity and the process of breaking free from structures. This approach is significant within Dynamic Relationality Theory (DRT) for its capacity to represent the continuous transformation and complex interplay between identity (becoming) and structural dynamics (lines of flight), offering a rich framework for understanding the fluidity of existence and systems.

Core concepts and processes: The theory is grounded in the concepts of continuous transformation and differentiation, utilizing the symmetry groups $U(1)$ for phase transformations (identity shifts) and $SU(2)$ for capturing the nuances of structural dynamics and breaking free from established norms. It aims to capture the dynamics of transformation, change, and liberation from established structures, with energy conceptualized as the combined potential for change, transformation, and innovation within systems.

Key groups and mathematical constructs: The $U(1)$ group represents continuous, cyclic transformations akin to phase shifts in identity, while $SU(2)$ captures the nonabelian, complex transformations associated with structural dynamics and lines of flight. Within DRT, these mathematical constructs metaphorically represent the fluidity of identity and the complexity of navigating and breaking free from rigid structures, aligning with Deleuze and Guattari's philosophical explorations.

Basis states and transformations: States in the $U(1) \times SU(2)$ combined group represent intricate interplays between identity ($U(1)$ transformations) and structure ($SU(2)$ dynamics), emphasizing the interconnected transformations within these domains. The combined symmetry group $U(1) \times SU(2)$ encapsulates both the identity transformations and structural dynamics, indicating how changes in one aspect can profoundly influence and be influenced by the other.

Lagrangian framework: The Lagrangian is structured to reflect the dynamics of transformation and liberation from established structures, with energy symbolizing the potential for change and innovation. The kinetic term (T) represents active processes of becoming and lines of flight, while the potential term (V) captures the tensions and interactions between identity and structure.

Specific interactions and transformations governed by the Lagrangian:

- **Continuous transformation of identity:** Transformations under $U(1)$ symbolize the fluid, continuous change in identity, mirroring the process of becoming.
- **Interactions with established structures:** Interactions between identity transformations (ψ) and structural dynamics (ϕ) lead to changes in the Lagrangian's potential energy term, signaling new configurations of tensions between identity and structure.
- **Inducing lines of flight:** Continuous transformations in identity may induce corresponding "lines of flight" in structures, reflected in the kinetic term of the Lagrangian, showcasing intertwined dynamics.
- **Stabilization or destabilization of identity:** Depending on context, transformations may lead to stabilization or destabilization, with the Lagrangian reflecting these dynamics through its structure.
- **Direct challenges and inversions (σ_1), subtle shifts (σ_2), dominance or bifurcation (σ_3):** These transformations under $SU(2)$ capture the varied dynamics of engaging with and breaking free from structures, highlighting direct challenges, subtle shifts, and dominance or bifurcation within the system.

Unified Gauge theory of differentiation

The Unified Gauge Theory of Differentiation employs the mathematical framework of $SU(2) \times U(1)$ to explore the nuanced interplay between differentiation and repetition within the fabric of reality. This theoretical construct is pivotal within Dynamic Relationality Theory

(DRT), providing a robust mechanism to conceptualize the processes of becoming and the cyclical nature of phenomena. It allows for a sophisticated understanding of how entities navigate the continuum between the virtual and the actual, and how repetition underpins and transforms these journeys, embodying the core of Deleuze and Guattari's philosophical inquiries into the nature of change and constancy.

Core concepts and processes: The theory is founded on the exploration of differentiation (through $SU(2)$) as the process of becoming, where the virtual differentiates into the actual, and repetition (through $U(1)$) as the cyclical, nuanced recurrence of phenomena. This dual focus captures the essence of transformative processes across various domains of existence. It captures the dialectic between the emergence of new realities (differentiation) and the persistence of patterns (repetition), aiming to elucidate the dynamic, ever-evolving landscape of complex systems.

Key groups and mathematical constructs: $SU(2)$ represents the dynamics of difference and the virtual-actual continuum, while $U(1)$ encapsulates the phases of repetition, emphasizing the recurring nature of processes with subtle variations. In DRT, these groups metaphorically articulate the fluidity and cyclicity of existence, aligning with philosophical contemplations on identity, structure, and change.

Basis states and transformations: The states represent the spectrum of virtuality and actuality ($SU(2)$) and the varied phases of repetition ($U(1)$), highlighting the complex interplay between becoming and persisting in systems. The interplay between differentiation and repetition is encapsulated in the combined group $SU(2) \times U(1)$, illustrating how transformations in one domain affect and are influenced by the other, showcasing a dynamic process of evolution and constancy.

Lagrangian framework: The Lagrangian, $L = T(\psi, \dot\psi, \phi, \dot\phi) - V(\psi, \phi)$, reflects the active processes of differentiation and repetition. The kinetic term (T) captures the "energy" of becoming and cyclical changes, while the potential term (V) signifies the tensions and potentials within the virtual-actual dynamics.

Specific interactions and transformations governed by the Lagrangian:

- **Inversion and new cycles of repetition (σ_1):** The inversion transformation underscores moments of transition, where new cycles of repetition are initiated, capturing the dynamic energy of transitioning from virtual to actual.
- **Hidden processes of differentiation and repetition (σ_2):** This transformation highlights the latent dynamics beneath surface appearances, emphasizing the intertwined nature of differentiation and nuanced phases of repetition.
- **Distinction and tension modulating repetition (σ_3):** The emphasis on distinction between virtual and actual states influences the frequency and nature of repetition, reflecting the system's evolving dynamics.
- **Phase shifts influencing differentiation:** Phase shifts in repetition can induce or modulate differentiation processes, revealing the interconnected dynamic between becoming and persisting, and illustrating the complex feedback loops that define and redefine entities and systems.

Unified Gauge theory of signification and transversality

The Unified Gauge Theory of Signification and Transversality articulates the intricate dynamics between the production and interpretation of meaning (signification) and the crossing of conceptual, disciplinary, or territorial boundaries (transversality). Within the framework of Dynamic Relationality Theory (DRT), this theory underscores the interconnected processes by which meaning is coded, decoded, and centralized (signification) and how these processes inherently involve transgressing established norms and boundaries (transversality). The significance of this theory lies in its ability to provide a mathematical and conceptual framework for understanding the fluid and dynamic nature of meaning-making and boundary-crossing, highlighting their roles in fostering resistance, liberation, and innovation.

Core concepts and processes: Signification encompasses the processes of coding, decoding, and overcoding, focusing on how meanings are constructed, deconstructed, and dominated. Transversality pertains to the crossing of boundaries, promoting the flow of ideas and intensities across established territories or domains. The theory delineates the interaction between the creation and interpretation of meaning and the dynamic, boundary-crossing actions that disrupt traditional categorizations and frameworks, offering a rich landscape for the exploration of resistance and liberation.

Key groups and mathematical constructs: The $GL(2,C)$ group, representing complex 2×2 invertible matrices, models the multifaceted processes of signification. The Braid Group $B3$, illustrating the intertwining and crossing of strands, symbolizes the essence of transversality. In DRT, $GL(2,C)$ metaphorically captures the complexity of signification processes, while $B3$ embodies the transformative potential of transversality in navigating and redefining boundaries.

Basis states and transformations: Basis States for $GL(2,C)$ might include specific configurations that represent coded, decoded, and overcoded states of signification. Braid states symbolize various degrees of boundary crossing, from established norms to innovative interdisciplinary engagements.

Lagrangian framework: The Lagrangian integrates the dynamics of signification and transversality, conceptualizing energy as the force generated by these processes and the potential for interdisciplinary innovation. It provides a mathematical representation of the tension, dynamism, and potential inherent in the interactions between signification and transversality.

Specific interactions and transformations governed by the Lagrangian:

- **Decoding and boundary crossings**: Highlights how acts of decoding can provoke boundary crossings, leading to interdisciplinary innovations and the challenge of established norms.
- **Overcoding and transversality limitation**: Examines how processes of overcoding can limit transversality, serving as a form of control over the generation and flow of meaning.
- **Transformation through transversality**: Explores how transversal actions can induce decoding or overcoding, transforming the landscape of signification.

- **Equilibrium and phase transitions**: Analyzes stable states and the potential for significant shifts that redefine the balance between signification and transversality, illustrating the theory's applicability to understanding dynamic ecosystems of meaning and interaction.

Unified Gauge theory of organization

The Unified Gauge Theory of Organization adapts the $SU(2) \times SU(2)$ symmetry groups to explore the dynamics of organization within the context of Dynamic Relationality Theory (DRT). It provides a sophisticated framework for understanding the fluid and dynamic transitions between chaos and order, embodied by the Body without Organs (BwO) and the Organized Body (OB) and the interplay between molecular and molar lines of segmentation. This theory is pivotal for decoding the complex processes of organization and scaling in various domains, reflecting the perpetual negotiation between stability and transformation, and individual desires versus collective structures.

Core concepts and processes: The theory integrates the dynamics of the BwO transitioning into the organized body, and the interplay between molecular and molar lines, using the mathematical richness of the $SU(2) \times SU(2)$ groups. This approach captures the essence of transitions from chaos to order and the nuances of organizational scaling. It aims to elucidate the continuous process of organization from the virtual plane of potentials (BwO) to actualized structures (OB), and the oscillation between micro (molecular) and macro (molar) organizational forms, reflecting the inherent tension and dynamics between chaos and structure.

Key groups and mathematical constructs: The first $SU(2)$ group symbolizes the transition between the BwO and OB, capturing states of pure potential and their actualization. The second $SU(2)$ group represents the scaling dynamics between molecular and molar lines, illustrating the shift between micro and macro organizational scales. In DRT, these groups metaphorically embody the fluid nature of organizational processes, the creative tension between chaos and order, and the dynamic scaling between individual desires and collective structures.

Basis states and transformations: The basis states ($|BwO\rangle$, $|OB\rangle$, $|Mol\rangle$, $|Mar\rangle$) and Pauli matrices (σ_1, σ_2, σ_3) for each $SU(2)$ group offer a rich vocabulary for articulating the transformations and tensions between chaos and order, and between molecular and molar organizational dynamics. The transitions represented by the Pauli matrices elucidate the mechanisms through which organizations navigate between chaos and structure (BwO and OB), and scale (molecular and molar lines), highlighting the constant evolution of organizational forms. Combined symmetry group $SU(2) \times SU(2)$ encapsulates the complex interplay between organization and scale, chaos and structure. It represents the dynamic, ongoing processes where organization is conceptualized as an evolving phenomenon, perpetually influenced by the tensions between the BwO and OB, and the scaling dynamics of molecular and molar lines.

Lagrangian framework: The Lagrangian L = T($\psi,\dot\psi,\phi,\dot\phi$)−V($\psi,\phi$) models the "energy" dynamics of organization and scaling, capturing the active processes and tensions inherent in the transition between chaos and order, and in organizational scaling. Kinetic terms represent the dynamism of organization, illustrating how entities navigate between chaos (BwO) and structure (OB), and engage in scaling processes between molecular and molar lines. Potential terms capture the tensions and potentials associated with the BwO's process of organization and the scaling dynamics, emphasizing the continuous negotiation and transformation within organizational structures.

Specific interactions and transformations governed by the Lagrangian:

- **BwO and molecular transformations**: The BwO's intensity can initiate shifts towards more fluid, molecular processes, challenging and transforming molar structures, highlighting the deterritorializing force of the BwO.
- **Organized body and molar structures**: The emergence of the OB can reinforce and stabilize molar structures, illustrating the negotiation between chaos and order.
- **Molecular disruptions and BwO emergence**: Intense molecular activity can destabilize the OB, reactivating the BwO's potentials, demonstrating the fluidity and transformative potential within organizations.
- **Molar suppression and organizational stability**: Strong molar structures may suppress BwO potentials, forcing a more organized state, reflecting the tension between stability and transformation.
- **Continuous negotiation for new formations**: The OB, in dialogue with molecular forces, engages in a continuous process of transformation, underlining the inherent dynamism and creativity in organizational processes.

Unified Gauge theory of folding and ecosophy

The Unified Gauge Theory of Folding and Ecosophy brings to light the intricate dynamics between the Deleuzian concepts of Folding/Unfolding and Guattari's extended ecosophy, articulated through the mathematical lens of $SU(2) \times SU(4)$. This theory underscores the profound relationship between the processes of internal complexity creation (Folding) and the expression of this complexity in external realities (Unfolding), as well as their interplay with the multifaceted dimensions of existence proposed by Guattari.

Core Concepts and Processes: This theory encapsulates the dual processes of Folding (the inward integration of external realities to create internal complexity) and unfolding (the externalization of internal complexities), interwoven with Guattari's four existential territories, energetic-signaletic flows, incorporeal universes, and abstract machinic phyla. It aims to capture the continuous interplay between internal and external states (folding/unfolding) and the transformative dynamics among existential territories, energetic flows, incorporeal universes, and abstract structures, highlighting the ecosystem's inherent dynamism and capacity for innovation.

Key groups and mathematical constructs: This theory utilizes $SU(2)$ for folding/unfolding dynamics and $SU(4)$ for the nuanced interrelations among Guattari's four domains, facilitating a complex model of existential and ecological transformations. These groups metaphorically represent the fluid interplay between internal and external states and the intricate relationships within Guattari's ecosophical model, reflecting the deep interconnectedness and transformative potential of beings and systems within DRT.

Basis states and transformations: Basis States for $SU(2)$ and $SU(4)$ represent the dynamic positions and potential transformations within the processes of folding/unfolding and across Guattari's four domains, encapsulating the continuous evolution of identity, structure, and ecological interrelations. The interplay between $SU(2)$ and $SU(4)$ captures the complex transformations between internal and external states and among the existential, energetic, incorporeal, and machinic dimensions, highlighting the system's fluid and transformative nature.

Lagrangian framework: Lagrangian approach constructs a function to model the energy dynamics of Folding/Unfolding and ecological interrelations, with kinetic terms representing the active processes of internal and external state dynamics and potential terms capturing the tensions and synergies among Guattari's domains.

Specific interactions and transformations governed by the Lagrangian:

- **Folding inducing existential reconfiguration**: Demonstrates how internalization processes (Folding) can lead to the reconfiguration of existential territories, creating new subjective spaces and experiences.
- **Existential crisis leading to unfolding**: Shows how crises within existential territories can trigger unfolding processes, externalizing internal complexities as adaptive strategies.
- **Unfolding stimulating energetic flows**: Illustrates how the act of Unfolding can initiate new energetic-signaletic flows, fostering novel exchanges and transformations.
- **Energetic flows inducing folding**: Explores how intense energetic flows can induce Folding, integrating external complexities into new internal landscapes.
- **Folding integrating incorporeal universes**: Reveals how Folding processes can incorporate new incorporeal universes into subjective experiences, enriching the internal landscape with new meanings.
- **Incorporeal universes influencing folding/unfolding**: Discusses how incorporeal universes can shape the dynamics of folding/unfolding, guiding the navigation of internal and external complexities.
- **Abstract machines guiding unfolding**: Explains how abstract machines can direct the unfolding process, influencing the expression of internal complexities in external forms.
- **Unfolding revealing new abstract machines**: Highlights how unfolding processes can unveil new abstract machines, guiding the evolution of existential territories and revealing underlying structures and potentials.

References

Aarras, N., Rönkä, M., Kamppinen, M., Tolvanen, H., Vihervaara, P., 2014. Environmental technology and regional sustainability—the role of life-based design. Technol. Soc. 36, 52—59.

Abramowitz, S., Stevens, L.A., Kyomba, G., Mayaka, S., Grépin, K.A., 2023. Data flows during public health emergencies in LMICs: a people-centered mapping of data flows during the 2018 Ebola epidemic in Equateur, DRC. Soc. Sci. Med. 318.

Adams, S.S., Arel, I., Bach, J., Coop, R., Furlan, R., Goertzel, B., Hall, J.S., Samsonovich, A., Scheutz, M., Schlesinger, M., Shapiro, S.C., Sowa, J.F., 2012. Mapping the landscape of human-level artificial general intelligence. AI Mag. 33 (1), 25—41.

Adamsone-Fiskovica, A., 2015. Technoscientific futures: public framing of science. Technol. Soc. 40, 43—52.

Ahn, S., 2016. Becoming a network beyond boundaries: brain-Machine Interfaces (BMIs) as the actor-networks after the internet of things. Technol. Soc. 47, 49—59.

Akin, D., Jakobsen, K.C., Floch, J., Hoff, E., 2021. Sharing with neighbours: insights from local practices of the sharing economy. Technol. Soc. 64.

Allal-Chérif, O., 2022. Intelligent cathedrals: using augmented reality, virtual reality, and artificial intelligence to provide an intense cultural, historical, and religious visitor experience. Technol. Forecast. Soc. Change 178.

Allen, E.J., St-Yves, G., Wu, Y.H., Breedlove, J.L., Prince, J.S., Dowdle, L.T., Nau, M., Caron, B., Pestilli, F., Charest, I., Hutchinson, J.B., Naselaris, T., Kay, K., 2022. A massive 7T fMRI dataset to bridge cognitive neuroscience and artificial intelligence. Nat. Neurosci. 25 (1), 116—+.

Anderson, C.W., Kreiss, D., 2013. Black boxes as capacities for and constraints on action: electoral politics, journalism, and devices of representation. Qual. Sociol. 36 (4), 365—382.

Aristotle, 2011. In: Bartlett, R.C., Collins, S.D. (Eds.), Aristotle's Nicomachean Ethics. University of Chicago Press, Chicago; London.

Atmanspacher, H., 2012. Dual-Aspect Monism a la Pauli and Jung. J. Conscious. Stud. 19 (9—10), 96—120.

Aw, E.C.X., Tan, G.W.H., Cham, T.H., Raman, R., Ooi, K.B., 2022. Alexa, what's on my shopping list? Transforming customer experience with digital voice assistants. Technol. Forecast. Soc. Change 180.

Baabdullah, A.M., Alalwan, A.A., Algharabat, R.S., Metri, B., Rana, N.P., 2022. Virtual agents and flow experience: an empirical examination of AI-powered chatbots. Technol. Forecast. Soc. Change 181.

Bachelard, G., 2002. The Formation of the Scientific Mind. Clinamen, Manchester.

Baez, J.C., Javier, P.M., 1994. Gauge Fields, Knots, and Gravity, vol 4. World Scientific, Singapore; River Edge, NJ.

Bagis, M., Kryeziu, L., Akbaba, Y., Ramadani, V., Karaguezel, E.S., Krasniqi, B.A., 2022. The micro-foundations of a dynamic technological capability in the automotive industry. Technol. Soc. 70.

Banich, M.T., Compton, R.J., 2023. Cognitive Neuroscience, fifth ed. Cambridge University Press, Cambridge, United Kingdom; New York, NY.

Barad, K.M., 2007. Meeting the Universe Halfway: Quantum Physics and the Entanglement of Matter and Meaning. Duke University Press, Durham.

Baskerville, R.L., Myers, M.D., Yoo, Y., 2020. Digital first: the ontological reversal and new challenges for information systems research. MIS Q. 44 (2), 509—523.

Batayeh, B.G., Artzberger, G.H., Williams, L.D.A., 2018. Socially responsible innovation in health care: cycles of actualization. Technol. Soc. 53, 14—22.

Bateson, G., 2000. Steps to an Ecology of Mind. University of Chicago Press, Chicago.

Beckmann, P., Köstner, G., Hipólito, I., 2023. An alternative to cognitivism: computational phenomenology for deep learning. Minds Mach. 33 (3), 397—427.

Benkler, Y., 2006. The Wealth of Networks: How Social Production Transforms Markets and Freedom. Yale University Press, New Haven.
Bennett, J., 2010. Vibrant Matter: A Political Ecology of Things. Duke University Press, Durham.
Bergson, H., 2004. Matter and Memory. Dover Publications, Mineola, NY.
Bostrom, N., 2014. Superintelligence : Paths, Dangers, Strategies, first ed. Oxford University Press, Oxford.
Bourdieu, P., 1977. Outline of a Theory of Practice. Cambridge University Press, Cambridge; New York.
Brachman, R.J., Levesque, H.J., 2022. Machines like Us : Toward AI with Common Sense. The MIT Press, Cambridge, Massachusetts.
Brey, P., 2005. The epistemology and ontology of human-computer interaction. Minds Mach. 15 (3–4), 383–398.
Briggle, A., Mitcham, C., 2009. Embedding and networking: conceptualizing experience in a technosociety. Technol. Soc. 31 (4), 374–383.
Buchanan, I., 2021. Assemblage Theory and Method : An Introduction and Guide. Bloomsbury Academic, London; New York.
Butler, J., 1990. Gender Trouble: Feminism and the Subversion of Identity. Routledge, New York.
Butler, J., 2021. Markets in the Making: Rethinking Competition, Goods, and Innovation. Zone Books, Brooklyn, New York.
Callon, M., 1986a. The sociology of an actor-network: the case of the electric vehicle. In: Mapping the Dynamics of Science and Technology. Springer, pp. 19–34.
Callon, M., 1986b. Some elements of a sociology of translation: domestication of the scallops and the fishermen of St. Brieuc Bay. In: Power, Action, and Belief: A New Sociology of Knowledge. Routledge & Kegan Paul, London, pp. 196–223.
Campbell, C., 2020. Social capital, social movements and global public health: fighting for health-enabling contexts in marginalised settings. Soc. Sci. Med. 257.
Capra, F., Luisi, P.L., 2014. The Systems View of Life: A Unifying Vision. Cambridge University Press, Cambridge.
Carhart-Harris, R.L., Friston, K.J., 2010. The default-mode, ego-functions and free-energy: a neurobiological account of Freudian ideas. Brain 133, 1265–1283.
Carroll, P., Witten, K., Duff, C., 2021. How can we make it work for you Enabling sporting assemblages for disabled young people. Soc. Sci. Med. 288.
Castells, M., 2010. The Rise of the Network Society, second ed. vol 1. Wiley-Blackwell, Chichester, UK.
Chang, A.Y., Ogbuoji, O., Atun, R., Verguet, S., 2017. Dynamic modeling approaches to characterize the functioning of health systems: a systematic review of the literature. Soc. Sci. Med. 194, 160–167.
Chesbrough, H.W., 2003. Open Innovation: The New Imperative for Creating and Profiting from Technology. Harvard Business School Press, Boston.
Chiang, A.H., Trimi, S., Lo, Y.J., 2022. Emotion and service quality of anthropomorphic robots. Technol. Forecast. Soc. Change 177.
Çipi, A., Fernandes, A., Ferreira, F.A.F., Ferreira, N., Meidute-Kavaliauskiene, I., 2023. Detecting and developing new business opportunities in society 5.0 contexts: a sociotechnical approach. Technol. Soc. 73.
Clark, A., 2016. Surfing Uncertainty: Prediction, Action, and the Embodied Mind. Still Image. Oxford University Press, Oxford.
Clarke, B., 2020. Gaian Systems: Lynn Margulis, Neocybernetics, and the End of the Anthropocene, vol 60. University of Minnesota Press, Minneapolis, MN; London, England.
Coleman, J.S., 1990. Foundations of Social Theory. Belknap Press of Harvard University Press, Cambridge, MA.
Cruz, T.M., 2022. The social life of biomedical data: capturing, obscuring, and envisioning care in the digital safety-net. Soc. Sci. Med. 294.
Dai, Y.X., Hao, S.T., 2018. Transcending the opposition between techno-utopianism and techno-dystopianism. Technol. Soc. 53, 9–13.
Daugherty, P.R., James Wilson, H., 2018. Human + Machine: Reimagining Work in the Age of AI. Harvard Business Review Press, Boston, MA.
de Boer, B., Te Molder, H., Verbeek, P.P., 2018. The perspective of the instruments: mediating collectivity. Found. Sci. 23 (4), 739–755.
De Filippi, P., Mannan, M., Reijers, W., 2020. Blockchain as a confidence machine: the problem of trust & challenges of governance. Technol. Soc. 62.
De Luca, G., 2021. The development of machine intelligence in a computational universe. Technol. Soc. 65.

de Neufville, R., Baum, S.D., 2021. Collective action on artificial intelligence: a primer and review. Technol. Soc. 66.
Deane, G., 2022. Machines that feel and think: the role of affective feelings and mental action in (artificial) general intelligence. Artif. Life 28 (3), 289–309.
DeLanda, M., 2002. Intensive Science and Virtual Philosophy. Continuum, London.
DeLanda, M., 2006. A New Philosophy of Society: Assemblage Theory and Social Complexity. Continuum, London; New York.
DeLanda, M., 2016. Assemblage Theory. Edinburgh University Press, Edinburgh, UK.
Deleuze, G., 1988. Bergsonism. Zone Books, New York.
Deleuze, G., 1990. The Logic of Sense. Columbia University Press.
Deleuze, G., 1993. The Fold: Leibniz and the Baroque. University of Minnesota Press, Minneapolis.
Deleuze, G., 1994. Difference and Repetition. Columbia University Press, New York.
Deleuze, G., 2001. Pure Immanence: Essays on a Life. Translated by Anne Boyman. Zone Books, New York.
Deleuze, G., Guattari, F., 1977. Anti-Oedipus: Capitalism and Schizophrenia. Viking Press, New York.
Deleuze, G., Guattari, F., 1987. A Thousand Plateaus: Capitalism and Schizophrenia. University of Minnesota Press, Minneapolis.
Desanctis, G., Poole, M.S., 1994. Capturing the complexity in advanced technology use: adaptive structuration theory. Organ. Sci. 5 (2), 121–147.
Di Paolo, E.A., Thomas, B., Barandiaran, X.E., 2017. Sensorimotor Life: An Enactive Proposal, first ed. Oxford University Press, Oxford, United Kingdom.
Dixon, J., Manyau, S., Kandiye, F., Kranzer, K., Chandler, C.I.R., 2021. Antibiotics, rational drug use and the architecture of global health in Zimbabwe. Soc. Sci. Med. 272.
Doloreux, D., Parto, S., 2005. Regional innovation systems: current discourse and unresolved issues. Technol. Soc. 27 (2), 133–153.
Duff, C., 2014. Assemblages of Health: Deleuze's Empiricism and the Ethology of Life. Springer, New York.
Duff, C., 2023. The ends of an assemblage of health. Soc. Sci. Med. 317.
Dundas, B.I., 2018. A Short Course in Differential Topology. Cambridge University Press, Cambridge, United Kingdom; New York, NY.
Dunjko, V., Briegel, H.J., 2018. Machine learning & artificial intelligence in the quantum domain: a review of recent progress. Rep. Prog. Phys. 81 (7).
Dwivedi, Y.K., Hughes, L., Baabdullah, A.M., Ribeiro-Navarrete, S., Giannakis, M., Al-Debei, M.M., Dennehy, D., Metri, B., Buhalis, D., Cheung, C.M.K., Conboy, K., Doyle, R., Dubey, R., Dutot, V., Felix, R., Goyal, D.P., Gustafsson, A., Hinsch, C., Jebabli, I., Janssen, M., Kim, Y.G., Kim, J., Koos, S., Kreps, D., Kshetri, N., Kumar, V., Ooi, K.B., Papagiannidis, S., Pappas, I.O., Polyviou, A., Park, S.M., Pandey, N., Queiroz, M.M., Raman, R., Rauschnabel, P.A., Shirish, A., Sigala, M., Spanaki, K., Tan, G.W.H., Tiwari, M.K., Viglia, G., Wamba, S.F., 2022. Metaverse beyond the hype: multidisciplinary perspectives on emerging challenges, opportunities, and agenda for research, practice and policy. Int. J. Inf. Manag. 66.
Fei, N.Y., Lu, Z.W., Gao, Y.Z., Yang, G.X., Huo, Y.Q., Wen, J.Y., Lu, H.Y., Song, R.H., Gao, X., Xiang, T., Sun, H., Wen, J.R., 2022. Towards artificial general intelligence via a multimodal foundation model. Nat. Commun. 13 (1).
Felin, T., Zenger, T.R., 2017. The theory-based view: economic actors as theorists. Strat. Sci. 2 (4), 258–271.
Felin, T., Foss, N.J., Heimeriks, K.H., Madsen, T.L., 2012. Microfoundations of routines and capabilities: individuals, processes, and structure. J. Manag. Stud. 49 (8), 1351–1374.
Fields, C., Friston, K., Glazebrook, J.F., Levin, M., Marcianò, A., 2022. The free energy principle induces neuromorphic development. Neuromorph. Comput. Eng. 2 (4).
Finkler, K., 2004. Biomedicine globalized and localized: western medical practices in an outpatient clinic of a Mexican hospital. Soc. Sci. Med. 59 (10), 2037–2051.
Fjeldstad, O.D., Snow, C.C., Miles, R.E., Lettl, C., 2012. The architecture of collaboration. Strat. Manag. J. 33 (6), 734–750.
Fjelland, R., 2020. Why general artificial intelligence will not be realized. Humanit. Soc. Sci. Commun. 7 (1).
Fligstein, N., McAdam, D., 2012. A Theory of Fields. Oxford University Press, New York.
Floridi, L., 2014. The 4th Revolution: How the Infosphere Is Reshaping Human Reality, first ed. Oxford University Press, New York; Oxford.
Flusser, V., 2000. Towards a Philosophy of Photography. Reaktion, London.
Foley, R., 2014. The Roman-Irish Bath: medical/health history as therapeutic assemblage. Soc. Sci. Med. 106, 10–19.

Fong, B., Spivak, D.I., 2019. An Invitation to Applied Category Theory: Seven Sketches in Compositionality. Cambridge University Press, Cambridge; New York, NY.

Fox, S., 2016. Open prosperity: how latent realities arising from virtual-social-physical convergence (VSP) increase opportunities for global prosperity. Technol. Soc. 44, 92—103.

Fox, N., Ward, K., O'Rourke, A., 2005. The birth of the e-clinic. Continuity or transformation in the UK governance of pharmaceutical consumption? Soc. Sci. Med. 61 (7), 1474—1484.

Friston, K.J., 2009. The free-energy principle: a rough guide to the brain? Trends Cognit. Sci. 13 (7), 293—301.

Friston, K.J., 2010. The free-energy principle: a unified brain theory? Nat. Rev. Neurosci. 11 (2), 127—138.

Friston, K.J., Ramstead, M.J.D., Kiefer, A.B., Tschantz, A., Buckley, C.L., Albarracin, M., Pitliya, R.J., Heins, C., Brennan, K., Millidge, B., Sakthivadivel, D.A.R., Smithe, T.S.C., Koudahl, M., Tremblay, S.E., Petersen, C., Fung, K., Fox, J.G., Swanson, S., Mapes, D., René, G., 2024. Designing ecosystems of intelligence from first principles. Collectiv. Intell. 3 (1).

Furlong, K., 2014. STS beyond the "modern infrastructure ideal": extending theory by engaging with infrastructure challenges in the South. Technol. Soc. 38, 139—147.

Galaz, V., Centeno, M.A., Callahan, P.W., Causevic, A., Patterson, T., Brass, I., Baum, S., Farber, D., Fischer, J., Garcia, D., McPhearson, T., Jimenez, D., King, B., Larcey, P., Levy, K., 2021. Artificial intelligence, systemic risks, and sustainability. Technol. Soc. 67.

Gangle, R., 2016. Diagrammatic Immanence: Category Theory and Philosophy. Edinburgh University Press, Edinburgh.

Geels, F., 2005. Co-evolution of technology and society: the transition in water supply and personal hygiene in The Netherlands (1850—1930)-a case study in multi-level perspective. Technol. Soc. 27 (3), 363—397.

Geels, F.W., Kemp, R., 2007. Dynamics in socio-technical systems: typology of change processes and contrasting case studies. Technol. Soc. 29 (4), 441—455.

Gibson, J.J., 1979. The Ecological Approach to Visual Perception. Houghton Mifflin, Boston.

Gkiouleka, A., Huijts, T., Beckfield, J., Bambra, C., 2018. Understanding the micro and macro politics of health: inequalities, intersectionality & institutions—a research agenda. Soc. Sci. Med. 200, 92—98.

Goulet, F., 2021. Characterizing alignments in socio-technical transitions. Lessons from agricultural bio-inputs in Brazil. Technol. Soc. 65.

Grudniewicz, A., Tenbensel, T., Evans, J.M., Gray, C.S., Baker, G.R., Wodchis, W.P., 2018. 'Complexity-compatible' policy for integrated care? Lessons from the implementation of Ontario's Health Links. Soc. Sci. Med. 198, 95—102.

Guattari, F., 1995. Chaosmosis: An Ethico-Aesthetic Paradigm. Indiana University Press, Bloomington.

Guattari, F., 2013. Schizoanalytic Cartographies. Bloomsbury, London; New York.

Gunderson, R., 2020. A materialist conception of the lifeworld: Enzo Paci's social phenomenology of technology and the environment. Technol. Soc. 63.

Haken, H., 2006. Information and Self-Organization : A Macroscopic Approach to Complex Systems, 3rd ed. Springer, Berlin; New York.

Hampel, H., Au, R., Mattke, S., van der Flier, W.M., Aisen, P., Apostolova, L., Chen, C., Cho, M., De Santi, S., Gao, P., Iwata, A., Kurzman, R., Saykin, A.J., Teipel, S., Vellas, B., Vergallo, A., Wang, H.L., Cummings, J., 2022. Designing the next-generation clinical care pathway for Alzheimer's disease. Nat. Aging 2 (8), 692—703.

Hannigan, B., 2013. Connections and consequences in complex systems: insights from a case study of the emergence and local impact of crisis resolution and home treatment services. Soc. Sci. Med. 93, 212—219.

Hansen, M.B.N., 2009. System-environment hybrids. In: Emergence and Embodiment: New Essays on Second-Order Systems Theory. Duke University Press, pp. 113—142.

Hansen, M.B.N., 2015. Feed-forward: On the Future of Twenty-First-Century Media. University of Chicago Press, Chicago; London.

Haraway, D.J., 1991. Simians, Cyborgs, and Women: The Reinvention of Nature. Routledge, New York.

Harvey, D., 2006. Spaces of Global Capitalism. Verso, New York, NY.

Haven, E., Khrennikov, A.I.U., 2013. Quantum Social Science. Cambridge University Press, Cambridge; New York.

Hayles, N.K., 1999. How We Became Posthuman: Virtual Bodies in Cybernetics, Literature, and Informatics. University of Chicago Press, Chicago, IL.

Hayles, N.K., 2017. Unthought: The Power of the Cognitive Nonconscious. The University of Chicago Press, Chicago; London.

Heidegger, M., 1996. In: Stambaugh, J. (Ed.), Being and Time. State University of New York Press, Albany, NY.

Helfat, C.E., Martin, J.A., 2015. Dynamic managerial capabilities: review and assessment of managerial impact on strategic change. J. Manag. 41 (5), 1281–1312.

Helfat, C.E., Winter, S.G., 2011. Untangling dynamic and operational capabilities: strategy for the (N) ever-changing world. Strat. Manag. J. 32 (11), 1243–1250.

Herrera-Vega, E., 2015. Relevance of N. Luhmann's theory of social systems to understand the essence of technology today. The Case of the Gulf of Mexico Oil Spill. Technol. Soc. 40, 25–42.

Heylighen, F., Lenartowicz, M., 2017. The Global Brain as a model of the future information society: an introduction to the special issue. Technol. Forecast. Soc. Change 114, 1–6.

Hopster, J., 2021. What are socially disruptive technologies? Technol. Soc. 67.

Huang, T.J., 2017. Imitating the brain with neurocomputer A "new" way towards artificial general intelligence. Int. J. Autom. Comput. 14 (5), 520–531.

Huang, J.Z., 2023. A break in the cloud: the local sociotechnical affordances underlying global internet infrastructures. Technol. Soc. 74.

Ihde, D., 1990. Technology and the Lifeworld: From Garden to Earth. Indiana University Press, Bloomington.

Ingold, T., 2015. The Life of Lines. Routledge, London; New York.

Jebari, K., Lundborg, J., 2021. Artificial superintelligence and its limits: why AlphaZero cannot become a general agent. AI Soc. 36 (3), 807–815.

Kanger, L., Sillak, S., 2020. Emergence, consolidation and dominance of meta-regimes: exploring the historical evolution of mass production (1765–1972) from the Deep Transitions perspective. Technol. Soc. 63.

Keding, C., Meissner, P., 2021. Managerial overreliance on AI-augmented decision-making processes: how the use of AI-based advisory systems shapes choice behavior in R&D investment decisions. Technol. Forecast. Soc. Change 171.

Kim, E.S., Oh, Y., Yun, G.W., 2023. Sociotechnical challenges to the technological accuracy of computer vision: the new materialism perspective. Technol. Soc. 75.

Kriegeskorte, N., Douglas, P.K., 2018. Cognitive computational neuroscience. Nat. Neurosci. 21 (9), 1148–1160.

Kundu, D.K., Gupta, A., Mol, A.P.J., Rahman, M.M., van Halem, D., 2018. Experimenting with a novel technology for provision of safe drinking water in rural Bangladesh: the case of sub-surface arsenic removal (SAR). Technol. Soc. 53, 161–172.

Kurzweil, R., 2005. The Singularity Is Near: When Humans Transcend Biology. Viking, New York.

Latour, B., 2005. Reassembling the Social: An Introduction to Actor-Network-Theory. Oxford University Press, Oxford; New York.

Latour, B., 2013. An Inquiry into Modes of Existence: An Anthropology of the Moderns. Harvard University Press, Cambridge, Massachusetts.

Lavie, D., Stettner, U., Tushman, M.L., 2010. Exploration and exploitation within and across organizations. Acad. Manag. Ann. 4, 109–155.

Law, J., 2009. Actor network theory and material semiotics. In: Turner, B.S. (Ed.), The New Blackwell Companion to Social Theory. Wiley-Blackwell, Oxford, pp. 141–158.

Lefkowitz, D., 2022. Black boxes and information pathways: an actor-network theory approach to breast cancer survivorship care. Soc. Sci. Med. 307.

Leinster, T., 2014. Basic Category Theory, vol 143. Cambridge University Press, Cambridge, United Kingdom.

Locock, L., Nettleton, S., Kirkpatrick, S., Ryan, S., Ziebland, S., 2016. I knew before I was told': breaches, cues and clues in the diagnostic assemblage. Soc. Sci. Med. 154, 85–92.

Löhr, G., 2023. Conceptual disruption and 21st century technologies: a framework. Technol. Soc. 74.

Luhmann, N., 1995. Social Systems. Stanford University Press, Stanford, CA.

Luna-Ochoa, S.M.A., Robles-Belmont, E., Suaste-Gomez, E., 2016. A profile of Mexico's technological agglomerations: the case of the aerospace and nanotechnology industry in Queretaro and Monterrey. Technol. Soc. 46, 120–125.

Mac Lane, S., 1998. Categories for the Working Mathematician, second ed. Vol 5. Springer, New York.

Maglio, P.P., Kieliszewski, C.A., Spohrer, J.C. (Eds.), 2010. Handbook of Service Science, Service Science Research and Innovation in the Service Economy. Springer, New York.

Malone, T.W., 2018. Superminds: The Surprising Power of People and Computers Thinking Together, first ed. Little, Brown and Company, New York.

Manovich, L., 2002. The language of new media. In: Leonardo. MIT Press, Cambridge, MA.

Maslen, S., Harris, A., 2021. Becoming a diagnostic agent: a collated ethnography of digital-sensory work in caregiving intra-actions. Soc. Sci. Med. 277.

Massumi, B., 2002. Parables for the Virtual: Movement, Affect, Sensation. Duke University Press, Durham [N.C.].

Mayhew, S.H., Balabanova, D., Vandi, A., Mokuwa, G.A., Hanson, T., Parker, M., Richards, P., 2022. (Re)arranging "systems of care" in the early Ebola response in Sierra Leone: an interdisciplinary analysis. Soc. Sci. Med. 300.

Merleau-Ponty, M., 1962. Phenomenology of Perception. Humanities Press, New York,.

Merleau-Ponty, M., 1968. The visible and the invisible. In: Lefort, C. (Ed.), Followed by Working Notes. Northwestern University Press, Evanston, IL.

Micheli, P., Wilner, S.J.S., Bhatti, S.H., Mura, M., Beverland, M.B., 2019. Doing design thinking: conceptual review, synthesis, and research agenda. J. Prod. Innovat. Manag. 36 (2), 124–148.

Misra, S., Stokols, D., 2012. A typology of people-environment relationships in the Digital Age. Technol. Soc. 34 (4), 311–325.

Mossabir, R., Milligan, C., Froggatt, K., 2021. Therapeutic landscape experiences of everyday geographies within the wider community: a scoping review. Soc. Sci. Med. 279.

Mukama, M., Musango, J.K., Smit, S., Ceschin, F., Petrulaityte, A., 2022. Development of living labs to support gendered energy technology innovation in poor urban environments. Technol. Soc. 68.

Mulder, K., Kaijser, A., 2014. The dynamics of technological systems integration: water management, electricity supply, railroads and industrialization at the Gota Alv. Technol. Soc. 39, 88–99.

Mwale, S., 2020. Becoming-with' a repeat healthy volunteer: managing and negotiating trust among repeat healthy volunteers in commercial clinical drug trials. Soc. Sci. Med. 245.

Naber, G.L., 2011a. Topology, Geometry and Gauge Fields: Foundations, second ed., vol 25. Springer Science+Business Media, New York.

Naber, G.L., 2011b. Topology, Geometry, and Gauge Fields: Interactions, second ed., vol 141. Springer, New York.

Niu, Z.Y., Zhong, G.Q., Yu, H., 2021. A review on the attention mechanism of deep learning. Neurocomputing 452, 48–62.

Northoff, G., 2012. Psychoanalysis and the brain—why did Freud abandon neuroscience? Front. Psychol. 3.

Nugus, P., Carroll, K., Hewett, D.G., Short, A., Forero, R., Braithwaite, J., 2010. Integrated care in the emergency department: a complex adaptive systems perspective. Soc. Sci. Med. 71 (11), 1997–2004.

Odling-Smee, F.J., Laland, K.N., Feldman, M.W., 2003. Niche Construction: The Neglected Process in Evolution, Vol 37. Princeton University Press, Princeton.

Ong, V., Skinner, K., Minaker, L.M., 2021. Life stories of food agency, health, and resilience in a rapidly gentrifying urban centre: building a multidimensional concept of food access. Soc. Sci. Med. 280.

Onno, J., Khan, F.A., Daftary, A., David, P.M., 2023. Artificial intelligence-based computer aided detection (AI-CAD) in the fight against tuberculosis: effects of moving health technologies in global health. Soc. Sci. Med. 327.

Orlikowski, W.J., 2000. Using technology and constituting structures: a practice lens for studying technology in organizations. Organ. Sci. 11 (4), 404–428.

Ostheimer, J., Chowdhury, S., Iqbal, S., 2021. An alliance of humans and machines for machine learning: hybrid intelligent systems and their design principles. Technol. Soc. 66.

Parr, T., Pezzulo, G., Friston, K.J., 2022. Active Inference: The Free Energy Principle in Mind, Brain, and Behavior. The MIT Press, Cambridge, Massachusetts.

Penrose, R., 2004. The Road to Reality: A Complete Guide to the Laws of the Universe. Jonathan Cape, London.

Pérez-Pérez, J.F., Parra, J.F., Serrano-García, J., 2021. A system dynamics model: transition to sustainable processes. Technol. Soc. 65.

Pezzulo, G., Parr, T., Cisek, P., Clark, A., Friston, K., 2023. Generating meaning: active inference and the scope and limits of passive AI. Trends Cognit. Sci. 28, 97–112.

Picard, R.W., 2000. Affective Computing. MIT Press, Cambridge, MA.

Pickering, A., 1995. The Mangle of Practice: Time, Agency, and Science. University of Chicago Press, Chicago.

Polk, J.B., Campbell, J., Drilon, A.E., Keating, P., Cambrosio, A., 2023. Organizing precision medicine: a case study of Memorial Sloan Kettering Cancer Center's engagement in/with genomics. Soc. Sci. Med. 324.

Possati, L.M., 2021. Freud and the algorithm: neuropsychoanalysis as a framework to understand artificial general intelligence. Humanit. Soc. Sci. Commun. 8 (1).

Prahalad, C.K., Venkat, R., 2004. The Future of Competition: Co-creating Unique Value with Customers. Harvard Business School Press, Boston.

Preece, J., Rogers, Y., Sharp, H., 2015. Interaction Design: Beyond Human-Computer Interaction, fourth ed. Wiley, Chichester, West Sussex, United Kingdom.

Priest, G., 2018. The Fifth Corner of Four: An Essay on Buddhist Metaphysics and the Catuṣkoṭi, first ed. Oxford University Press, Oxford.

Quitzau, M.B., 2007. Water-flushing toilets: systemic development and path-dependent characteristics and their bearing on technological alternatives. Technol. Soc. 29 (3), 351–360.

Ramaswamy, V., Ozcan, K., 2014. The Co-creation Paradigm. Stanford University Press, Stanford, CA.

Ramaswamy, V., Ozcan, K., 2018a. Offerings as digitalized interactive platforms: a conceptual framework and implications. J. Market. 82 (4), 19–31.

Ramaswamy, V., Ozcan, K., 2018b. What is co-creation? An interactional creation framework and its implications for value creation. J. Bus. Res. 84 (March), 196–205.

Ramaswamy, V., Ozcan, K., 2022. Brands as co-creational lived experience ecosystems: an integrative theoretical framework of interactional creation. In: Markovic, S., Gyrd-Jones, R., von Wallpach, S., Lindgreen, A. (Eds.), Research Handbook on Brand Co-creation: Theory, Practice, and Ethical Implications. Edward Elgar, pp. 47–64.

Ramstead, M.J.D., Badcock, P.B., Friston, K.J., 2018. Answering Schrodinger's question: a free-energy formulation. Phys. Life Rev. 24, 1–16.

Ramstead, M.J.D., Seth, A.K., Hesp, C., Sandved-Smith, L., Mago, J., Lifshitz, M., Pagnoni, G., Smith, R., Dumas, G., Lutz, A., Friston, K., Constant, A., 2022. From generative models to generative passages: a computational approach to (neuro) phenomenology. Rev. Philos. Psychol. 13 (4), 829–857.

Ramstead, M.J.D., Sakthivadivel, D.A.R., Heins, C., Koudahl, M., Millidge, B., Da Costa, L., Klein, B., Friston, K.J., 2023. On Bayesian mechanics: a physics of and by beliefs. Interface Focus 13 (3).

Raven, R.P.J.M., Verbong, G.P.J., 2009. Boundary crossing innovations: case studies from the energy domain. Technol. Soc. 31 (1), 85–93.

Reale, F., 2019. Governing innovation systems: a Parsonian social systems perspective. Technol. Soc. 59.

Rhodes, T., 2018. The becoming of methadone in Kenya: how an intervention's implementation constitutes recovery potential. Soc. Sci. Med. 201, 71–79.

Rodríguez, D., Busco, C., Flores, R., 2015. Information technology within society's evolution. Technol. Soc. 40, 64–72.

Roitblat, H.L., 2020. Algorithms Are Not Enough: Creating General Artificial Intelligence. The MIT Press, Cambridge, Massachusetts.

Roli, A., Jaeger, J., Kauffman, S.A., 2022. How organisms come to know the world: fundamental limits on artificial general intelligence. Front. Ecol. Evol. 9.

Roman, S., 2017. An introduction to the language of category theory. In: Compact Textbooks in Mathematics. Springer International Publishing Imprint: Birkhäuser, Cham.

Romele, A., 2024. Digital Habitus: A Critique of the Imaginaries of Artificial Intelligence. Routledge, New York, NY.

Rosen, R., 1991. Life Itself: A Comprehensive Inquiry into the Nature, Origin, and Fabrication of Life. Columbia University Press, New York.

Rosenberger, R., 2014. Multistability and the agency of mundane artifacts: from speed bumps to subway benches. Hum. Stud. 37 (3), 369–392.

Rosiak, D., 2022. Sheaf Theory through Examples. The MIT Press, Cambridge, Massachusetts.

Sanders, E.B.N., Jan Stappers, P., 2008. Co-creation and the new landscapes of design. CoDesign 4 (1), 5–18.

Sanne, J.M., 2012. Learning from adverse events in the nuclear power industry: organizational learning, policy making and normalization. Technol. Soc. 34 (3), 239–250.

Sassen, S., 2006. Territory, Authority, Rights: From Medieval to Global Assemblages. Princeton University Press, Princeton, NJ.

Schatzki, T.R., 2002. The Site of the Social : A Philosophical Account of the Constitution of Social Life and Change. Pennsylvania State University Press, University Park, PA [Great Britain].

Schön, D.A., 1983. The Reflective Practitioner: How Professionals Think in Action. Basic Books, New York.

Schubert, C., 2015. Situating technological and societal futures. Pragmatist engagements with computer simulations and social dynamics. Technol. Soc. 40, 4–13.

Sengupta, B., Tozzi, A., Cooray, G.K., Douglas, P.K., Friston, K.J., 2016. Towards a neuronal gauge theory. PLoS Biol. 14 (3).

Sha, X.W., 2013. Poiesis and Enchantment in Topological Matter. MIT Press, Cambridge, Massachusetts.

Simondon, G., 2020. Individuation in light of notions of form and information. In: Translated by Taylor Adkins, vol 57. University of Minnesota Press, Minneapolis.

Sirmon, D.G., Hitt, M.A., Ireland, R.D., Gilbert, B.A., 2011. Resource orchestration to create competitive advantage: breadth, depth, and life cycle effects. J. Manag. 37 (5), 1390–1412.

Smart, P., 2018. Emerging digital technologies: implications for extended conceptions of cognition and knowledge. In: Adam Carter, J., Clark, A., Kallestrup, J., Orestis Palermos, S., Pritchard, D. (Eds.), Extended Epistemology. Oxford University Press, Oxford, pp. 266–304.

Sneltvedt, O., 2018. Experience the future in full-scale: technological background relations and visions of the good society at the World's Columbian Exposition. Technol. Soc. 52, 46–53.

Solms, M., 2020. New project for a scientific psychology: general scheme. Neuro-psychoanalysis 22 (1–2), 5–35.

Sony, M., Naik, S., 2020. Industry 4.0 integration with socio-technical systems theory: a systematic review and proposed theoretical model. Technol. Soc. 61.

Spicer, N., Bhattacharya, D., Dimka, R., Fanta, F., Mangham-Jefferies, L., Schellenberg, J., Tamire-Woldemariam, A., Walt, G., Wickremasinghe, D., 2014. Scaling-up is a craft not a science': catalysing scale-up of health innovations in Ethiopia, India and Nigeria. Soc. Sci. Med. 121, 30–38.

Stewart, M., Brown, J.B., Wayne Weston, W., McWhinney, I.R., McWilliam, C.L., Freeman, T.R., 2014. Patient-centered Medicine: Transforming the Clinical Method, third ed. Radcliffe Publishing, London.

Stiegler, B., 1998. *Technics and Time*. 2 Vols. Stanford University Press, Stanford, CA.

Tabarés, R., 2021. HTML5 and the evolution of HTML; tracing the origins of digital platforms. Technol. Soc. 65.

Tania, F.G., Vicente, L., Blanca, P.G., Fernando, R.M., 2022. Measuring the territorial effort in research, development, and innovation from a multiple criteria approach: application to the Spanish regions case. Technol. Soc. 70.

Teece, D.J., 2007. Explicating dynamic capabilities: the nature and microfoundations of (sustainable) enterprise performance. Strat. Manag. J. 28 (13), 1319–1350.

Teece, D., Peteraf, M., Leih, S., 2016. Dynamic capabilities and organizational agility: risk, uncertainty, and strategy in the innovation economy. Calif. Manag. Rev. 58 (4), 13–35.

Tekic, Z., Füller, J., 2023. Managing innovation in the era of AI. Technol. Soc. 73.

Thorndike, A.N., 2020. Healthy choice architecture in the supermarket: does it work? Soc. Sci. Med. 266.

Tijmes, P., 1999. Philosophy in the service of people. Technol. Soc. 21 (2), 175–189.

Tilson, D., Lyytinen, K., Sorensen, C., 2010. Digital infrastructures: the missing IS research agenda. Inf. Syst. Res. 21 (4), 748–759.

Tironi, M., Lisboa, D.I.R., 2023. Artificial intelligence in the new forms of environmental governance in the Chilean State: towards an eco-algorithmic governance. Technol. Soc. 74.

Tiwana, A., 2014. Platform Ecosystems: Aligning Architecture, Governance, and Strategy. Morgan Kaufmann, Waltham, MA.

Tiwana, A., Konsynski, B., Bush, A.A., 2010. Platform evolution: coevolution of platform architecture, governance, and environmental dynamics. Inf. Syst. Res. 21 (4), 675–687.

Tonkonoff, S., 2017. From Tarde to Deleuze and Foucault: the infinitesimal revolution. In: Palgrave Studies in Relational Sociology. Palgrave Macmillan, Cham, Switzerland.

Trnka, S., 2021. Multi-sited therapeutic assemblages: virtual and real-life emplacement of youth mental health support. Soc. Sci. Med. 278.

Umbrello, S., Bernstein, M.J., Vermaas, P.E., Resseguier, A., Gonzalez, G., Porcari, A., Grinbaum, A., Adomaitis, L., 2023. From speculation to reality: enhancing anticipatory ethics for emerging technologies (ATE) in practice. Technol. Soc. 74.

Varela, F.J., Thompson, E., Rosch, E., 2016. The Embodied Mind: Cognitive Science and Human Experience, revised edition. MIT Press, Cambridge, Massachusetts; London England.

Veatch, R.M., 2009. Patient, Heal Thyself: How the New Medicine Puts the Patient in Charge. Oxford University Press, Oxford.

Voinea, C., 2018. Designing for conviviality. Technol. Soc. 52, 70–78.

Wall, C.T.C., 2016. Differential Topology, vol 156. Cambridge University Press, Cambridge, United Kingdom.

Wang, V., Tucker, J.V., 2016. Phatic systems in digital society. Technol. Soc. 46, 140–148.

Watanabe, C., Naveed, K., Neittaanmäki, P., Fox, B., 2017. Consolidated challenge to social demand for resilient platforms—lessons from Uber's global expansion. Technol. Soc. 48, 33–53.

Watanabe, C., Naveed, N., Neittaanmäki, P., 2018. Digital solutions transform the forest-based bioeconomy into a digital platform industry—a suggestion for a disruptive business model in the digital economy. Technol. Soc. 54, 168–188.

Weick, K.E., Sutcliffe, K.M., Obstfeld, D., 2005. Organizing and the process of sensemaking. Organ. Sci. 16 (4), 409–421.

Wendt, A., 2015. Quantum Mind and Social Science: Unifying Physical and Social Ontology. Cambridge University Press, Cambridge, United Kingdom; New York.

Wu, C.L., 2012. IVF policy and global/local politics: the making of multiple-embryo transfer regulation in Taiwan. Soc. Sci. Med. 75 (4), 725–732.

Xiang, Y.S., Zhang, X.L., Wu, W., 2021. Coupling or lock-in? Co-evolution of cultural embeddness and cluster innovation-exploratory case study of Shaoxing textile cluster. Technol. Soc. 67.

Yamakawa, H., 2021. The whole brain architecture approach: accelerating the development of artificial general intelligence by referring to the brain. Neural Network. 144, 478–495.

Yannakakis, G.N., Togelius, J., 2011. Experience-driven procedural content generation. IEEE Trans. Affect. Comput. 2 (3), 147–161.

Yoo, Y., 2010. Computing in everyday life: a call for research on experiential computing. MIS Q. 34 (2), 213–231.

Zhang, Z.W., Yoo, Y., Lyytinen, K., Lindberg, A., 2021. The unknowability of autonomous tools and the liminal experience of their use. Inf. Syst. Res. 32 (4), 1192–1213.

Zhu, M.L., He, T.Y.Y., Lee, C.K., 2020. Technologies toward next generation human machine interfaces: from machine learning enhanced tactile sensing to neuromorphic sensory systems. Appl. Phys. Rev. 7 (3).

Zittrain, J.L., 2006. The generative Internet. Harv. Law Rev. 119 (7), 1974–2040.

Zott, C., Amit, R., Massa, L., 2011. The business model: recent developments and future research. J. Manag. 37 (4), 1019–1042.

Glossary of terms

Note: The bold-type numbers (**1**—**12**) and letters (**P**rologue, **E**pilogue) at the end of each entry denote the chapters of this book, where more detailed discussion of that term can be found.

Abstract Machinic Phyla (M) conceptual framework encompassing the organizational, technological, and interactional dimensions of ecosystems; pathways and infrastructures that enable connectivity, movement, and complex interactions within systems, highlighting the potential for innovation and the formation of intricate relational networks among human and technological entities. **1, 2, 11, 12, E**

Active inference cognitive process where organisms and systems actively engage with their environment to reduce uncertainty and enhance predictive accuracy; dynamic, self-organizing mechanism fundamental to decision-making and intelligence, continuously refining models through new experiences to make sense of and interact with the world effectively. **4, 5, 7**

Actual realm of tangible, realized manifestations of ideas and processes, where potentialities from the virtual dimension take concrete form; emphasizing adaptability and foresight in shaping lived reality and experiences, and the physical embodiment of theoretical constructs in navigating real-world challenges. **P, 1**

Actualization dynamic transition from the realm of unmanifested potentialities (virtual) to tangible realities (actual), involving the process of creative differentiation and "becoming"; influenced by the virtual's possibilities, playing a crucial role in the evolution of concepts, technologies, or policies within ecosystems. **1, 12**

Adaptive structuration process where social structures and practices dynamically evolve in response to technological innovations and changes; the interaction between social actors, technology, and organizational practices, emphasizing the co-evolution and adaptive capacity of ecosystems to integrate new technologies, thereby reflecting the emergent, transformative nature of living systems. **2, 3, 4, 6, 7**

Agenc*ed* corporation These are adaptable, innovative, and entrepreneurial organizations with a variable functor set and limit object focus. They constantly experiment with new strategies and models to align their activities towards a specific "ideal" future state. **9**

Agenc*ing* collective disruptive and transformative organizations focused on expansion and diversification with a variable functor set and colimit object focus; adapt their processes to navigate their ecosystem, aiming to create new relationships and interactions that extend beyond the current state. **9**

Anthropocene a geological epoch characterized by the significant impact of human activity on the Earth's geology and ecosystems; the era where human actions have started to have a global impact on Earth's environment, marking a departure from the Holocene epoch. **E**

Artificial general intelligence (AGI) development stage of artificial intelligence where systems can perform any intellectual task that a human being can; convergence of cognitive processes and machine capability, key for understanding future technological evolutions. **5**

Assemblage (agencement) networks of heterogeneous components within ecosystems, emphasizing dynamic, fluid interactions that enable emergent and recurrent changes; the capacity of diverse elements to adapt and transform through morphisms, reflecting the complex, evolving nature of relational phenomena and the contingent properties of interconnected entities. **P, 3, 4, 8, 11**

Base sets foundational territories within ecosystems, representing various fields of practice that form the grounding elements of the topology; core areas within the plane of immanence, essential for analyzing and understanding the transformations and dynamics within MLXEs. **3, 4**

Becoming perpetual transformation and evolution of identities and structures beyond static states; fluid and dynamic interplay between entities, facilitated by counter-actualization and natural transformations, leading to the emergence of new possibilities and transformative changes across virtual and actual realms. **2, 4, 5, 6**

Black boxes complex systems or elements within ecosystems whose internal mechanisms are not fully transparent or understood. **3, 7**

Body without organs (BwO) the potentiality within systems, denoting spaces of pure possibility and innovation; in organizational contexts, areas for growth and evolution, emphasizing adaptability and the exploration of uncharted territories. 10

Boundary-spanning creating connections that cross-established boundaries or territories, fostering innovation, adaptation, and change by bridging gaps between different domains; see **Transversality**. 4

CARE (complex adaptive relational event-sense) architecture framework that views organizations as dynamic assemblages, emphasizing complexity, relational thinking, and adaptivity. 8

Category theory mathematical framework focusing on the structure, dynamics, and transformations of entities within complex systems; abstractly modeling relationships (morphisms) and entities (objects) within assemblages, providing a robust tool for understanding and analyzing the intricate interconnections and evolutionary processes within and across ecosystems. P, 3, 8, 9, 10

Cocreative power synergistic potential that emerges from the collaboration between humans and machines (AI); emphasizes the collective intelligence and creative capabilities that are unlocked when diverse intelligences work together in co-intelligent creation, leading to innovative solutions and transformations. P

Cocreative transformation dynamic process by which different agents, including humans and machines, collaborate to generate innovative solutions and transformations in cointelligent creation; driving changes in machinic ecosystems grounded in life-experiences. E

Coinnovation process wherein technological and human domains mutually evolve, leading to the creation of value-oriented platforms deeply resonant with individuals as creative-experiencers, bridging virtual and actual experiences. 2

Coinnovation platforms strategic platforms within organizations that facilitate collaborative innovation, aligning with the overarching goals and dynamics of ecosystems. 8

Cognitive metamorphosis transformation of cognitive processes as influenced by the development and integration of AGI into experiential computing; deterritorialization and reconfiguration of traditional perceptions and cognitive frameworks in light of advanced AI capabilities. 5

Colimit object represents the potential for growth, exploration, and diversification within ecosystems, embodying organizations' orientation toward expanding strategic and operational boundaries to proactively navigate and thrive in complex environments by identifying emergent opportunities and challenges. 8, 9, 10

Collective assemblage of enunciation (CAE) networks of diverse components that collectively contribute to the production and emergence of meaning within ecosystems; underscores the collaborative nature of expressing and navigating complexities, facilitating entities' enhanced agency in internal and external environments. 1, 12

Collective individuation process through which collective identities and relations are formed and transformed in interaction with technology, highlighting the social aspect of technological engagement. 2

Complexion intricate interplay of factors shaping the present state of energetic-signaletic ecosystem flows (E), reflecting the tangible forces and interactions within a system. 1

Constellation assemblages of abstract ideas forming significant patterns within Incorporeal Experience Universes (X), guiding transformations across domains. 1

Coshaping technology and organizational practices reciprocal shaping and adaptation between technological advancements and organizational changes within ecosystems; continuous transformation and integration of new technologies into social practices. 4, 6

Counter-actualization dynamic process where actualized entities influence and enrich the virtual realm; ensures that reality remains in a state of perpetual flux, allowing for continuous novelty and becoming. 2

Creative emergence process through which new forms, structures, and systems arise within an MLXE; differentiation from a virtual multiplicity into actualized entities, driven by the dynamics of transformation. 11

Creative transformation perpetual process of change and innovation within ecosystems, propelled by the synergistic interactions among diverse elements, including the virtual and actual realms, human and artificial intelligence, and cross-sectoral engagements; underscores the adaptability and evolutionary potential of living systems toward sustainable futures. P, E

Cutouts specific configurations within existential life territories (L) that define the structure of life's experiential fabric, capturing moments or durations of potentiality. 1

De Rham Cohomology method for evaluating the evolution of categories, using differential forms to study the topology of spaces; classifies spaces based on their topology, offering insights into the structural and dynamic properties of categories. 6

Desiring machines productive potential of desire within ecosystems, driving entities to connect, produce, and create within their territories; fundamental relational force, fostering the generation of new combinations, assemblages, and transversal connections among diverse entities. **3, 4**

Deterritorialization process that disrupts established identities and structures within assemblages, emphasizing fluidity and the potential for transformation; entities breaking free from traditional territories, driven by machinic desire and socio-political forces. **1, 3**

Deterritorialization and reterritorialization disruption and recreation of ecosystem structures, with morphisms as forces of change, symbolizing the ecosystem's perpetual state of flux and the constant negotiation between stability and transformation. **4**

(Shared) Digitalized infrastructure foundational systems and platforms of digital infrastructures enabling digitalized experiences and interactions, pivotal for fostering coinnovation and supporting engagement of individuals in ecosystems. **2**

Diagrammatic method use of abstract diagrams as nonrepresentational maps to capture and analyze the intricate relationships, potentials, and dynamics within assemblages and ecosystems; foundational approach for understanding complex networks, facilitating strategic planning, and navigating organizations through current states toward envisioned growth. **8, 10**

Difference and repetition the mechanism by which actualization occurs, generating variations and ensuring that the transition from virtual to actual brings forth new and unexpected outcomes. **12**

Differential forms quantify the orientation-dependent rates of transformation and flow within complex systems; measure the intensity of changes across categories, integral to the gauge-invariant analysis of transformations. **6, 12**

Differential topology provides insights into the continuous, dynamic nature of transformations and relational dynamics within systems; complements Category Theory by focusing on the nuances of morphisms, smooth manifolds, and the adaptive processes of entities, for a comprehensive understanding of complex system evolution. **P, 3, 6, 12**

Differentiation process by which a virtual multiplicity actualizes into diverse, distinct entities within an MLXE; driven by the dynamics of creative emergence and the monadic structure. **11**

Digital niche construction co-evolutionary relationship between technology and organizations, where digital technologies are strategically implemented and integrated into ecosystems to create new environments and enhance adaptability as well as the ecosystem's dynamic response and the scaling of technologies to meet evolving demands and improve survivability. **3, 4, 6, 7**

Discursivity production and intersection of meanings across various platforms; beyond formal language to include semiotic systems, bodily expressions, and material flows. **1**

Double articulation foundational process that integrates the material (machinic assemblages of bodies) and semiotic (collective assemblages of enunciation) dimensions within ecosystems; the reciprocal influence between physical—material components and social-communicative aspects, mediated by morphisms, shaping entity trajectories and ecosystem dynamics. **1, 4, 11**

Dynamic relationalities evolving, interconnected interactions among entities within and across categories; focus on nonlinear, complex dynamics that drive system transformations and evolution, utilizing tools like functors, smooth manifolds, and vector fields to model continuous intra- and inter-category transformations. **5, 6**

Dynamic relationality theory (DRT) comprehensive framework that synthesizes ethical, epistemological, and ontological perspectives to conceptualize systems as fluid networks of interrelated entities; explores the essence and identity of entities through evolving relationships, emphasizing the transformative, interconnected local and global dynamics within digitalized ecosystems and strategic organizational architecting, guided by category theory for navigating complex landscapes. **1, 7, 10**

Dynamic **relationality (pragmatic core)** relations and interactions as inherently dynamic, leading to continuous change, evolution, and the creation of new possibilities; transformative potential of practices and relations as the pragmatic tenet of DRT. **P**

Endofunctor maps a category to itself, transforming objects and morphisms within that category to illustrate the internal dynamics and continuous transformation processes of systems; plays a crucial role in understanding and modeling the evolutionary changes and state behaviors within ecosystems. **6, 11**

Energetic-signaletic ecosystem flows (E) dynamic processes and currents driving transformations within entities and ecosystems, characterized by the exchange of energy and information; the actual real, molding existential territories and emphasizing the continuous, nonstatic nature of entities across multiple sectors. 1, 2, 11, 12, E

Engagement and enactment processes highlighting the transition from virtual tendencies to actual interactions, underscoring the co-evolutionary dynamics of technology and human practices. 2

Enrollment and agencing process by which entities become part of larger assemblages or organizations within an ecosystem, embedded inherently through their roles and interactions as defined by category theory. 8

Epistemic alignment conceptual consistency and coherence between epistemic structures and outcomes, facilitating a comprehensive understanding of the interconnected nature of cognitive processes, AGI, and experiential computing. 5

Events transformative moments where shifts from macroscopic structures to microscopic transformations occur, catalyzing the intersection of virtual potentialities and tangible dynamics within ecosystems; dynamic evolution through disruption and creation, bridging actual mechanisms and virtual principles, and recalibrating relationships for continuous ecosystem evolution. 2, 4

Existential life territories (L) map the dynamic interplay between personal subjective realms and specific socio-behavioral conditions within ecosystems, encompassing both physical and symbolic spaces; foundational for identity formation, cultural evolution, and the interaction with technology, serving as the basis from which entities' transformations and machinic desires emerge and actualize. 1, 2, 3, 11, 12, E

Experience-verse an assemblage focused on sense-making processes, encompassing elements like "series and structure" and "poiesis and enchantment"; characterized by its dynamic and evolving nature, shaping our experiences through continuous interaction and transformation. 4

Experiencial computing use of computing technologies to engender emergent experiences and enhance lived experiences through interactivity, immersion, and personalization; explores how digitalized environments can be designed to more intuitively, responsively, and personally engage with individuals, shaping cognitive and experiential realities. 5

Fiber bundles topological structure used in the Abstract Machinic Phyla domain to model the complex interconnections between entities within MLXEs, representing the intricate organization of system. 11

Folding/unfolding processes of internalization and externalization that entities undergo in their continuous transformation: enrichment of an entity's internal landscape by integrating external complexities and externalization of internal complexities in response to shifts in existential territories, respectively. 12

Free Energy Principle systems' efforts to minimize discrepancies between expected and actual sensory inputs by minimizing the free energy of their internal states; adaptive, predictive capabilities of living systems in maintaining stability and coherence within dynamic environments, crucial for understanding cognitive processes and transformations within MLXEs. 4, 5, 7

Functor maps objects and their relationships from one category to another, preserving structural relationships; essential for modeling transformations and interactions within and across ecosystems, enabling the visualization of complex systems' dynamics and the structured interplay of assemblages and their manifold interconnections. P, 4, 5, 6, 8, 10, 11

Functor set stability consistency of the processes or transformations an organization uses to interact with its environment; "stable functor sets" indicate organizations with established, consistent processes, reflecting traditional hierarchical structures, while "variable functor sets" characterize organizations that are flexible and innovative, adapting their operational processes in response to changing business environments. 9

Gauge theory elucidates the complex interactions and transformations within ecosystems, aligning the structure of existential territories, energetic-signaletic flows, incorporeal experiences, and abstract machinic phyla; employs Lagrangian dynamics to analyze kinetic and potential energies, facilitating a rigorous understanding of the dynamics governing MLXEs. 1, 11, 12

Gauge transformations mechanics of transformations between different states or configurations of the system, embodying the symmetries of the Lagrangian. 12

Gauge-invariant Lagrangian describes the system dynamics, remaining constant under gauge transformations; encapsulates the "energy" or "action" that governs the motion equations for fields involved, offering a mathematical framework to quantify and represent the transformation dynamics and evolutionary processes of entities over time. 9, 12

Generative AI Artificial intelligence (AI), which by generating new data, insights, or content from learned datasets, redefines relational structures and enhances creative, adaptive capabilities, significantly transforming assemblage dynamics and operational essence; bridging artificial neural networks and human intelligence (HI), evolving toward machinic generalized intelligence (MGI). **5, 6**

Generativity capacity of technology to enable new forms of value creation, innovation, and novel user experiences, signifying the formative impact of technological interactions. **2**

Gluing process method of integrating diverse components or data points within an ecosystem to form a coherent whole; essential for ensuring that the integrated system remains comprehensive and customized. **10**

Higher-order categories conceptual layers above basic categories and functors to include objects, morphisms, and 2-morphisms or higher transformations; capture intricate relationships and dynamic transformations between different system components, addressing the multifaceted nature of ecosystems. **8, 10**

Homotopy theory dynamic perspective on natural transformations, offering continuous paths between functions and capturing the essence of "becoming" in dynamic systems. **6**

Human-artificial intelligence (HAI) interactivity collaborative and interactive relationship between humans and artificial intelligence systems; foundational to creative transformation, leveraging the strengths of both human creativity and AI's analytical prowess to drive innovation and adaptability in digitalized ecosystems. **P**

Immanence potentialities and actualities as internal dynamics, not external possibilities; inherent or embedded nature of forces within MLXEs driving transformation and evolution. **11**

Incorporeal experience universes (X) realms of principles, goals, and values that guide and influence the transformational trajectories within ecosystems; embodying the Virtual Possible, play a pivotal role in shaping subjective life-experiences, steering entities' paths, and influencing the differentiation and emergence within dynamic systems. **1, 2, 11, 12, E**

Individual-technology-environment relations systemic, dynamic interdependencies among individuals, technology, and environmental contexts within MLXEs; complex web of interactions, fundamental to understanding the intricate mechanics and transformative dynamics that define the co-evolutionary processes within these ecosystems. **3, 7**

Induction process by which global potentials or overarching patterns are inferred and actualized into structured individual entities or local processes, reflecting the dynamic relationality within ecosystems. **7**

Interactional creation emergent pattern of potential futures derived from present realities, shaped by the multifaceted interplay of human engagement, technological capabilities, and emergent outcomes. **2**

Kinetic and potential terms the momentum of transformation (kinetic) and the latent capacities driving change (potential) in ecosystems, derived from Lagrangian dynamics; dynamic forces and tension guiding the evolution of entities and systems. **3, 12**

Lie derivatives measure changes in vector fields, capturing the essence of symmetries and transformations within categories; quantitative assessment of dynamic shifts and evolutions, providing a detailed understanding of intra-category transformations. **6**

Life-experiences multidimensional, cumulative encounters shaped by dynamic interactions within natural, technological, societal, and economic realms, and influenced by memories, desires, and external forces; constitute both tangible and intangible aspects of existence, central to assessing the impacts of MLXEs on individual and community well-being, wealth, empowerment, and welfare. **P, 1, E**

Limit object future states or "ideal" configurations organizations aim to achieve, representing goals that guide strategic decisions and optimize interactions; pivotal for aligning organizational activities and relationships towards a coherent, predefined outcome, serving as benchmarks for development and strategic alignment within ecosystems. **8, 9, 10**

Limit/colimit construction universal methods for summarizing and combining structures across categories, capturing the essence and dynamics of architectural frameworks or ecosystemic assemblages; representing the most generalized or specified forms of organizational configurations. **8, 10**

Lines of flight transformative pathways that enable escape from rigid structures, facilitating the creation of new assemblages and systems; forces of deterritorialization and innovation within ecosystems, challenging established configurations and enabling systemic evolution and change. **1, 4, 5, 11**

MAB-CAE Nexus conjunction of machinic assemblages of bodies (MABs) and collective assemblages of enunciation (CAEs), representing the intertwined physical and communicative elements within an MLXE. **11**

Machinic assemblage of bodies (MAB) entities as dynamic combinations of components within deterritorialization, highlighting their interconnected and productive nature. 1

Machinic ecosystems dynamic networks where technology, humans, and their environments interact and co-evolve, characterized by continuous transformation; synthesis of forces—economic, biological, and experiential—that shape lived realities and potential pathways, with technology central to influencing interactions and evolutionary outcomes. P, 1, 2

Machinic generalized intelligence (MGI) emergent intelligence from the synergistic interaction between human intelligence (HI) and artificial intelligence (AI); amplifies transformative processes and outcomes across co-intelligent ecosystems, integrating and converging diverse intelligences into a novel, dynamic form of intelligence in various domains. P, 5

Machinic life-experience ecosystem (MLXE) dynamic, interconnected networks where humans, machines, and environments co-evolve, emphasizing the transformative interactions that shape life-experiences; characterized by their adaptability and co-creative processes, MLXEs highlight the reciprocal influence of global and local dynamics. 2, 7, 8, 10, 11, E

Mixed reality (MR) blending of physical and digital worlds to create new environments where physical and digital objects co-exist and interact in real time; connects to phenomenotechniques, altering human perception and experience. 5

Molar lines of segmentation large-scale structures influenced by socio-political forces; delineate the global systems shaping collective identity and overarching structures within ecosystems, highlighting the role of codification and socio-political dynamics. 7

Molar vs. molecular dual nature of systems and organizations, with molar representing structured, uniform aspects that confer stability, and molecular denoting fluid, diverse elements that enable adaptability; balance between innovation and established practices, crucial for navigating the dynamics of sustainable development and evolution within ecosystems. 1, 10

Molecular lines of segmentation micro-transformations within ecosystems, emphasizing the fluid and dynamic nature of cognitive processes and experiences; nuances of individual and localized sense-making processes, affecting global systems subtly. 7

Monadic structure an endofunctor (for internal dynamics), a unit transformation (for emergence of new entities) and a multiplication transformation (for interaction and merging of entities); models the differentiation and integration processes within ecosystems and the ongoing metamorphosis of entities, illustrating the complex interconnections and potential for transformation within and across categories, effectively guiding strategic analysis and development within dynamic relational systems. 6, 11, 12

Morphisms functions or arrows signifying relationships and transformations between objects within categories; dynamic influences, interactions, and transformative processes driving the evolution of assemblages, enabling a comprehensive understanding of the systemic dynamics and relational intricacies within and across ecosystems. P, 3, 4, 8

Multiplication transformation merging, interaction, and consolidation of entities within ecosystems, facilitating the development of new structures or coherent wholes; dynamic process of organization formation and ensuring coherence in transformations, akin to the function of Lie derivatives in measuring systemic changes. 6, 11

Natural transformation describes the evolution between functors, acting as structured pathways for change within ecosystems; ensures the coherent structural integrity of categories amidst transformations, serving as a bridge that maintains systemic coherence and facilitates the alignment and development of complex relational dynamics across assemblages. 5, 6, 10

NEST complex arrangements (assemblages) of natural, societal, economic, and technological elements in co-intelligent ecosystems. E

Neuromorphic computing emulates the neural structures and processing methods of the human brain; provides insights into active inference, enabling a deeper understanding of cognitive processes and their potential replication in artificial systems. 5

Open set represents "neighboring" or "close" elements within a topological space, indicating groups of similar interactions within the MLXE. 3, 4

Organic integration seamless integration of human and machine capabilities, emphasizing a harmonious, symbiotic relationship between organic and inorganic elements; a fluid and integrated co-existence that leverages the strengths of both entities. 5

Organizational morphogenesis process by which organizations evolve over time; interplay between meaning-making (signification) and boundary-crossing (transversality), providing insights into the dynamics of organizational change and development. 9

Organizational morphology form and structure of organizations; how organizations evolve, adapt, and interact within their ecosystems, utilizing concepts from category theory to provide insight into their operational consistency, adaptability, and strategic orientation. 9

Organized body (OB) actualized, structured aspects of an organization, encompassing established protocols and standardized practices; crystallization of potential into tangible forms, focusing on stability and consistency. 10

Paths or patches trajectory or segments within an ecosystem that an organization navigates through a sequence of transformations or processes, depicted by functors. 8

Perception-cognition mental processes involved in interpreting sensory information and making sense of the environment; fundamental to both human cognitive science and the development of AGI, highlighting the intersection of perception, cognition, and artificial intelligence in shaping experiences. 5

Phenomenotechnicity manifestation of phenomena through technological means; critical role of technology in constructing, mediating, and shaping experiences, interactions, and realities in digitalized ecosystems. P, 4, 5, 7

Plane of consistency/immanence (P) virtual space of pure becoming and interconnectedness, where potential states of being and interactions pre-exist, serving as the fertile ground for the emergence and evolution of actual entities through differentiation; all entities and their transformative potentials within MLXEs. 1, 3, 11, 12

Platform(ization) (strategic development and integration of) shared digitalized infrastructures within ecosystems, crucial for facilitating seamless data flow, collaboration, and innovation; alignment and steering of ecosystems' dynamics, enabling diverse engagements and optimizing the strategic architecturing of entities towards achieving coherent and innovative outcomes. 2, 10

Poiesis and enchantment creative and transformative aspects of experiences, focusing on the generation of meaningful and captivating experiences within ecosystems. 7

Pragmatic semiotics of diagrams interpreting diagrams to bring forth new categories, objects, and morphisms, akin to creating a new language or system of meaning within ecosystems; instrumental in designing life-experience ecosystems that are engaging and immersive. 10

Presheaf functor representing the organization of objects within a topological space, showing how local entities relate and restrict across different contexts; underpins the sheaf analysis by tracking interactions and transformations within ecosystems. 7

Quantum computing leveraging quantum mechanics principles to enhance information processing capabilities; promising transformative advancements in computational efficiency, AI systems, human cognition, and the development of artificial general intelligence (AGI). 5, 6

Relation entities exist and derive their essence through relations with others; challenging traditional notions of isolated existence, interconnectedness and interdependence of components within systems as the ontological tenet of DRT. P

Relation-*al* knowledge and relations as not static but continuously crafted, emphasizing the importance of interactional knowledge creation in understanding complex systems; generative, evolving nature of knowledge and relations as the epistemological tenet of DRT. P

Relational-*ity* (axiological core) relations and interactions as not superficial but deeply rooted in value systems which also inform the assessment and valorization of outcomes and actions; every relational dynamic being imbued with inherent worth and significance as the axiological tenet of DRT. P

Reterritorialization dynamic reconfiguration or creation of new structures, identities, or territories within assemblages following deterritorialization; adaptive mechanisms by which systems channel machinic desire and integrate new elements to stabilize and evolve, marking a continuous negotiation and transformation within ecosystems. 3, 5

Rhizome decentralized, nonlinear, and nonhierarchical networks of connections within Abstract Machinic Phyla (M), fostering flexibility, multiplicity, and spontaneous growth; interconnectedness and potential for free connections among components challenging traditional structures. P, 1, 2, 3

Sense-making process by which individuals interpret and give meaning to their experiences, highlighting the cognitive and narrative construction of reality within ecosystems. 7

Sentio-experientiality emotional, affective, and sensory dimensions of experiences generated through interactions with digitalized ecosystems; subjective, felt experiences being central to engagement with technology, calls for designing emotionally resonant digitalized environments. 5

Series and structure organization and sequencing of experiences or processes, emphasizing the structured nature of cognitive and social dynamics. 7

Sheaf extension of the presheaf that satisfies certain conditions, ensuring that global data can be restricted to local data and vice versa; encapsulates the coherence and integration of local and global information within ecosystems. 7

Sheaf morphisms mappings between sheaves that respect the sheaf structure, transforming one sheaf into another while preserving the relationships between objects and morphisms; used to model the processes of induction, transduction, and translation within ecosystems. 7

Sheaf theory as a tool to bridge local and global perspectives within systems, facilitates understanding of how local processes, whether cognitive or functional, interact with overarching dynamics. 6, 7

Singularity transformative point where AI systems potentially achieve recursive self-improvement, leading to exponential technological growth; contemporary technologies driving categorical transformations and advanced states within perception-cognition and experiential computing. 5, 6

Smooth manifolds spaces where local behaviors and changes are continuously trackable, applied to categories to facilitate the analysis of dynamic relationalities. 3

Social actor-network intricate web of dynamic interactions among human and nonhuman actors within ecosystems, forming a networked complex that underpins energetic-signaletic flows. 2, 3, 7

Stack (sheaf) analysis extends the concept of sheaves to model the integration and "gluing" of system components into a coherent whole, accommodating complex, multilayered relationships within ecosystems; facilitates understanding and strategic architecturing by tracking local data aggregation and managing intricate interactions among diverse entities across various domains. 8, 10, 11

Stalks collections of local information at specific points within a topological space, representing the direct limit of sheaf sections; provide a comprehensive view of local processes and their contribution to global dynamics. 7

Strata (stacked) hierarchical or nested arrangement of layers or levels within a system or ecosystem, where each stratum may contain entities or assemblages with distinct properties. 5

Strategic architecturing designing and adjusting the macro-architecture of an ecosystem to align with envisioned goals, using diagrams, categories, and functors; integration of diverse components and the reconciliation of potential conflicts within the enterprise. 10

Structur*ed* corporation organizations that operate with a stable set of processes and aim toward a specific "ideal" state; characterized by centralized control mechanisms and a focus on optimization and alignment of activities to achieve an "ideal" configuration. 9

Structur*ing* collective communal and decentralized organizations with a stable functor set and colimit object focus; aim to create new relationships and interactions that extend beyond their current configuration, emphasizing expansion and diversification. 9

Substance and form of content actual processes and structures within a domain of MLXEs and how these are organized and structured. 11

Substance and form of expression narratives, policies, and discourses that articulate and guide the content within MLXEs, shaping understanding and representation. 11

Techno-epistemology how technology influences our ways of knowing, interpreting, and understanding the world; cognitive and conceptual frameworks that evolve in response to technological advancements, reshaping human epistemology. 5

Technological affordances potential actions made possible by technology, dictating how users can engage and navigate within ecosystems, leading to unforeseen innovations and emphasizing the ecosystem's adaptive nature. 2

Territorial capture process by which new structures or norms are formed within assemblages, channeling machinic desire in specific directions. 3

Territorial escape the movement away from established norms or structures within assemblages, driven by the deterritorializing forces of machinic desire. 3

Territorial Life Spacetime (L-Spacetime) defines the conditions and states of entities within existential life territories, extending beyond physical dimensions to include influences from past experiences, desires, and external factors. **1, 2**

Topological space conceptual framework used to map and understand the positions, interactions, and potential configurations of elements within a system, identifying all possible assemblages within a given context; facilitates the exploration of the system's structure, dynamics, and interaction states, crucial for analyzing complex ecosystems and their constituent entities. **3, 4, 7**

Trans-epistemic shift transition from traditional sense-making frameworks to more technologically integrated, organic, and immersive approaches to cognition, intelligence, and experience. **5**

Transduction structuration process where changes in one part of a system lead to transformations in other parts, creating a new structured whole. **7**

Translation shift in scale or context, where local entities or insights are expanded, reinterpreted, and applied to broader global contexts, taking on new meanings or functions. **7**

Transversality dynamic, transformative intersections across categories, domains, and levels of organization within ecosystems; nontangential meeting points to enhance understanding of complex relational dynamics, facilitating multifaceted interactions and the emergence of novel connections and possibilities within various systems. **3, 4, 6**

Twisted gluing conditions complex integration scenarios within ecosystems where simple aggregation is insufficient and necessitate additional layers or transformations for integration, highlighting the need for nuanced understanding and adaptability in managing overlaps and intersections. **10**

Unit transformation embedding of category objects into transformative contexts, facilitating the emergence of new entities and potentials in monadic structures; illustrates the transition from virtual spaces to actualized forms within ecosystems. **6, 11**

Universal constructions/mapping property characterize objects or morphisms with universal properties (including limits and colimits), offering a framework to understand the generalizable and optimal structure and dynamics of ecosystems. **8, 10**

Vector fields assign vectors to each point in space or category manifold to represent forces or directions of change; model the dynamics of transformation, providing insights into the flow, direction, and guiding forces influencing system evolution within MLXEs. **3, 6, 11**

Virtual dimension realm of potentialities, ideas, and multiplicities before their actualization; embodies the space of conceptual possibilities, serving as a fertile ground for innovation, the emergence of new forms, and transformative experiences. **P, 1**

Wellbeing, wealth, empowerment, and welfare meta-level outcomes as holistic impacts of transformations within existential life territories, energetic-signaletic flows, incorporeal universes, and abstract machinic phyla; measure health, happiness, prosperity, resource richness, skill acquisition, and societal participation, influenced by access to food, healthcare, education, and mobility, driving societal evolution. **12, E**

Index

'Note: Page numbers followed by "f" indicate figures, "t" indicate tables and "b" indicate boxes.'

A

Abstract machinic phyla (M), 59, 253–254, 256–257, 271, 281–283
Abstract problem solving, 239
Active participation, 49
Actual dimensions, 3–6
Actual dynamics, 13–20
Actualization, 6–9, 38–41
    processes, 40–41
Adaptive structuration, 140–142, 144
Adjunctions, 237
Agenced corporations, 202–204, 203f, 230
Agencing collective, 205–212, 206f
Anthropocene, 293, 299–304
Anticipatory Ethics for Emerging Technologies (ATE), 11–12
Application programming interfaces (APIs), 48
Architectural transformation
    2-categories/higher categories, 234–236
    diagrammatic alignment, 240
    diagrams or limit/colimit objects, 229–231
    functors and natural transformations, 231–234
    healthcare application, of strategic architecturing, 239–241
    higher order category, 241
    life-experience ecosystems, 236–239
    multilayered global to local analysis, 216–228
    stack perspective, 239–240
    strategic architecturing, 228–236
Artificial General Intelligence (AGI), 111, 114, 138, 152–155
Artificial intelligence (AI), 115
    mammography interpretation, 99–100
    personalized treatment plans, 100
    predictive analytics, for breast cancer risk, 100
Assemblage, dynamic relationalities in, 70–72
    plane of consistency, 83–87
    smooth manifolds and submanifolds, 76–78
    territories of assemblage, 83–87
    vector fields, 79–80

B

Balancing adaptability (BwO), 245
    active push toward transformations, 219b
    dissolving molar structures, 219
    emergence, 219b
    inducing molecular transformations, 219b
    molar suppression, 219b
Blockchain technology, 10
Boundary state, 213–214
Breast cancer care, 14, 234
Breast Cancer Research Foundation, 204

C

Cardiology department, 239–240
Category theory, 187
Centralized adaptive structuration, 99
Clinical decision-making community, 174
Cocreative transformation, 65–66
Co-innovation
    life-experience platforms, 42, 49–50
    platform strata, 191
Colimit construction, 189
Collaborative evolution, 103b
Collaborative research network, 206
Collaborative social-actor networks, 84
Collective assemblages of enunciation (CAE), 19–20, 27, 260, 270
Complementary goals, 103b
Complex Adaptive Relational Event-Sense (CARE) architecture, 184
Cone, 237
Constellation, 17–18
Continuous innovation, 49
Continuous transformation of identity, 64
Corporation communicates, 201–202
Counteractualization, 34–41
Countersignifying regime, 214
COVID-19, 254–258, 265–268
Creative plane, of immanence, 10, 15, 23–24, 27
Creative transformation, 295–296

CRISPR, 236
Cross stacked strata, 95–96
Curvature, 80

D

Decentralized adaptive structuration, 99
Deepened user engagement, 51
Deeper digital integration, 47
De Rham cohomology, 64–65, 139, 141–143
Deterritorialization processes, 18, 162–163
Diagrammatic relationality, in ecosystem dynamics, 65
Differential topology, 77, 99–100, 137–139, 252–253, 288, 291, 293–295
Differential transformation, 138–140
Digital health startups (DHS), 180–182
Digital niche construction, 79, 140–142, 144
Digital voice assistants (DVAs), 121
Direct induction, 171
Dominance/bifurcation operator, 219b
Double articulation, 258
Dynamic adaptation, 44
Dynamic relationalities, on strata
    assemblages across stacked strata, 95–96
    functorial transformations across stacked strata, 96–97
    machinic and collective assemblages, 93–94
    territorialization, of strata between assemblages, 89–93
    transversality, 98–110
Dynamic relationality, 64–65, 70–72, 89–93, 115–119, 185–187, 228–236, 287–289
Dynamic relationality theory (DRT), 3, 11f, 26f, 27–29, 63t, 65, 89, 112, 159, 269
    transformative potential, 29–31

E

Ecosystem sectors, via double articulation, 258–265
Electronic health record (EHR), 141, 143–144, 146–147
Embodied machines, 51
Emergency department (ED), 80
Empowerment, 274–279
Endofunctor, 143, 250, 255, 276
Endofunctor T, 153–154
Energetic-signaletic ecosystem flows, 11–12, 16, 21–24
Energetic-signaletic flows (E), 57, 250, 253, 255–256, 270, 276, 281, 290
Enriched reality transformation, 51
Environmental and developmental policy, 31
Environmental policies, 31
Euler-Lagrangian framework, 103b

Existential life territories (L), 33
Existential territories (L), 57, 253, 255–256, 270, 275–276
Experiencial computing, 114

F

Feedback loops, 103b
Fiber bundles, 250–251, 272
Free energy principle, 118
Functorial domains, 23–26
Functorial mapping, 100–102
Functorial transformations, 96–97
Functors, 237
    from E to X, 251
    from L to E, 251
    from M to L, 251–252
    and natural transformations, 228
    set, stability of, 197–199
    from X to M, 251

G

Gauge-invariant Lagrangian, 273
Gauge theoretic interpretation, 26f
Gauge theory, 23–24, 269–273
Generative artificial intelligence, 112–115, 153
Global and local transformations, 175f
Global immanence, 278
Global transformations, 161–166, 162f
Grounding machinic ecosystem
    actualization and counter-actualization, 38–41
    counter-actualization, 34–38
    machinic life-experience ecosystems, 55–62
    virtual
        experience ecosystems, 41–48
        machinic life, 48–55

H

Harmonize internal models, 37
Healthcare and technology management, 30–31
Healthcare application, of strategic architecturing, 239–241
Healthcare ecosystems, 194–196
Healthcare organization, 229, 231
Healthcare professional desire (HPD), 103b
Healthcare system, 230, 254–258, 265–268, 271
Healthcare technology innovators field, 85, 170–171
Health ecosystems, 27–29
Health policy advocacy groups (HPAGs), 180–182
Hierarchical social actor networks, 94
Higher-order categories, 190–191
Homotopies, 150–152
Homotopy theory, 137, 150–151
    for natural transformation dynamics, 154–155

Human-centric designs, 54
Human cognition, 30
Human experiences and emotions, 49
Human intelligence (HI), 137
Hybrid intelligent systems, 22–23
Hyper-intelligent systems, 116–117

I

ICT-driven business model, 17–18
Immanence, creative plane of, 10–13
Incorporeal universes (X), 251, 253–254, 256–257, 264, 271, 276, 281
Individual technology environment relations, 170–172
Infrastructural tools, 48–49
Intangible interactions, 42
Integrated care networks (ICN), 180–182
Intense molecular activity, 233–234
Interdisciplinary state, 213–214
Interruption, 103b
Inversion operator, 219b

L

Lagrangian, 214
  dynamics, 33
Latent operator, 219b
Latour's Actor-Network Theory, 57
Lie derivative, 139, 147–148
Life-experience ecosystems, 236–239
Life-experience platforms, 52
Limit/colimit objects, 229–231, 240
Limit construction, 188–189
Limited digitalized engagement, 42

M

Machinic assemblages of bodies (MAB), 18–19, 259, 270, 292
Machinic ecosystem co-innovation, 45
Machinic generalized intelligence (MGI), 58, 111, 123–129, 137
Machinic life, 48–55
Machinic life-experience ecosystem (MLXE), 31, 55–62, 69, 70f, 159, 179, 215, 245–248, 247f
  ecosystem sectors via double articulation, 258–265
  global immanence to well-being, wealth, empowerment, and welfare, 274–279
  grounding social-behavioral phenomena, 269–273
  healthcare system and COVID-19, 254–258, 265–268
  immanence of emergence and evolution, 287–289
  immanence within, 246–248
  monadic structure, 248–253
  strategy development, 289–292
  vaccination, 257
  virus
    emergence, 255–256
    global spread, 256–257
Machinic phyla (M), 11–12, 14–15, 25, 291–292
Mammography technology, 161–162
Molar/molecular lines, 159–160
Molecular activity, 233–234
Molecular processes, 232–233
Monadic structure, 249–251
Multiplication transformation, 143–144, 151, 154, 246–247, 250–251, 255

N

Natural transformations, 117, 127, 140–143, 150, 231–234, 232f, 237
"NEST-networked" systems, 274
Neuromorphic computing, 153
Nontangential intersection, 101

O

Open set, 62, 72, 78, 83–85, 93–94, 96–97, 168, 170, 174, 216–217, 309
Organic integration, 123
Organizational morphogenesis, 207b
Organizational morphology, 197–201, 200t
Organizations
  colimit object focus, 205–206, 212–213
  ecosystems, 183–185
  limit object focus, 202–203, 213
  with stable functor sets, 204–205
  with variable functor sets, 205–212
Organized body (OB), 230, 233–234
  negotiating with molecular forces, 219b
  stabilizing molar structures, 219b
Overcoding state, 213

P

Patient-centered care community, 172–173
Patient-centered health, 97
Patient's desire (PD), 103b
Perception cognition, 114, 128
Phenomenotechnics, 122–123
Plane of consistency, 83–87
Postsignifying regime, 214
Pragmatic semiotics of diagrams, 236–238
Predictive analytics community, 170–171
Presignifying regime, 214
Preventive Healthcare Organizations (PHO), 180–182

Q

Qualitative transformation, 12
Quantum computing, 113, 153

R

Rapid technological iteration, 38–39
Remote consultation tools, 195
Resource allocation, 103b
Reterritorialization processes, 64, 119–123, 162–163

S

Sense-making activities, 120
Sentio experientiality, 125
Shared spaces, 103b
Sheaf theory, 137–138, 172
Signification gauge transformations, 213
Signifying regime, 214
Smooth manifolds, 76–80, 138, 146
Social actor networks, 71, 83, 172–173
Social-behavioral phenomena, 269–273
Societal backlash, 40–41
Socio-behavioral ecosystem, 270
Socio-technological environment, 41–42
Stable functor sets, 201–202, 212–213
Stable/variable functor, 200–201
Stack analysis, 240
Stack gluing conditions, 239–240
Stack perspective, 239–240
Stakeholder ecosystems, 52, 58, 95–96
Strata via assemblages, 20–23
Strategic architecture, 245
    organizational forms, 240
    transformations of diagrams, 228–236
Strategy development, 289–292
Stratification, 93–94
Structured corporations, 201
Structuring collective, 204–205
Submanifolds, 76–78
Surjective derivative, 101
Sustainable innovations, 54
Systemic constraints, 103b
System's desire (SD), 103b

T

Technological determinism, 38
Technological evolution, 103b
Technological innovations, 30, 97
Technology-dependent interactions, 94
Technology-independent individual-technology-environment relations, 94
Topological spaces, 161, 234, 250
Transduction, 166–167, 174–178
Translation process, 166–167
Transversality, 98–111, 139
    gauge transformations, 213–214
    inducing decoding, 214

U

Unit transformation, 143, 151, 154, 250, 255, 275–276
Universal constructions, 237
User-machine interactions, 49

V

Vaccination, 257
Value-oriented platforms, 47
Variable functor set, 202–203, 205–206, 212–213
Vector fields, 79–80, 139, 146–147
Virtual dimension, 3–6
Virtual dynamics, 13–20
Virtual realities, 9–10
Virtual space, 247–248
Virus
    emergence, 255–256
    global spread, 256–257

W

Wealth, 274–279
Welfare, 274–279
Well-being, 274–279

Y

Yoneda Lemma, 237–238

Printed in the United States
by Baker & Taylor Publisher Services